NEW PATHWAYS IN HIGH-ENERGY PHYSICS I

Magnetic Charge and Other Fundamental Approaches

Studies in the Natural Sciences

A Series from the Center for Theoretical Studies

University of Miami, Coral Gables, Florida

Recent Volumes in this Series

Volume 4 – **QUANTUM STATISTICAL MECHANICS IN THE NATURAL SCIENCES**
Edited by Behram Kursunoglu, Stephan L. Mintz, Susan M. Widmayer, Mou-Shan Chen, C. S. Hui, Joseph Hubbard, Joseph F. Malerba, Laurence Mittag, and George Soukup • 1974

Volume 5 – **FUNDAMENTAL THEORIES IN PHYSICS**
Edited by Stephan L. Mintz, Laurence Mittag, Susan M. Widmayer, Chui-Shuen Hui, and Joseph F. Malerba • 1974

Volume 6 – **PROGRESS IN THE NEUROSCIENCES AND RELATED FIELDS**
Edited by Behram Kursunoglu, Stephan L. Mintz, Susan M. Widmayer, Chui-Shuen Hui, Joseph Hubbard, and Joseph F. Malerba • 1974

Volume 7 – **TOPICS IN ENERGY AND RESOURCES**
Edited by Behram Kursunoglu, Stephan L. Mintz, Susan M. Widmayer, Chui-Shuen Hui, Joseph Hubbard, Joseph Malerba, and George Soukup • 1974

Volume 8 – **PROGRESS IN LASERS AND LASER FUSION**
Edited by Behram Kursunoglu, Arnold Perlmutter, Susan M. Widmayer, Uri Bernstein, Joseph Hubbard, Christian Le Monnier de Gouville, Laurence Mittag, Donald Pettengill, George Soukup, and M. Y. Wang • 1975

Volume 9 – **THEORIES AND EXPERIMENTS IN HIGH-ENERGY PHYSICS**
Edited by Behram Kursunoglu, Arnold Perlmutter, Susan M. Widmayer, Uri Bernstein, Joseph Hubbard, Christian Le Monnier de Gouville, Laurence Mittag, Donald Pettengill, George Soukup, and M. Y. Wang • 1975

Volume 10 – **NEW PATHWAYS IN HIGH-ENERGY PHYSICS I**
Magnetic Charge and Other Fundamental Approaches
Edited by Arnold Perlmutter • 1976

Volume 11 – **NEW PATHWAYS IN HIGH-ENERGY PHYSICS II**
New Particles – Theories and Experiments
Edited by Arnold Perlmutter • 1976

A Continuation Order Plan is available for this series. A continuation order will bring delivery of each new volume immediately upon publication. Volumes are billed only upon actual shipment. For further information please contact the publisher.

Orbis Scientiae, University of Miami, 1976.

NEW PATHWAYS IN HIGH-ENERGY PHYSICS I

Magnetic Charge and Other Fundamental Approaches

Edited by

Arnold Perlmutter

Center for Theoretical Studies
University of Miami
Coral Gables, Florida

PLENUM PRESS • NEW YORK AND LONDON

Library of Congress Cataloging in Publication Data

Orbis Scientiae, University of Miami, 1976.
New pathways in high-energy physics.

(Studies in the natural sciences; v. 10-11)
Includes indexes.
1. Particles (Nuclear physics)—Congresses. I. Perlmutter, Arnold, 1928- II. Miami, University of, Coral Gables, Fla. Center for Theoretical Studies. III. Title. IV. Series.
QC793.07 1976 539.7'6 76-20476
ISBN 0-306-36910-9 (v. 1)

A part of the Proceedings of Orbis Scientiae 1976 held by the
Center for Theoretical Studies, University of Miami, Coral Gables, Florida,
January 19-22, 1976

A Division of Plenum Publishing Corporation
227 West 17th Street, New York, N.Y. 10011

Printed in the United States of America

Preface

This year, Orbis Scientiae 1976, dedicated to the Bicentennial of the United States of America, was devoted entirely to recent developments in high energy physics. These proceedings contain nearly all of the papers presented at Orbis, held at the Center for Theoretical Studies, University of Miami, during January 19-22, 1976.

The organization of Orbis this year was due mainly to the moderators of the sessions, principally Sydney Meshkov, Murray Gell-Mann, Yoichiro Nambu, Glennys Farrar, Fred Zachariasen and Behram Kursunoglu, who was also chairman of the conference. The coherence of the various sessions is due to their efforts, and special thanks are due to Sydney Meshkov who was responsible for coordinating many of the efforts of the moderators and for including essentially all of the frontier developments in high energy physics.

Because of the number of papers and their integrated length, it has been necessary to divide these proceedings into two volumes. An effort has been made to divide the material in the two volumes into fundamental questions (including the appearance of magnetic charge in particle physics) and recent high energy results and attendant phenomenology.

These volumes were prepared by Mrs. Helga Billings, Mrs. Elva Brady and Ms. Yvonne Leber, and their dedication and skill are gratefully acknowledged. Their efforts during Orbis were supplemented by those of Mrs. Jacquelyn Zagursky, with our appreciation. The photographs were taken by Ms. Shirley Busch.

Orbis Scientiae 1976 received some support from the National Science Foundation Office of International Programs and Energy Research and Development Administration.

The Editor

Contents of Volume 10

Contents of Volume 11

Participants of the Orbis Scientiae 1976, January 19-22, 1976

New Pathways in High Energy Physics

Some of the other participants of the Orbis Scientiae

Professor P.A.M. Dirac

THEORY OF MAGNETIC MONOPOLES

P. A. M. Dirac

Florida State University

Tallahassee, Florida 32306

1. CHARGES AND MONOPOLES

The Maxwell equations for empty space are symmetric between the electric and magnetic fields E and H. One can replace E by H and H by -E and the equations are unchanged. One can add further terms to the field equations to represent the effects of electric charges and currents. If one then brings in the Lorentz equation for the motion of a charged particle

$$m\frac{d^2v_\mu}{ds^2} + e\ F_{\mu\nu}v^\nu = 0\ , \qquad (1.1)$$

one has the complete scheme of equations for the ordinary classical electrodynamics.

Another way to generalize the vacuum equations would be to bring in terms corresponding to magnetic charges and currents. One could set up an equation analogous to Lorentz's for the motion of a particle carrying a magnetic monopole

$$m\frac{d^2v_\mu}{ds^2} + g\,\tilde{F}_{\mu\nu}\,v^\nu = 0, \tag{1.2}$$

where g is the strength of the monopole and the sign ~ denotes the dual of a 6-vector. The resulting scheme of equations is mathematically just as reasonable as the usual one, but it is not so useful because, whereas electric charges are continually being observed, magnetic monopoles have never (or almost never) been observed.

The theory allows one to have both charges and monopoles co-existing and interacting with one another. If one wants to quantize such a theory one runs into a difficulty because, in the quantum theory for the motion of a charged particle, one does not use directly the Lorentz equation (1.1) involving the field quantities $F_{\mu\nu}$, but one has a Schrödinger equation involving the potentials A_μ. The vector potentials satisfy

$$H = \text{curl } A, \tag{1.3}$$

which requires

$$\text{div } H = 0. \tag{1.4}$$

Now if there is a magnetic monopole, equation (1.4) cannot hold all round it. There must be (at least) one point on any surface enclosing the monopole where (1.4) fails, and where consequently the potentials cannot be introduced to satisfy (1.3). The points where (1.3) fail form, at any time, a line extending from the monopole to infinity, or terminating on an equal and opposite monopole. We call such a line a <u>string</u>.

The position of the string is arbitrary, subject only to the condition that it has the correct end point, or end points. When we change from one position to another, we are changing the potentials without changing the field quantities, so we are making a gauge transformation.

2. A CHANGE OF GAUGE

Take as an example the monopole g at the origin. The magnetic field is

$$H_x = g\, x_s r^{-3} \quad (s = 1,2,3). \qquad)2.1)$$

A possible choice of the potentials is

$$A_1 = \frac{-gx_2}{r(r+x_3)}, \quad A_2 = \frac{gx_1}{r(r+x_3)}, \quad A_3 = 0.$$

One easily checks that this leads to the field (2.1), except where $r + x_3 = 0$. This means $x_1 = x_2 = 0$, $x_3 \leq 0$. Thus the string extends along the axis $x_1 = x_2 = 0$ from the origin to $x_3 = -\infty$.

A second possible choice of the potentials is

$$A_1 = \frac{gx_2}{r(r-x_3)}, A_2 = \frac{-gx_1}{r(r-x_3)}, A_3 = 0,$$

for which the string extends along the axis $x_1 = x_2 = 0$ from the origin to $x_3 = \infty$. The change of the string from one position to the other corresponds to the change of the potentials

$$\Delta A_1 = \frac{gx_2}{r}\left[\frac{1}{r-x_3} + \frac{1}{r+x_3}\right] = \frac{2gx_2}{r^2-x_3^2} ,$$

$$\Delta A_2 = \frac{-2gx_1}{r^2-x_3^2}, \ \Delta A_3 = 0.$$

This is a gauge transformation

$$\Delta A_s = \partial S/\partial x_s , \tag{2.2}$$

with

$$S = 2g\tan^{-1}\frac{x_1}{x_2} = 2g\phi , \tag{2.3}$$

where ϕ is the azimuthal angle of the point (x_1, x_2, x_3) about the axis $x_1 = x_2 = 0$.

We see that the function S producing the gauge transformation is not single-valued. If we go round a closed loop passing between the two positions of the string, the change in S is $4\pi g$. This result is characteristic of any change in the position of the string. The gauge transformation is always generated by a multivalued function S with the multiplicity $4\pi g$.

In the Schrödinger theory, when we make a gauge transformation like (2.2) with the function S, the wave function gets multiplied by $e^{ieS/\hbar}$. This comes from the momentum variables $p_s = -i\hbar\partial/\partial x_s$ being added onto eA_s in the Schrödinger equation. In order that the wave function may be single-valued with S having the value (2.3), containing the uncertainty $4\pi g$, it is necessary that

$$4\pi eg = 2\pi n\hbar ,$$

where n is some integer. We can conclude that, if a monopole of strength g interacts with a charge e according to quantum mechanics then

$$eg = \frac{1}{2}n\hbar. \tag{2.4}$$

For a given e, there can only be quantized values of g, and for a given g there can only be quantized values of e.

One may discuss the problem without making a gauge transformation, as was done by the author.* The result is the same, the quantum condition (2.4) connecting e and g.

If one uses the experimental value of e for an electron

$$e^2 = \hbar/137 ,$$

one gets for the smallest g with n = 1

$$g = \frac{137}{2} e.$$

The smallest monopole value is thus much stronger than the electronic charge. This may account for why monopoles are so rare. With these values for e and g there is, of course, no symmetry between monopoles and charges.

It is desirable to have a field theory for electric charges and monopoles in interaction. With the theory in the form of an action principle one can apply standard rules for quantization.

---

*P. A. M. Dirac, Proc. Roy. Soc. A vol 133, p. 60 (1931).

If one works with point particles one meets the difficulty that the field quantities are not well-defined at the points where the particles are situated. The right-hand sides of equations (1.1) and (1.2) are thus not well-defined. One can avoid the difficulty by working with extended particles.

Consider a particle with a continuous charge distribution, with each element of charge moving in accordance with Lorentz's equation (1.1) with the local value of the $F_{\mu\nu}$. The elements will then be moving apart under the influence of their Coulomb repulsion, so the particle will be exploding. However, it lasts a short time, long enough for one to be able to discuss its equations of motion. The resulting theory is mathematically more satisfactory than any theory involving ill-defined quantities.

The monopole must be handled in the same way. It must be considered as an extended distribution of magnetic charge with each element moving according to (1.2). It will likewise be exploding. When we introduce the potentials we shall have equation (1.3) failing along a fiber stretching out from each element of magnetic charge. The string extending out from the monopole now has a definite thickness and is composed of a bundle of fibers, each fiber being a mathematical line.

The method of setting up an action principle for an extended particle has been given by the author.* It was there applied to a charged particle moving in the electromagnetic and gravitational fields. We shall here ignore the gravitational field and shall generalize the treatment to bring in monopoles.

---

*Dirac, GRG vol. 5, p. 741, 1974.

3. KINEMATICS

We shall be considering a continuous distribution of something in space-time and shall determine how it is changed when each bit of it is shifted from the point x^μ to $x^\mu + b^\mu$, the b^μ being small. The problem involves only kinematical relationships.

Let us first take a distribution of points in 4-dimensions, with density ρ, a scalar. With the shift we get, for the change in ρ at a particular place x^μ,

$$\delta\rho = -(\rho b^\nu)_{,\nu} . \qquad (3.1)$$

Now consider a dust in 3-dimensional space and suppose each speck to move along a world-line. We get a distribution of world-lines in 4-dimensional space. Their density is specified by a contravariant vector u^μ lying in the direction of the world-lines, such that the number intersecting an element of volume $dx^1dx^2dx^3$ at a certain time, equal to the number of specks in the volume, is

$$u^0dx^1dx^2dx^3. \qquad (3.2)$$

If each speck is provided with an element of mass m, we shall have mu^μ as the density of energy and momentum. Similarly if each speck is provided with an element of electric charge e, we shall have eu^μ as the charge density and current. The quantities m and e here are very small (mathematically infinitesimal) and are much smaller than the mass and charge of a physical particle.

We represent an extended physical particle by such a distribution of world-lines, with u^μ vanishing except within the volume of the particle.

With the dust shifted from x^μ to $x^\mu + b^\mu$, we get

$$\delta u^{\mu} = -(u^{\mu} b^{\nu})_{,\nu} + u^{\nu} b^{\mu}{}_{,\nu}. \qquad (3.3)$$

Here the first term corresponds to (3.1) and the second term is the extra effect coming from the rotation of the vectors u^{μ}.

If the dust is conserved, so that the world-lines are endless, we have

$$u^{\mu}{}_{,\mu} = 0. \qquad (3.4)$$

The formula (3.3) now reduces to

$$\delta u^{\mu} = (u^{\nu} b^{\mu} - u^{\mu} b^{\nu})_{,\nu}. \qquad (3.5)$$

This agrees with the result (2.3) of the earlier GRG paper.

The monopoles are also to be considered as extended particles. The string stretching out from a monopole then becomes a bundle of fibers, with one fiber stretching out from each element of magnetic charge. The fibers form a distribution of 2-dimensional sheets in 4-dimensions. Their density is specified by a 6-vector $w^{\mu\nu} = -w^{\nu\mu}$. Corresponding to (3.2), the sheets intersect an element of volume $dx^1 dx^2 dx^3$ at a certain time in segments of fiber whose total is the contravariant vector (in 3 dimensions)

$$w^{s0} dx^1 dx^2 dx^3 \ (s=1,2,3) \quad .$$

The connection between the electromagnetic potentials and the field quantities is

$$F_{\mu\nu} = A_{\mu,\nu} - A_{\nu,\mu} + g\tilde{w}_{\mu\nu}, \quad , \qquad (3.6)$$

where the symbol ~ denotes the dual of a 6-vector and g is the element of magnetic charge associated with a fiber. One easily checks the relativistic formula (3.6) by applying it to one of the components of the magnetic field

$$H_{12}=A_{1,2}-A_{2,1}+gw^{03},$$

which shows the effect of a fiber in disturbing equation (1.3).

When the sheets are shifted from x^μ to $x^\mu+b^\mu$, the change in $w^{\mu\nu}$ is

$$\delta w^{\mu\nu}=-(w^{\mu\nu}b^\lambda)_{,\lambda}+w^{\lambda\nu}b^\mu{}_{,\lambda}+w^{\mu\lambda}b^\nu{}_{,\lambda}. \tag{3.7}$$

This is the natural generalization of (3.1) and (3.3).

The condition for the sheets to have no edges is

$$w^{\mu\nu}{}_{,\nu}=0 \tag{3.8}$$

corresponding to (3.4). This enables one to write the variation (3.7) as

$$\delta w^{\mu\nu}=-(w^{\mu\nu}b^\lambda+w^{\nu\lambda}b^\mu+w^{\lambda\mu}b^\nu)_{,\lambda}. \tag{3.9}$$

Note that the quantity $w^{\mu\nu}b^\lambda+w^{\nu\lambda}b^\mu+w^{\lambda\mu}b^\nu$ is antisymmetrical between λ, μ and ν.

When condition (3.8) does not hold, we put

$$w^{\mu\nu}{}_{,\nu}=q^\mu. \tag{3.10}$$

A non-vanishing q^μ occurs at the ends of the fibers, which is where there is magnetic charge. The q^μ defined by

(3.10) determine the magnetic charge and current, in a similar way to u^μ for the electric charge and current. The definition (3.10) leads to

$$q^\mu{}_{,\mu} = 0,$$

the conservation law for magnetic charge, which now follows automatically from the magnetic charge being associated with the ends of fibers.

When (3.8) does not hold, (3.9) must be amended to

$$\delta w^{\mu\nu} = -(w^{\mu\nu} b^\lambda + w^{\nu\lambda} b^\mu + w^{\lambda\mu} b^\nu)_{,\lambda} + q^\nu b^\mu - q^\mu b^\nu . \qquad (3.11)$$

This leads to

$$\delta q^\mu = (q^\nu b^\mu - q^\mu b^\nu)_{,\nu}, \qquad (3.12)$$

which corresponds to the formula (3.5) for δu^μ.

4. THE ACTION PRINCIPLE

The contribution of the inertia of a particle to the action is

$$I_m = -m \int (u^\mu u_\mu)^{\frac{1}{2}} d^4 x . \qquad (4.1)$$

On varying the u^μ we get

$$\delta I_m = -m \int (u^\nu u_\nu)^{-\frac{1}{2}} u_\mu \delta u^\mu d^4 x . \qquad (4.2)$$

Let us put

$$(u^\nu u_\nu)^{-\frac{1}{2}} u_\mu = v_\mu \ .$$

Here v_μ is the velocity vector, satisfying

$$v^\mu v_\mu = 1. \tag{4.3}$$

It is defined, of course, only where $u^\nu \neq 0$, that is, inside the particle. Equation (4.3) leads to

$$v^\mu v_{\mu,\nu} = 0. \tag{4.4}$$

Using the value for δu^μ given by (3.5) in (4.2), we get

$$\begin{aligned}\delta I_m &= -m\int v_\mu (u^\nu b^\mu - u^\mu b^\nu)_{,\nu} d^4x \\ &= m\int v_{\mu,\nu}(u^\nu b^\mu - u^\mu b^\nu) d^4x \\ &= m\int u^\nu v_{\mu,\nu} b^\mu d^4x \end{aligned} \tag{4.5}$$

with the help of (4.4).

If the particle has an electric charge, it contributes to the action

$$I_e = -e\int A_\mu u^\mu d^4x. \tag{4.6}$$

Thus

$$\begin{aligned}\delta I_e &= -e\int \delta A_\mu u^\mu d^4x - e\int A_\mu (u^\nu b^\mu - u^\mu b^\nu)_{,\nu} d^4x \\ &= -e\int \delta A_\mu u^\mu d^4x + e\int (A_{\mu,\nu} - A_{\nu,\mu}) u^\nu b^\mu d^4x. \end{aligned} \tag{4.7}$$

Let us assume that a string and a charged particle never overlap, so that $w^{\mu\nu} = 0$ inside a charged particle.

Then from (3.6) the second term of (4.7) becomes

$$e\int F_{\mu\nu}u^{\nu}b^{\mu}d^{4}x. \tag{4.8}$$

We get the equations of motion for the particle by adding this term to (4.5) and equating the total coefficient of b^{μ} to zero. The result is

$$mu^{\nu}v_{\mu,\nu}+eF_{\mu\nu}u^{\nu}=0. \tag{4.9}$$

Here we may replace u^{ν} in each term by v^{ν}, which is proportional to it, and then use

$$d^2v_{\mu}/ds^2 = v_{\mu,\nu}v^{\nu}.$$

The result is of the form of Lorentz's equation (1.1), and shows that each element of charge in the particle moves correctly.

We take the action for the electromagnetic field to be the usual expression

$$I_F=-\tfrac{1}{4}\int F_{\mu\nu}F^{\mu\nu}d^4x,$$

with neglect of a factor $(4\pi)^{-1}$. Thus

$$\begin{aligned}\delta I_F&=-\tfrac{1}{2}\int F^{\mu\nu}\delta F_{\mu\nu}d^4x\\ &=-\int F^{\mu\nu}\delta A_{\mu,\nu}d^4x+\tfrac{1}{2}g\int \tilde{F}_{\mu\nu}\delta w^{\mu\nu}d^4x\\ &=\int F^{\mu\nu}{}_{,\nu}\delta A_{\mu}d^4x+\tfrac{1}{2}g\int \tilde{F}_{\mu\nu}\{-(w^{\mu\nu}b^{\lambda}+w^{\nu\lambda}b^{\mu}+w^{\lambda\mu}b^{\nu})_{,\lambda}\\ &\qquad +q^{\nu}b^{\mu}-q^{\mu}b^{\nu}\}d^4x\end{aligned} \tag{4.10}$$

with the help of (3.11). We get the electromagnetic field equations by equating to zero the coefficient of δA_μ in $\delta(I_e+I_F)$. This gives

$$F^{\mu\nu}{}_{,\nu}-eu^\mu = 0, \qquad (4.11)$$

which is in agreement with the Maxwell theory for the charge-current density eu^μ.

The first term in the {} in (4.10) gives

$$-\tfrac{1}{2}g\int \tilde{F}_{\mu\nu,\lambda}(w^{\mu\nu}b^\lambda+w^{\nu\lambda}b^\mu+w^{\lambda\mu}b^\nu)d^4x$$

$$=-\tfrac{1}{2}g\int(\tilde{F}_{\mu\nu,\lambda}+\tilde{F}_{\lambda\mu,\nu}+\tilde{F}_{\nu\lambda,\mu})w^{\mu\nu}b^\lambda d^4x. \qquad (4.12)$$

The quantity $(\tilde{F}_{\mu\nu,\lambda}+\tilde{F}_{\lambda\mu,\nu}+\tilde{F}_{\nu\lambda,\mu})$ evidently vanishes when two of the suffixes μ,ν,λ are equal. To see its value when μ,ν,λ are all different, take the case when they equal 1,2,3. Then

$$\tilde{F}_{12,3}+\tilde{F}_{23,1}+\tilde{F}_{31,2}=F^{03}{}_{,3}+F^{01}{}_{,1}+F^{02}{}_{,2}.$$

This equals the charge density, according to the field equation (4.11), and vanishes from the condition that the string must not overlap a charged particle. The other components of (4.12) vanish similarly.

The vanishing of this part of δI_F means that there are no equations of motion for the strings. The strings can move arbitrarily, subject to their ends being anchored at the monopoles, and subject to the condition that they must not move through charges.

The remainder of δI_F is

$$g\int \tilde{F}_{\mu\nu}q^\nu b^\mu d^4x. \qquad (4.13)$$

Note the similarity of this expression to (4.8).

For a monopole with an element of mass m' attached to the end of each fiber, there will be an inertial term like (4.1),

$$I_{m'} = -m' \int (q^{\mu} q_{\mu})^{\frac{1}{2}} d^4x. \tag{4.14}$$

We must vary this, add on the term (4.13), and equate the total coefficient of b^{μ} to zero. The calculation goes just the same as for an electric charge, and the result is the same as Lorentz's equation (1.1) with m' for m, g for e, and $\tilde{F}_{\mu\nu}$ for $F_{\mu\nu}$. This checks that the action principle gives correctly the motion of each element of magnetic charge.

It should be noted that the action does not contain any term giving directly the interaction of the monopole and the electromagnetic field, corresponding to the term (4.6) for the electric charge. The force on the monopole comes simply from the monopole being constrained to lie at the end of a string and the string variables occuring in the expression for the action of the field.

When one has an action principle one can introduce a Hamiltonian and can pass to the quantum theory. The Hamiltonian provides a definite expression for the total energy. People have sometimes doubted whether a monopole that has lost its kinetic energy will really be trapped by a ferromagnetic, but the existence of a total energy shows that it must be trapped, provided its rest-mass is sufficient, so that its zero-point motion does not take it beyond the range of the ferromagnetic.

ELEMENTARY PARTICLES IN THE GENERALIZED THEORY OF GRAVITATION

Behram Kurşunoğlu

Center for Theoretical Studies

University of Miami, Coral Gables, Fla. 33124

ABSTRACT

This paper contains, for the time independent spherically symmetric fields, various regular solutions of the field equations. An elementary particle structure consists of the entire spectrum of magnetic charges g_n, $n=0,1,2,\ldots$, with alternating signs where $g_n \to 0$ for $n \to \infty$ and where $\sum_{n=0}^{\infty} g_n = 0$. The screening caused by the stratified distribution generates short range magnetic forces. The strength of the coupling between the field and particle is described by $e^2+g_n^2$ where $n=\infty$ corresponds to the distances of the order of a Compton wave length $\frac{\hbar}{Mc}$. The observed mass M of particle or antiparticle is obtained, as a consequence of the equations of motion, in the form $Mc^2 = \frac{1}{2} mc^2 + 2E_s$, where E_s is the finite selfenergy of particle (or antiparticle) and where m and E_s have opposite signs. The "bare gravitational mass" m, obtained as a constant of integration, is estimated to be of the order of 10^{21} Mev. The spectrum of fundamental lengths r_{on} $[= \frac{\sqrt{(2G_0)}}{c^2} \sqrt{(e^2+g_n^2)} \sim 10^{-33}\text{cm}]$ measure the deviation of the theory from general relativity. The selfenergy E_s in the limit $r_{on} = 0$ tends to infinity and the solutions reduce to the corresponding spherically symmetric solutions in general relativity. The spin $\frac{1}{2}$ of an elementary particle is found to be the result of its neutral magnetic structure and the latter exists only for nonvanishing $\Gamma^{\rho}_{[\mu\nu]}$, the antisymmetric part of the affine connection. The two states of spin correlate with the two possible sequences of signs of g_n, i.e g_n and $- g_n$, $n=0,1,2,....$.

For the solutions where e=0 the symmetries of charge conjugation and parity are not conserved. The latter lead to the assumption of small masses for the two neutrinos ν_e, ν_μ. Conservation of the electric charge multiplicity i.e the existence of -1, +1, 0 units of electric charge, is found to be the basis for the existence of four massive fundamental particles p, e, ν_e, ν_μ and the corresponding antiparticles $\bar{p}, e^+, \bar{\nu}_e, \bar{\nu}_\mu$. Based on a new concept of "vacuum" predicted by the theory it may be possible to construct all other elementary particles as bound or resonance states of the "fundamental quartet" p, e, ν_e, ν_μ and the "antiquartet" $\bar{p}, e^+, \bar{\nu}_e, \bar{\nu}_\mu$.

1. INTRODUCTION

This paper is a sequel to an earlier one[1] hereafter to be referred to as (I). The paper (I) contains the author's version (versus the Einstein[2] and Schrödinger[3] theories) of the nonsymmetric generalization of general relativity, proposed over 20 years ago. In (I) we were able, for the first time, to obtain some of the spherically symmetric solutions of the field equations. These solutions are regular everywhere. It was furthermore found that all fundamental particles, in addition to their usual electromagnetic properties, carry a neutral magnetic core (with different distributions for different particles) associated with a short range field. The short range character of the field is due to magnetic charge screening arising from the distribution of magnetic charge density in an infinite sequence of stratified layers of magnetic charge densities, the signs of which alternate (see Fig. 1). The magnetic charge predicted by this theory is in no way related to the magnetic monopole theory of Dirac[4] producing a long range field and quantized according to $ge = \frac{1}{2}n\hbar e$, or to the magnetic charge theories of Schwinger[5] for the structure of elementary particles. The new magnetic charge g is carried by all elementary particles and anti-particles as described above. The charge g has an infinite spectrum of values g_n, n=0,1,2,3,.... which correspond to the amounts of magnetic charge contained in each of the stratified layers along with their alternating signs.

In (I) it was shown that for $g = 0$ or $g_n = 0$ the solutions, at distances large compared to the fundamental length r_{on}, reduce to the solutions of general relativity and classical electrodynamics and that they are no longer regular everywhere. Thus using a "magnetic" basis of matter in terms of the new magnetic charge g with the properties described above and in (I) we can hope to establish a new approach to elementary particle physics.

In this series of papers we shall discuss these interesting consequences for the fundamental interactions of elementary particles associated with the finiteness of the self-energy. It is pleasing to see that a theory without any infinities does, in fact, provide a unification of all fundamental interactions. The magnetic charge of this theory is the novel idea which was

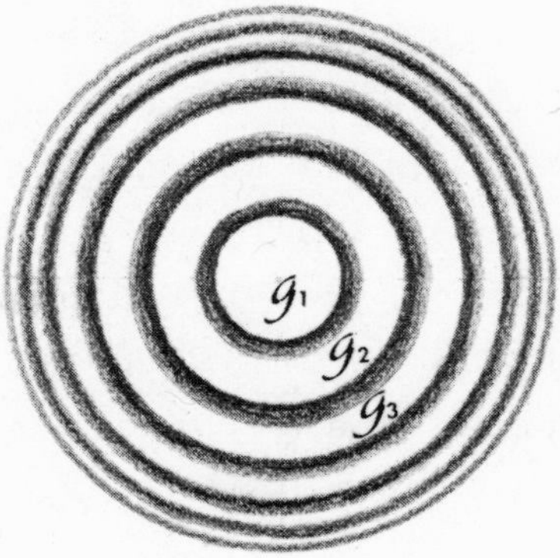

Figure 1

Fig. 1. The partial magnetic charges $|g_{n-1}| > |g_n|$ are distributed with alternating signs in the stratified layers n=0,1,2,... which have the total magnetic charge value $(-1)^{s+\tau} \sum_{n=0}^{\infty} g_n = 0$. The values s=0, s=1, or the sequences of positive and negative signs of g_n correlate with spin up (or down) and spind down (or up) states, respectively. The shaded bands describe the uncertainties in the partial charges g_n or the uncertainties in the location of the neutral magnetic surfaces. The distribution extends to distances of the order of $r_c \sim \frac{\hbar}{Mc}$, where r_c is the indeterminate distance of the "magnetic horizon" from the particle's center.

missing in conventional theories. However the present theory deviates from Einstein's 1916 general relativity in the way quantum theory deviates from classical theory. The degree of deviation of this theory from general relativity plus classical electrodynamics is measured by the size of a fundamental length r_o which is a function of the electric and magnetic charges. In (I) the length r_o

was shown to be an infinite spectrum of lengths given by

$$r_{on}^2 = \sqrt{(\ell_{on}^4 + \lambda_{on}^4)} \quad , \tag{1.1}$$

where

$$\ell_{on}^2 = \frac{2G_o}{c^4} |g_n| \sqrt{(e^2+g_n^2)} \quad , \quad n=0,1,2,3,\ldots, \tag{1.2}$$

$$\lambda_{on}^2 = \frac{2G_o}{c^4} |e| \sqrt{(e^2+g_n^2)} \quad , \tag{1.3}$$

G_o is the gravitational constant, e and c represent the unit of electric charge and speed of light, respectively. As n tends to infinity g_n tends to zero, r_{on} tends to a smaller limiting value

$$r_{on} \to \frac{2G_o}{c^4} e^2 \quad ,$$

and the field equations approach the field equations of general relativity. In fact for distances $r \gg r_{on}$ the generalized theory of gravitation reduces to the field equations of general relativity and classical electrodynamics of free fields.

Both the lengths ℓ_o and λ_o in (1.1) were obtained in (I) as constants of integration. Thus the existence of a correspondence principle (i.e. the $r_o=0$ limit yields general relativity plus the electromagnetic fields) provides a powerful basis for the unique and unambigous physical interpretation of the theory. The four fundamental conserved currents j_e^μ (charged electromagnetic current vector density), j_o^μ (neutral electromagnetic vacuum current vector density), s^μ (neutral intrinsic magnetic axial current density), ζ^μ (neutral intrinsic magnetic axial vacuum current density) play a crucial role in the physical interpretation of the theory. These currents are derived from the field equations and can be determined only if we know the solutions of the field equations. The currents cannot be prescribed arbitrarily. The neutral electric and magnetic currents j_o^μ and ζ^μ and their corresponding fields describe the vacuum pairs associated with the field. The theory predicts only four massive particles (and their antiparticles) which hopefully will be identified as proton, electron and two chargeless (i.e. e=0) particles. The latter two have the same symmetry properties as the two neutrinos ν_e and ν_μ. The current experimental measurements have provided upper limits for ν_e and ν_μ masses but did not rule out the possibility of the neutrinos having finite rest masses. The restrictions on the four currents mentioned above applies also to the total energy density of the field since it is also determined in terms of the solutions of the field equations. A further important result concerns the concept of the spin of an elementary particle. We shall demonstrate in this paper that the $\frac{1}{2}$ and $-\frac{1}{2}$ values of spin is a consequence of stratified distribution and symmetry properties of the magnetic charge in an elementary particle.

We shall also derive the equations of motion of particles from an action principle of the field. The motion of the particles results from varying the extremum action function obtained by substituting the field equations in the action function of the field. This new way of obtaining the equations of motion from varying the extremum action function is a consequence of the regularity of the fields everywhere since the latter property leads after the volume integration to the appearance of the particle mass in the action function. However, because of the indeterminacy or spread in the particle boundary the resulting trajectory cannot be defined as sharply as in classical theory. This spread in the particle trajectory is a quantum like behavior though in the present case it results from effect of the requirement of relativistic invariance on an object extended in space and time. In general relativity and in electrodynamics the particles move along the points where the field assumes an infinite value.

An exact treatment of the field equations, at least in the spherically symmetric case, leads to various interesting results. A linearization of any form would be quite useless since none of the results of this paper and the previous paper (I) could have been obtained without an exact treatment of the field equations. Spin, charge, mass, structure, multiplicity and the laws of motion of elementary particles are consequences of a nonlinear field theory. In this paper the concept of nonlinearity will be regarded as a basic principle of physics.

2. SYMMETRIES OF THE FIELD EQUATIONS

The special case of time independent spherically symmetric field equations contains a wealth of information on the physics of the generalized theory of gravitation. They were derived in (I) in the form of equations (4.13)-(4.16). These equations for the four functions $\exp(\rho)$, $\exp(u)$, υ and Φ can readily be further simplified into the form

$$\frac{1}{2}r_o^2 f[fS\exp(\rho)\Phi']' = R^2\cos\Phi + (-1)^s \ell_o^2 \sin\Phi \qquad (2.1)$$

$$\frac{1}{2}r_o^2 f[fS\ \exp(\rho)\rho']' = -R^2\sin\Phi + (-1)^s \ell_o^2\cos\Phi + \exp(\rho)\ , \qquad (2.2)$$

$$\frac{1}{2}r_o^2 f[f\ \exp(\rho)S']' = \left(1 - \frac{\sin\Phi}{\cosh\Gamma}\right)\exp(\rho)\ , \qquad (2.3)$$

$$\rho''+\rho'\frac{f'}{f} + \frac{1}{2}(\rho'^2+\Phi'^2) = 0\ , \qquad (2.4)$$

where prime indicates differentiation with respect to r and where

$$f = \upsilon\cosh\Gamma\ ,\quad S = \frac{\exp(u)}{\cosh^2\Gamma}\ , \qquad (2.5)$$

$$\cosh\Gamma = (R^2+r_o^2)\exp(-\rho)\ ,\quad R^2+r_o^2 = \sqrt{[\exp(2\rho)+\lambda_o^4]}\ , \qquad (2.6)$$

$$\ell_o^2 = q^{-1}|g|\ ,\quad \lambda_o^2 = q^{-1}|e|\ .$$

The spherically symmetric form of the nonsymmetric field variables

$$\hat{g}_{\mu\nu} = g_{\mu\nu} + q^{-1}\Phi_{\mu\nu}$$

is, as derived in (I), given by

$$[\hat{g}^s_{\mu\nu}] = \begin{bmatrix} -\frac{e^{-u}}{\upsilon^2} & 0 & 0 & \frac{\tanh\Gamma}{\upsilon} \\ 0 & -e^{\rho}\sin\Phi & e^{\rho}\cos\Phi\sin\theta & 0 \\ 0 & -e^{\rho}\cos\Phi\sin\theta & -e^{\rho}\sin\Phi\sin^2\theta & 0 \\ -\frac{\tanh\Gamma}{\upsilon} & 0 & 0 & e^{u} \end{bmatrix}$$

The fundamental length r_o was calculated in (I) in terms of the constant λ_o and ℓ_o as

$$r_o^2 = \sqrt{(\ell_o^4+\lambda_o^4)} = \frac{2G_o}{c^4}(e^2+g^2) \quad , \quad r_o^2 q^2 = \frac{c^4}{2G} \qquad (2.7)$$

where

$$\ell_o^2 = q^{-1}|g| = \frac{2G_o}{c^4}|g|\sqrt{(e^2+g^2)} \quad , \qquad (2.8)$$

$$\lambda_o^2 = q^{-1}|e| = \frac{2G_o}{c^4}|e|\sqrt{(e^2+g^2)} \quad . \qquad (2.9)$$

We thus have four field equations (2.1)-(2.4) for the four unknown functions $\rho(r)$, $u(r)$, $\upsilon(r)$, and $\Phi(r)$. It is interesting to observe that the equation (2.4) does not, explicitly, depend on the constants r_o, ℓ_o, λ_o. However, the relation of (2.4) to the first three equations can be seen by differentiating the equation (2.2), using (2.1) and (2.3), and by deriving the equation

$$\frac{d}{dr}\left[S^2\exp(2\rho)(2\rho''+2\rho'\frac{f'}{f}+\rho'^2+\Phi'^2)\right] = 0 \quad . \qquad (2.10)$$

The result (2.10) together with the equation (2.4) is related to the compatibility of the field equations (2.1)-(2.4).

The field equations (2.1)-(2.2), through the factor $(-1)^s\ell_o^2$, depend on the sign of the magnetic charge g. Now, by using the results (4.18), (4.19) of (I) the explicit dependence of the field equations on $(-1)^s$ can be represented as a new degree of freedom of the field. Thus, by using the new set of angle functions

$$\Phi_{ns}(r) = \pm\,(2n+s)\pi+(-1)^s\Phi(r) \quad , \qquad (2.11)$$

where now

$$0 \leqq \Phi(r) \leqq \frac{\pi}{2} \quad , \tag{2.12}$$

and where s=0,1 and n=0,1,2,... . , we can rewrite the first three of the field equations in the form

$$\frac{1}{2}r_o^2 f[fS\exp(\rho)\Phi'_{ns}]' = R^2\cos\Phi_{ns} + \ell_o^2\sin\Phi_{ns} , \tag{2.13}$$

$$\frac{1}{2}r_o^2 f[fS\exp(\rho)\rho']' = -R^2\sin\Phi_{ns} + \ell_o^2\cos\Phi_{ns} + \exp(\rho), \tag{2.14}$$

$$\frac{1}{2}r_o^2 f[f\exp(\rho)S']' = [1 - \exp(\rho)\frac{\sin\Phi_{ns}}{R^2+r_o^2}]\exp(\rho) , \tag{2.15}$$

where Φ_{ns} represent the ∞^2 distinct solutions of the field equations and where we have used the relations

$$\sin\Phi = \sin\Phi_{ns} , \cos\Phi=(-1)^s\cos\Phi_{ns} , \Phi'=(-1)^s\Phi'_{ns} .$$

It was pointed out in (I) that compatibility with the light cone partition of space-time events requires that all the solutions of the field equations must fulfill the condition

$$\sin\Phi \geqq 0 , \tag{2.16}$$

in order to insure positivity of the coefficients

$$-g_{22} = \exp(\rho)\sin\Phi , \quad g_{33} = g_{22}\sin^2\theta , \tag{2.17}$$

of the metric tensor $g_{\mu\nu}$. The fulfillment of the condition (2.16) confines all the solutions to the upper half of the Φ-plane where Φ is measured in the counterclockwise direction. Indeed, as was shown in (I), all the solutions satisfy (2.16) and there exist no solutions violating condition (2.16) provided only that the possibility of the solutions like $-\exp(\rho)$ are excluded. However, if negative solutions of the form $-\exp(\rho)$ are included then the condition (2.16) can be replaced by

$$\exp(\rho)\sin\Phi \geqq 0 \quad , \tag{2.18}$$

and in the Φ-plane, for the solution $-\exp(\rho)$, the angle functions $\Phi_{ns\tau}$ can also be measured in the clockwise direction. In this case we obtain ∞^4 distinct solutions satisfying (2.18), and no others exist. Thus the new sets of solutions are of the form

$$\Phi_{ns\tau}=\pm(2n+s+\tau)\pi+(-1)^s\Phi \ , \ \rho\pm(2n+\tau)i\pi=\rho_{n\tau} \ , \ 0\leqq\Phi(r)\leqq \frac{\pi}{2} \ , \tag{2.19}$$

where $\qquad \tau = 0,1 \quad ,$

and where now

$$\sin\Phi=(-1)^\tau\sin\Phi_{ns\tau} \ , \ \cos\Phi=(-1)^{s+\tau}\cos_{ns\tau} \ ,$$

$$\exp(\rho)=(-1)^\tau\exp(\rho_{n\tau}), \tag{2.20}$$

so that

$$\exp(\rho)\sin\Phi = \exp(\rho_{n\tau})\sin\Phi_{ns\tau} \quad ,$$

remains positive and unchanged. The field equations (2.1)-(2.4), in terms of the new solutions (2.19), can be replaced by

$$\frac{1}{2}r_o^2 f[fS\ \exp(\rho_{n\tau})\Phi'_{ns\tau}]' = R^2\cos\Phi_{ns\tau} + \ell_o^2\sin\Phi_{ns\tau}\ , \qquad (2.1')$$

$$\frac{1}{2}r_o^2 f[fS\ \exp(\rho_{n\tau})\rho'_{n\tau}]' = -R^2\sin\Phi_{ns\tau} + \ell_o^2\cos\Phi_{ns\tau} + \exp(\rho_{n\tau}), \qquad (2.2')$$

$$\frac{1}{2}r_o^2 f[f\ \exp(\rho_{n\tau})S']' = \exp(\rho_{n\tau})[1 - \frac{\exp(\rho_{n\tau})\sin\Phi_{ns\tau}}{R^2+r_o^2}]\ , \qquad (2.3')$$

$$2\rho''_{n\tau} + 2\rho'_{n\tau}\frac{f'}{f} + \rho'^2_{n\tau} + \Phi'^2_{n\tau} = 0\ . \qquad (2.4')$$

A further simplification of the field equations (2.1)-(2.4) can be achieved by introducing a new variable β by

$$dr = f\ d\beta\ . \qquad (2.21)$$

In terms of the new variable β the field equations can now be written as

$$\frac{1}{2}r_o^2\frac{d}{d\beta}[S\ \exp(\rho_{n\tau})\dot{\Phi}_{ns\tau}] = R^2\cos\Phi_{ns\tau} + \ell_o^2\sin\Phi_{ns\tau}, \qquad (2.22)$$

$$\frac{1}{2}r_o^2\frac{d}{d\beta}[S\ \exp(\rho_{n\tau})\dot{\rho}_{n\tau}] = -R^2\sin\Phi_{ns\tau} + \ell_o^2\cos\Phi_{ns\tau} + \exp(\rho_{n\tau}), \qquad (2.23)$$

$$\frac{1}{2}r_o^2\frac{d}{d\beta}[\exp(\rho_{n\tau})\ \dot{S}] = \exp(\rho_{n\tau})[1 - \frac{\exp(\rho_{n\tau})\sin\Phi_{ns\tau}}{R^2+r_o^2}]\ , \qquad (2.24)$$

$$2\ddot{\rho}_{n\tau} + \dot{\rho}^2_{n\tau} + \dot{\Phi}^2_{ns\tau} = 0\ , \qquad (2.25)$$

where

$$\dot{\rho} = \frac{d\rho}{d\beta} \quad ,$$

and where now S, ρ and Φ can be regarded as functions of the new variable β. In view of the invariance of the field equations (2.1)-(2.4) under the transformation $f \to - f$, the definition (2.21) implies that the range of the new variable β extends from $-\infty$ to ∞. It is interesting to note that the equations resulting from (2.1)-(2.4) by taking $f = \pm 1$ are, formally, the same (except being functions of r whose range extends from 0 to ∞) as the equations (2.22)-(2.25).

The new forms (2.22)-(2.25) of the field equations involve only three unknown functions S, ρ, Φ satisfying four field equations. This kind of overdetermination of the field variables S, ρ, Φ might imply a spurious character for the eliminated field variable f. Actually this interesting property of the field equations is, as will be demonstrated, a virtue since it establishes relations between the constants r_o, ℓ_o and λ_o. The field equations are compatible only through definite relationships between these constants.

Now, the field equations (2.1')-(2.4') are independent of the sign of the electric charge and therefore the solutions (2.19) refer to both positive and negative electric charge irrespective of the positive and the negative energy states. There are thus two signs for electric charges for a particle as well as for an antiparticle. Furthermore, as seen from (9.2) of (I) the energy density has a linear dependence on f and therefore it changes sign under the transformation

$$f \to - f \quad , \tag{2.26}$$

under which the field equations are unchanged. Hence the field equations have both positive and negative energy solutions for particles with positive as well as with negative sign of electric charge. The role of the symmetry (2.26) for the field equations (2.1')-(2.4') is taken over in the new form (2.22)-(2.25) of the field equations by the invariance under $\beta \to -\beta$.

The spherically symmetric fields and the corresponding electric and magnetic charge densities are given by

$$\mathcal{E}_e = \frac{q}{\upsilon} \tanh\Gamma = \frac{q\sinh\Gamma}{f} = \frac{\pm e}{\upsilon(R^2+r_o^2)} , \qquad (2.27)$$

for the charged electric field,

$$\mathcal{E}_o = q\, r_o^2\, f \sinh\Gamma\, \rho' S' \qquad (2.28)$$

$$= q\, r_o^2 \frac{1}{f} \sinh\Gamma\, \dot{\rho}\, \dot{S} = r_o^2\, E_e\, \dot{\rho}\, \dot{S} ,$$

for the neutral electric field, and

$$\mathcal{B} = q \exp(\rho)\cos\Phi \sin\theta , \qquad (2.29)$$

for the neutral magnetic field,

$$\mathcal{H}_o = \frac{q}{\upsilon} \frac{\cos\Phi}{\cosh\Gamma} = \frac{q\cos\Phi}{f} , \qquad (2.30)$$

for the neutral vacuum magnetic field. The corresponding charge densities are

$$j_e^4 = \frac{q}{4\pi} (\exp(\rho)\tanh\Gamma\sin\Phi)'\sin\theta = \frac{\pm e}{4\pi} \frac{1}{f} \frac{d}{d\beta} \left(\frac{\sin\Phi}{\cosh\Gamma}\right) \sin\theta , \qquad (2.31)$$

$$j_o^4 = \frac{\pm e r_o^2}{4\pi} \frac{1}{f} \frac{d}{d\beta} \left(\frac{\sin\Phi}{\cosh\Gamma} \dot{\rho} \dot{S}\right) \sin\theta \quad , \qquad (2.32)$$

for the electric charge densities, and

$$\mathcal{s}^4 = \frac{q}{4\pi} (\exp(\rho)\cos\Phi)' \sin\theta = \frac{q}{4\pi} \frac{1}{f} \frac{d}{d\beta} [\exp(\rho)\cos\Phi]\sin\theta \ , \qquad (2.33)$$

$$\zeta^4 = \frac{q}{4\pi} \left(\frac{\exp(\rho)\cos\Phi\sin\Phi}{\cosh\Gamma}\right)' \sin\theta = \frac{q}{4\pi} \frac{1}{f} \frac{d}{d\beta} \left[\frac{\exp(\rho)\cos\Phi\sin\Phi}{\cosh\Gamma}\right]\sin\theta \qquad (2.34)$$

for the magnetic charge densities, where

$$\int j_e^4 \, dr d\theta d\phi = \pm e \ , \quad \int j_o^4 \, dr d\theta d\phi = 0 \ , \qquad (2.35)$$

and

$$\int \mathcal{s}^4 dr d\theta d\phi = 0 \quad , \quad \int \zeta^4 dr d\theta d\phi = 0 \quad . \qquad (2.36)$$

We note that the r-integrations in (2.36) are carried out over the interval $(0,r_c)$ where r_c represents the indeterminate distance of the <u>magnetic horizon</u> (where $g=0$) from the origin. In this case, for the functions $\Phi(r)$ and $\rho(r)$, we have the relations

$$\Phi_{ns\tau}(r_c) = \tfrac{1}{2}\pi(-1)^s \pm (2n+s+\tau)\pi \quad , \qquad (2.37)$$

$$\rho_{n\tau}(r_c) = \mathit{Ln}\ r_c^2 \pm i(2n+\tau)\pi \ , \ n=0,1,2,\ldots \ , \qquad (2.38)$$

where $\Phi(r_c) = \frac{\pi}{2}$ refers to the critical value of the angle function pertaining to the exterior solutions of the field equations.

The invariance of the field equations under the

transformation

$$\Gamma \to -\Gamma \quad , \tag{2.39}$$

implies electric charge multiplicity invariance (i.e existence of +e, −e, 0 where the latter occurs for $\Gamma=0$) which leaves $\mathcal{B}$, $\mathcal{H}_o$, s^4, and ζ^4 unchanged and reverses the signs of $\mathcal{E}_e$, $\mathcal{E}_o$, j_e^4, j_o^4. The sign change of energy under (2.26) leads to the change of signs of $\mathcal{E}_e$ and $\mathcal{E}_o$. Hence, because of $f \to -f$ invariance, it follows that particles and antiparticles carry equal and opposite signs of electric charge. However, under (2.26) the neutral magnetic field $\mathcal{B}$ and magnetic charge density s^4 remain unchanged, and therefore particles and antiparticles can have the same magnetic field and magnetic charge density.

Now, if the transformations of (2.26) (i.e$(-1)^{\varepsilon}$, $\varepsilon=0,1$) and (2.39) are followed by a change of magnetic charge sign by spin inversion (i.e$(-1)^{s}$) or by parity inversion (i.e$(-1)^{\tau}$) then we obtain the results

$$\mathcal{E}_e \to (-1)^{J+\varepsilon+\tau}\,\mathcal{E}_e \;,\quad \mathcal{E}_o \to (-1)^{J+\varepsilon+\tau}\,\mathcal{E}_o \;, \tag{2.40}$$

$$\mathcal{B} \to (-1)^{s}\,\mathcal{B} \;,\quad \mathcal{H}_o \to (-1)^{s}\,\mathcal{H}_o \;,$$

and

$$j_e^4 \to (-1)^{J+\varepsilon}\,j_e^4 \;,\quad j_o^4 \to (-1)^{J+\varepsilon}\,j_o^4 \;, \tag{2.41}$$

$$s^4 \to (-1)^{s}\,s^4 \;,\quad \zeta^4 \to (-1)^{s}\,\zeta^4 \;, \tag{2.42}$$

where J=0, 1 select positive and negative charge. The fields $\mathcal{E}_o$ and $\mathcal{H}_o$ represent the electric and magnetic fields of the vacuum pairs. The vacuum charge densities j_o^4 and ζ_o^4 change signs under charge conjugation and spin inversion, respectively. The latter statement applies also to the charge densities j_e^4 and s^4. The relations of $(-1)^s$ and $(-1)^\tau$ to spin and parity inversions, respectively, are discussed in sections 3 and 4 of the paper.

3. SPIN AND MAGNETIC CHARGE

The nonlinearity of the field equations does not preclude their invariance under the 2-dimensional linear unitary transformations of the group SU(2).

The right hand sides of the field equations (2.22) -(2.25) contain a rotation by an angle $\Phi_{ns\tau}$ around the "3-direction" of the "vector"

$$|V\rangle = \begin{bmatrix} R^2 \\ \ell_o^2 \\ \exp(\rho_{n\tau}) \end{bmatrix} , \tag{3.1}$$

where the rotation matrix is given by

$$A = \begin{bmatrix} \cos\Phi_{ns\tau} & \sin\Phi_{ns\tau} & 0 \\ -\sin\Phi_{ns\tau} & \cos\Phi_{ns\tau} & 0 \\ 0 & 0 & 1 \end{bmatrix} . \tag{3.2}$$

The real orthogonal transformations

$$|V'\rangle = A|V\rangle , \tag{3.3}$$

can be represented by a stereographic projection of the unit sphere on to the equatorial plane, the south pole of the sphere being the Center of projection. Thus if the point (a',b',o) is the image on the plane of the point (a,b,d) on the sphere and we write $\xi = a'+i\,b'$, then the relations for the projection are

$$a+ib = \frac{2\xi}{1+\bar{\xi}\xi}\ ,\quad a-ib = \frac{2\bar{\xi}}{1+\bar{\xi}\xi}\ ,\quad d = \frac{1-\bar{\xi}\xi}{1+\bar{\xi}\xi}\ , \qquad (3.4)$$

where

$$a^2+b^2+d^2 = 1$$

and

$$a = \frac{R^2}{V^2}\ ,\quad b = \frac{\ell_o^2}{V^2}\ ,\quad d = \frac{\exp(\rho_{n\tau})}{V^2}\ , \qquad (3.5)$$

$$V^2 = \sqrt{[R^4+\ell_o^4+\exp(2\rho)]}\quad . \qquad (3.6)$$

In order to include the south pole of the sphere in the stereographic projection we use the homogeneous representation

$$\xi = \frac{u_1}{u_2}\quad .$$

Hence the relations (3.4) yield for the right hand sides of equations (2.22)-(2.24) the results

$$R^2\cos\Phi_{ns\tau}+\ell_o^2\sin\Phi_{ns\tau} = \langle v_{ns\tau}|\sigma_1|v_{ns\tau}\rangle\ , \qquad (3.7)$$

$$-R^2\sin\Phi_{ns\tau}+\ell_o^2\cos\Phi_{ns\tau} = \langle v_{ns\tau}|\sigma_2|v_{ns\tau}\rangle\ , \qquad (3.8)$$

$$\exp(\rho_{n\tau}) = \langle v_{ns\tau}|\sigma_3|v_{ns\tau}\rangle\ , \qquad (3.9)$$

where $\sigma_j (j=1,2,3)$ are the usual Pauli spin matrices, and where the spinor

$$|v_{ns\tau}\rangle = V \exp(-\tfrac{1}{2} i\sigma_3 \Phi_{ns\tau})|u\rangle , \qquad (3.10)$$

$$\langle u|u\rangle = 1 , \quad |u\rangle = \begin{bmatrix} u_1 \\ u_2 \end{bmatrix} .$$

Thus the rotations of the "sphere" of radius V are, for given s and τ, completely represented by the linear unitary transformations (3.10) which cover the double-valued representations of the proper rotation group. The equations (3.7)-(3.9) are invariant under the gauge transformation

$$|u'\rangle = \exp(i\alpha)|u\rangle , \qquad (3.11)$$

where α is a constant. The above properties of the field equations will be interpreted as a <u>spin degree of freedom</u> due to the magnetic structure of an elementary particle.

From the definition of $\Phi_{ns\tau}$ by (2.19) it follows that the unitary transformations (3.10) can be expressed as

$$|v_n{\uparrow}^+\rangle = (-1)^n V \exp(-\tfrac{1}{2} i\sigma_3 \Phi)|u\rangle , \qquad (3.12)$$

for spin up (s=0), even parity (τ=0) ,

$$|v_n{\downarrow}^-\rangle = -(-1)^n V \exp(\tfrac{1}{2} i\sigma_3 \Phi)|u\rangle , \qquad (3.13)$$

for spin down (s=1), odd parity (τ=1) ,

$$|v_n\uparrow^{-}\rangle = -i(-1)^n \sigma_3 \; V \exp(-\tfrac{1}{2} i\sigma_3 \Phi)|u\rangle \;, \qquad (3.14)$$

for spin up $(s=0)$, odd parity $(\tau=1)$,

$$|v_n\downarrow^{+}\rangle = i(-1)^n \; \sigma_3 \; V \exp(\tfrac{1}{2} i\sigma_3 \Phi)|u\rangle \;, \qquad (3.15)$$

for spin down $(s=1)$, even parity $(\tau=0)$ states, respectively. The phase factor $(-1)^n$ describes change of sign in the transition through the stratified layers of magnetic charges. A possible dependence of r_o on $\hbar$ may arise from a relationship where

$$g_n^2 = \gamma_n \; \hbar c \;, \qquad (3.16)$$

and where the numerical coefficients γ_n which satisfy

$$\lim_{n\to\infty} \gamma_n = 0 \;, \qquad (3.17)$$

may correspond to the solutions of the field equations (2.22)-(2.25) in the magnetic layers. Thus with the assumption (3.16) we obtain

$$r_{on}^2 = \hbar \frac{2G}{c^3} \left(\frac{e^2}{\hbar c} + \gamma_n\right) \;, \quad \frac{Q_n^2}{\hbar c} = \frac{e^2}{\hbar c} + \frac{g_n^2}{\hbar c} \;. \qquad (3.18)$$

The spin degree of freedom of an elementary particle is due to its magnetic charge structure as described here and in (I). Thus under spin flip (or time reversal) and also under parity transformation, the neutral magnetic charge content transforms as

$$0 = \sum_0^\infty g_n \rightarrow (-1)^{s+\tau} \sum_0^\infty g_n = 0 \;, \qquad (3.19)$$

and changes its sign. From the above results we see that the spin of an elementary particle arising from its magnetic content assumes a new physical status. The direction of the spin and the signs of the partial magnetic charges g_n are correlated. This result implies that in the coupling of particles and antiparticles at high energies with parallel spins and opposite parities, because of the equality of their partial magnetic charges, the annihilation process must slow down. In the case of anti-parallel spins, because of the opposite signs of their partial magnetic charges, the annihilation process is faster than in the previous case. Thus, the strongly bound magnetic layers or magnetic levels of an elementary particle with its antiparticle in a parallel spin state (i.e. same signs of their partial magnetic charges) results in bound states of a new particle of spin 1 and negative parity. The energy levels of the new particle are determined by electromagnetic, strong, weak (and even gravitational) interactions at short distances. For such systems (e.g. proton + anti-proton) the slow annihilation could lead to a discrete spectrum of photons. In fact the observed Ψ_n (n=1,2,3 so far) or J particles[6] could well be due to the formation of such bound states of particles and antiparticles.

From (3.18) we see that the term $g_n^2/\hbar c$ represents the magnetic coupling between the n-th layer of the particle and the field. Thus in the range of $(0,r_c)$ corresponding to each magnetic layer (n=0,1,2,...) there exists an infinite number of couplings between the field and the particle, the strength of which decreases as $n\to\infty$. Beyond $n\to\infty$ (i.e. beyond $r=r_c$) the coupling between the field and the particle is measured by the fine structure constant alone.

We have thus established that the field equations (2.22)-(2.25) are invariant under the following symmetry operations:

(i) $\Gamma \to -\Gamma$ corresponds to electric charge multiplicity C_m $(\equiv(-1)^J)$,

(ii) $\Phi \to \pi - \Phi$, $\ell_o^2 \to -\ell_o^2$ correspond to magnetic charge conjugation C_m,

(iii) $\Phi \to \pi + \Phi$, $e^\rho \to -e^\rho$ correspond to parity operation P $(\equiv(-1)^\tau)$,

(iv) $\Gamma \to -\Gamma$, $q \to -q$ correspond to time reflection operation T $(\equiv(-1)^s)$,

(v) $q \to -q$ corresponds to electric and magnetic charge conjugation $C_e\ C_m$,

(vi) $\upsilon \to -\upsilon$ corresponds to reversal of the sign of mass (and energy) $[\equiv(-1)^\varepsilon]$.

Some of these symmetries do, in a special situation, break down and are, therefore, not conserved. For example, if the constant of integration λ_o^2 vanishes (i.e if e=0) then the symmetries C_m and P are not conserved. This can be seen by noting that the field equations (2.1)-(2.4) in the limit $\lambda_o^2 = 0$ reduce to

$$\tfrac{1}{2}\ell_o^2\upsilon[\upsilon\exp(u+\rho)\Phi']' = [\exp(\rho)-\ell_o^2]\cos\Phi+(-1)^s\ell_o^2\sin\Phi \quad ,(3.20)$$

$$\tfrac{1}{2}\ell_o^2\upsilon[\upsilon\exp(u+\rho)\rho']'=-[\exp(\rho)-\ell_o^2]\sin\Phi+(-1)^s\ell_o^2\cos\Phi+\exp(\rho), \quad (3.21)$$

$$\tfrac{1}{2}\ell_o^2\upsilon[\upsilon\exp(u+\rho)u']' = \exp(\rho)(1 - \sin\Phi) \quad , \quad (3.22)$$

$$\rho'' + \rho'\frac{\upsilon'}{\upsilon} + \frac{1}{2}(\rho'^2+\Phi'^2) = 0 \quad . \quad (3.23)$$

In this case the solutions (2.19) involving both spin and parity do not satisfy the field equations (3.20)-(3.23). However, the solutions (2.11) without the parity quantum number do satisfy the field equations (3.20)-(3.23) and therefore they can be replaced by

$$\frac{1}{2}\ell_o^2 \upsilon[\upsilon \exp(u+\rho)\Phi'_{ns}]' = [\exp(\rho)-\ell_o^2]\cos\Phi_{ns} + \ell_o^2 \sin\Phi_{ns} , \tag{3.24}$$

$$\frac{1}{2}\ell_o^2 \upsilon[\upsilon \exp(u+\rho)\rho']' = -[\exp(\rho)-\ell_o^2]\sin\Phi_{ns} + \ell_o^2\cos\Phi_{ns} + \exp(\rho), \tag{3.25}$$

$$\frac{1}{2}\ell_o^2 \upsilon[\upsilon \exp(u+\rho)u']' = \exp(\rho)(1-\sin\Phi_{ns}) , \tag{3.26}$$

where, as before, the discrete indices n,s for the functions υ, u, ρ and for the ℓ_o have been suppressed and where

$$\Phi_{ns} = \pm (2n+s)\pi + (-1)^s\Phi, \; n=0,1,2,\ldots \; . \tag{3.27}$$

Thus the solutions of the equations (3.24)-(3.26) are not invariant under parity and magnetic charge conjugation operations. In this case we have only ∞^2 distinct solutions. Because of the invariance under $\upsilon \to -\upsilon$ we still have particle-antiparticle solutions, each with two spin states s=0 and 1 (i.e $(-1)^s$). These particles have no electromagnetic interactions and they couple through the magnetic charge alone. The absence of parity and charge conjugation symmetries for e=0 imply that these symmetries are of electromagnetic origin. Conversely, intrinsic parity and charge conjugation are space-time symmetries induced by electric charge. The chargeless particles have no continuum solutions occuring beyond the magnetic horizon since for $g_n = 0$ (ℓ_0=0) the equations (3.24)-(3.26), as a consequence of the relations $\lim_{\ell_o\to 0} \cos\Phi_{ns} = 0$, $\lim_{\ell_o\to 0} \sin\Phi_{ns} = 1$, are empty. Thus chargeless particles have short range interactions only, where the range of the force is $g_n^2/M_\nu c^2$ which, of course, has an indeterminacy specified by g_n. The chargeless massive particles predicted by this theory have the same symmetry properties as the two neutrinos ν_e and ν_μ.

4. EXACT SOLUTIONS

From (3.18) of (I) we obtain the relations

$$\cot\Phi = \pm\sqrt{(\tfrac{1}{2}\Omega+I)} \quad , \quad \tanh\Gamma = -\Lambda\tan\Phi \quad , \tag{4.1}$$

where

$$\Omega = \frac{1}{2}\Phi^{\mu\nu}\Phi_{\mu\nu}, \ \Lambda = \frac{1}{4} f^{\mu\nu}\Phi_{\mu\nu}, \ f^{\mu\nu} = \frac{1}{2\sqrt{(-g)}}\varepsilon^{\mu\nu\rho\sigma}\Phi_{\rho\sigma} \tag{4.2}$$

and where

$$I^2 = \frac{1}{4}\Omega^2 + \Lambda^2 \quad .$$

Hence we see that for spherically symmetric fields the functions Φ and Γ are scalar and pseudo-scalar quantities, respectively. These properties of Φ and Γ can be utilized in the solutions of the field equations (2.22) -(2.25). In fact all the solutions of the field equations can be classified in terms of the invariant function $\Phi(\beta)$. We shall first discuss the simple case where Φ is a constant. For the constant Φ the field equations

(2.22)-(2.25) become

$$R^2\cos\Phi_{ns\tau} + \ell_o^2\sin\Phi_{ns\tau} = 0 \quad , \tag{4.3}$$

$$\frac{1}{2}r_o^2\frac{d}{d\beta}[S\ \exp(\rho_{n\tau})\dot{\rho}_{n\tau}]=-R^2\sin\Phi_{ns\tau}+\ell_o^2\cos\Phi_{ns\tau}+\exp(\rho_{n\tau}), \tag{4.4}$$

$$\frac{1}{2}\ r_o^2\ \frac{d}{d\beta}\ [\exp(\rho_{n\tau})\dot{S}] = [1 - \frac{\exp(\rho_{n\tau})\sin\Phi_{ns\tau}}{R^2+r_o^2}]\exp(\rho_{n\tau}) \quad , \tag{4.5}$$

$$\frac{d^2}{d\beta^2}\ [\exp(\tfrac{1}{2}\rho_{n\tau})] = 0 \quad . \tag{4.6}$$

The solutions of the field equations (4.3)-(4.6) can further be classified according to the nature of ℓ_o^2. There are two possibilities where (i) $\ell_o = 0$, (ii) $\ell_o \neq 0$. For the case where $\ell_o = 0$ the equation (4.3) reduces to

$$R^2\cos\Phi = 0 \quad , \tag{4.7}$$

which will provide two classes of solutions subject to either one of the conditions:

$$R^2 = 0 \quad , \tag{4.8}$$

or

$$\cos\Phi = 0 \quad . \tag{4.9}$$

For the case (4.8) the equation (4.6) is solved by

$$\exp(\tfrac{1}{2}\rho) = \pm\ \beta \quad . \tag{4.10}$$

The equation (4.8) and the definition (2.6) of R^2 implies that $\exp(\rho)$ is independent of β. From equations (4.3), (4.4) and (4.8) we obtain

$$\exp(\rho) = 0 \tag{4.11}$$

and hence $\beta=0$. The definitions (2.5) and (2.6) together with the result (4.11) yields the solution

$$S = 0 \quad . \tag{4.12}$$

The equations (4.8), (4.10) and (4.11) imply that for the case (i) where $\ell_o^2 = 0$ (or $g = 0$), the magnetic charge g at the origin assumes the "eigenvalue" 0 and that

$$r_o^2 = \lambda_o^2 = \frac{2G}{c^4} e^2 \quad . \tag{4.13}$$

Thus for the case $\ell_o = 0$ where the conditions (4.8) applies the field equations (4.3)-(4.6) are solved by

$$\Phi = A = 0, \ \exp(\rho) = 0, \ S = 0, \ r_o^2 = \lambda_o^2, \ g = 0, \tag{4.14}$$

where A is a constant, and the solutions consist of the values of Φ, $\exp(\rho)$ and S at the origin $\beta = 0$. The choice of the constant A as zero is a consequence of the electric charge conservation. From the above results and the definitions (2.27)-(2.30) it follows that at $\beta=0$ (or $r=0$) the fields E_e, E_o, B_g, H_o vanish.

The field equations (4.4) and (4.5) for the case (i), as follows from the solution

$$\Phi(\beta) = \frac{\pi}{2} \quad , \tag{4.15}$$

of the equation (4.9), and the solutions

$$e^{\rho} = \beta^2 \ , \ f = \pm 1 \ , \ \beta = \pm r \qquad (4.16)$$

of the equation (4.6), become

$$r_o^2 \frac{d}{d\beta} (\beta S) = \beta^2 + r_o^2 - \sqrt{(\beta^4+\lambda_o^4)} \ , \qquad (4.17)$$

$$\frac{1}{2} r_o^2 \frac{d}{d\beta} (\beta^2 \dot{S}) = \beta^2 - \frac{\beta^4}{\sqrt{(\beta^4+\lambda_o^4)}} \ . \qquad (4.18)$$

As can easily be seen, the equation (4.18) is derivable from equation (4.17). The equation (4.17) can be solved in the form

$$S = 1 + \frac{A'}{\beta} + \frac{1}{3} \frac{\beta^2}{r_o^2} - \frac{1}{\beta r_o^2} \int \sqrt{(\beta^4+\lambda_o^4)} d\beta \ , \ \beta = \pm r \qquad (4.19)$$

where because of the definition of β by (2.21) and invariance of the field equations under the operation of sign reversal of mass (or energy) as defined by (2.26), the constant A' and the variable β must have the same sign. Thus the constant of integration A' can be positive or negative at the same time that β becomes positive or negative, respectively. For large β we obtain

$$S \to \exp(u) = 1 + \frac{A'}{\beta} + \frac{\lambda_o^4}{2r_o^2\beta^2} = 1 - \frac{2mG}{c^2\beta} + \frac{Ge^2}{c^4\beta^2} \ , \qquad (4.20)$$

where we used the definitions $\lambda_o^2 = q^{-1}e$, $r_o^2 q^2 = \frac{c^4}{2G}$, and where the constant of integration A' is identified, from the correspondence principle of the theory, as $\frac{2mG}{c^2}$. One basic difference of the solution (4.20) from Nordström's solution in general relativity arises from the fact that in the present case both signs of mass m are allowed. The integration in (4.19) can easily be carried out and one obtains the exact solution

$$\exp(u) = \frac{\beta^4+\lambda_o^4}{\beta^4}\left[1 - \frac{2mG}{c^2\beta} + \frac{1}{3}\frac{\beta^2}{\lambda_o^2} - \frac{1}{3}\frac{\sqrt{(\beta^4+\lambda_o^4)}}{\lambda_o^2} - \frac{2}{3}\frac{\lambda_o^2}{\beta}K(\beta)\right], \tag{4.21}$$

where

$$r_o^2 = \lambda_o^2 = \frac{2G}{c^4}e^2 \tag{4.22}$$

and where

$$K(\beta) = \int \frac{d\beta}{\sqrt{(\beta^4+\lambda_o^4)}} = \frac{1}{2\lambda_o}\int \frac{d\gamma}{\sqrt{(1-\frac{1}{2}\sin^2\gamma)}}$$

with

$$\gamma = \cos^{-1}\left[\frac{\lambda_o^2-\beta^2}{\lambda_o^2+\beta^2}\right] ,$$

is an elliptic integral of the first kind. For this solution also the "eigenvalue" of g corresponding to β is zero. However in this case as seen from <u>the invariant equation</u> (4.15) the distance β from the origin is <u>indeterminate</u>. Because of the invariance of the equation (4.15), and also of the field equations, it is not possible to sharply define the boundaries of an extended elementary particle of this theory. Thus the <u>indeterminacy</u> is a necessary consequence of the invariance principle. We may rewrite the equation (4.15) in the form

$$\Phi(\beta_c) = \Phi(r_c) = \frac{\pi}{2} , \tag{4.23}$$

where r_c is the indeterminate distance of the magnetic horizon from the origin. At and beyond the magnetic horizon we have the result $g = 0$. By using the above solutions we can obtain, using (2.27)-(2.30), the electric and magnetic fields in the form

$$E_e = \frac{\pm e(\pm 1)}{\beta^2} \ , \ E_o = \frac{4r_o^2}{\beta^2} E_e \ [\frac{mG}{c^2\beta} + \frac{\beta^3+\lambda_o^4 K(\beta)-\beta\surd(\beta^4+\lambda_o^4)}{3\lambda_o^2\beta}], \tag{4.24}$$

$$B = 0 \ , \ H_o = 0 \ . \tag{4.25}$$

For both of the above solutions corresponding to (4.8) and (4.9) the magnetic charge density s_4 vanishes. It is interesting to observe the mass dependence of the neutral electric field E_o and furthermore the fact that at and beyond the magnetic horizon the electric field is the usual Coulomb field. Both the neutral magnetic field B and the neutral vacuum field H_o vanish at magnetic horizon and beyond. The latter are due to the short range nature of the fields B and H_o.

The case where $\ell_o \neq 0$, and constant Φ is described by the field equations

$$R^2\cos\Phi_{ns\tau} + \ell_o^2\sin\Phi_{ns\tau} = 0 \ , \tag{4.26}$$

$$- R^2\sin\Phi_{ns\tau} + \ell_o^2\cos\Phi_{ns\tau} + \exp(\rho_{n\tau}) = 0 \ , \tag{4.27}$$

$$\frac{1}{2} r_o^2 \frac{d}{d\beta} [\exp(\rho_{n\tau})\dot{S}] = [1 - \frac{\exp(\rho_{n\tau})\sin\Phi_{ns\tau}}{R^2+r_o^2}] \exp(\rho_{n\tau}), \tag{4.28}$$

$$\frac{d^2}{d\beta^2} [\exp(\tfrac{1}{2}\rho_{n\tau})] = 0 \ , \tag{4.29}$$

where the equation (4.27) is a consequence of the field equations (4.26), (4.4) and the relation

$$R^2 = \surd\,[\exp(2\rho)+\lambda_o^4] - r_o^2 \quad , \tag{4.30}$$

where exp(ρ) is independent of β but at the same time the equation (4.29) implies

$$\exp(\rho) = \beta^2 \quad . \tag{4.31}$$

The field equations (4.26) and (4.27) yield the results

$$R^2 = \exp(\rho)\sin\Phi \;, \quad (-1)^{s+1}\,\ell_o^2 = \exp(\rho)\cos\Phi \;, \tag{4.32}$$

and

$$R^4 + \ell_o^4 = \exp(2\rho) \quad . \tag{4.33}$$

From the definition of coshΓ by

$$(R^2+r_o^2)\,\exp(-\rho) = \cosh\Gamma$$

and from the first relation of (4.32) we obtain the result

$$R^2 = \frac{r_o^2\,\sin\Phi}{\cosh\Gamma - \sin\Phi} \quad , \tag{4.34}$$

where, as follows from (4.1), (4.2), the right hand side is an invariant function but R^2 does not possess this invariance property. Thus we must set

$$R^2 = 0 \quad , \tag{4.35}$$

or

$$\sin\Phi = 0 \quad , \tag{4.36}$$

as the condition of compatibility of the equations (4.26) and (4.27). Hence from (4.36) and (4.32) we obtain

$$\exp(\rho) = \ell_o^2 \quad , \tag{4.37}$$

where now

$$\Phi = \pm(s+1)\pi \quad , \quad s = 0,1 \quad . \tag{4.38}$$

The field equation (4.28) reduces to

$$\frac{1}{2} r_o^2 \ddot{S} = 1 \quad , \tag{4.39}$$

and is solved by

$$\dot{S} = \frac{2\beta}{r_o^2} + \frac{2mG_o}{c^2 \ell_o^2} \quad , \quad S = \frac{\beta^2}{r_o^2} + \frac{2mG_o}{c^2 \ell_o^2} \beta + A'' \quad , \tag{4.40}$$

where A" is a constant of integration. Using (4.31) and (4.37) we see that

$$\beta = \pm \ell_o \quad . \tag{4.41}$$

From (4.39) and the first equation of (4.40) it follows that the function S has its minima at the points

$$\pm \ell_o = - \frac{mG_o r_o^2}{c^2 \ell_o^2} \quad . \tag{4.42}$$

Furthermore the equations (4.30), (4.35) and (4.37) yield the result

$$r_o^2 = \sqrt{(\ell_o^4 + \lambda_o^4)} \quad , \tag{4.43}$$

and hence

$$r_o^2 = \frac{2G_o}{c^4}(e^2 + g_o^2) \quad . \tag{4.44}$$

By using the definitions $\ell_o^2 = q^{-1} g_o$, $\lambda_o^2 = q^{-1}e$, we further obtain the relations

$$\ell_o^2 = \frac{2G_o}{c^4} g\sqrt{(e^2+g_o^2)} \quad , \quad \lambda_o^2 = \frac{2G_o}{c^4} e\sqrt{(e^2+g_o^2)} . \tag{4.45}$$

In this case g_o is the maximum "eigenvalue" of g which corresponds to the distance $\beta = \pm \ell_o$. The value of g_o can be estimated from the equation (4.42) which, using (4.44) and (4.45) becomes a cubic equation

$$(g_o^2)^3 - \frac{1}{4} m^4 G_o^2 g_o^2 - \frac{1}{4} m^4 G_o^2 e^2 = 0 \quad . \tag{4.46}$$

This equation has only one real root given by

$$g_o^2 = \frac{1}{2} (m^4 G_o^2 e^2)^{\frac{1}{3}} [(1+a)^{\frac{1}{3}} + (1-a)^{\frac{1}{3}}] \quad , \tag{4.47}$$

where

$$a = \sqrt{[1 - \frac{1}{27} (\frac{m^2 G_o^2}{e^2})]} \quad . \tag{4.48}$$

From the derivation of the equations of motion it will be seen that the mass m obtained as a constant of integration does not correspond to the inertial mass of a particle. The actual inertial mass is obtained as the difference between m and the self energy. In this theory m is interpreted as the "bare mass" of the particle and is of the order of the Planck mass, viz.,

$$m \sim \surd\left(\frac{\hbar c}{G_o}\right) \quad . \tag{4.49}$$

In fact for an order of magnitude estimate of g_o we may identify it with the maximum value of (4.47). The latter occurs at a = 0 or at

$$m^2 = \frac{e^2}{G_o} \surd 27 \quad , \tag{4.50}$$

and yields the value

$$g_o^2 = 3\ e^2 \quad , \tag{4.51}$$

which is not far from the more likely value $\hbar c$ of g_o^2.

For e = 0 the equation (4.46) yields the values

$$g_o^2 = 0 \quad , \quad g_o^2 = \frac{1}{2} m^2 G_o \quad , \tag{4.52}$$

which may be representing magnetic charge contents of the photon and neutrinos, respectively.

5. SPECTRUM OF MAGNETIC CHARGES

One common characteristic of the three classes of exact solutions of the field equations (2.22)-(2.25) discussed in the previous section is the fact that for all of them magnetic charge density vanishes. We must therefore seek all other solutions for which magnetic charge density vanishes. From the definition (2.33) it follows that we must impose the condition

$$\dot{\rho}_{n\tau} = \dot{\Phi}_{ns\tau} \tan\Phi_{ns\tau} \quad , \tag{5.1}$$

or its integrated form

$$\exp(\rho_{n\tau})\cos\Phi_{ns\tau} = (-1)^s \ell_o^2 \quad , \tag{5.2}$$

where the constant of integration $(-1)^s \ell_o^2$ is determined from the solutions of section 4 and where

$$\Phi_{ns\tau} \neq n\pi \quad , \quad \Phi_{ns\tau} \neq \frac{\pi}{2} \quad . \tag{5.3}$$

On substituting (5.1) in (2.25) and integrating once we obtain

$$\frac{1}{4} \lambda_1^2 \dot{\Phi}_{ns\tau}^2 \tan^3\Phi_{ns\tau} = (-1)^s \quad , \tag{5.4}$$

where $\frac{1}{4} \lambda_1^2$ is a constant of integration. By using the substitution

$$\tan\Phi_{ns\tau} = (-1)^s t \quad , \quad t > 0 \quad , \tag{5.5}$$

and the equation (5.4) we obtain

$$d\Phi_{ns\tau} = \frac{2}{\lambda_1}(-1)^{s+\varepsilon} t^{-\frac{3}{2}} d\beta = \frac{(-1)^s}{1+t^2} dt \quad , \tag{5.6}$$

where

$$\cos\Phi_{ns\tau} = \frac{(-1)^{s+\tau}}{\surd(1+t^2)} \quad , \quad \sin\Phi_{ns\tau} = \frac{(-1)^{\tau} t}{(1+t^2)} \quad , \tag{5.7}$$

$$\exp(\rho_{n\tau}) = (-1)^{\tau} \ell_o^2 \surd(1+t^2) \quad . \tag{5.8}$$

From the asymptotic limits of the equations (2.23) and (2.24) it follows that

$$t = \frac{r^2}{\ell_o^2} \quad , \tag{5.9}$$

and hence in the regions where magnetic charge density vanishes we have the solution

$$\exp(\rho) = \sqrt{(r^4+\ell_o^4)} \quad . \tag{5.10}$$

From (5.6) we obtain

$$\beta=(-1)^{\varepsilon}\ \lambda_1[\sqrt{(t)} - \frac{1}{2\sqrt{2}}\left(\tan^{-1}(\frac{\sqrt{(2t)}}{1-t}) + \tanh^{-1}(\frac{\sqrt{(2t)}}{1+t})\right)] \ , \tag{5.11}$$

where we observe the relation

$$\frac{d}{dr}\tan^{-1}(\frac{\ell_o\sqrt{2r}}{\ell_o^2-r^2}) = \frac{d}{dr}\tan^{-1}(\frac{r^2-\ell_o^2}{\ell_o\sqrt{2r}}) \quad , \tag{5.12}$$

which may be found useful in the discussion of various asymptotic limits.

Now, using the above results in the field equations (2.22)-(2.24) we obtain them in a directly integrable form

$$\frac{d}{dt}[S\ \frac{\sqrt{(1+t^2)}}{t^{\frac{3}{2}}}] = \frac{1}{2}\ \frac{m_1\ \ t^{\frac{3}{2}}}{(1+t^2)^{\frac{3}{2}}}\ [\sqrt{(1+m^2t^2)}-1 + (-1)^s tm] \ , \tag{5.13}$$

$$\frac{d}{dt}[S\frac{\sqrt{(1+t^2)}}{t^{\frac{1}{2}}}]=\frac{1}{2}\ \frac{m_1\ \ t^{\frac{3}{2}}}{(1+t^2)^{\frac{3}{2}}}\ [t-t\sqrt{(1+m^2t^2)}+m(1+t^2)+(-1)^s m], \tag{5.14}$$

$$\frac{d}{dt}\ [\frac{dS}{dt}\ \frac{(1+t^2)^{\frac{3}{2}}}{t^{\frac{3}{2}}}] = \frac{1}{2}\ m\ \frac{m_1\ \ t^{\frac{3}{2}}}{\sqrt{(1+t^2)}}\ [1 - \frac{mt}{\sqrt{(1+m^2t^2)}}] \quad , \tag{5.15}$$

where the "eigenvalues" m and m_1 are defined by

$$m^2 = \frac{\ell_o^4}{r_o^4} = \frac{g^2}{e^2+g^2} \quad , \quad m_1 = \frac{\lambda_1^2}{\ell_o^2} \quad . \tag{5.16}$$

For the points of zero magnetic charge density the field equations, as seen from (5.13)-(5.15), are decoupled and are not compatible for every t. In fact if we multiply (5.13) by t and add it to (5.14) the resulting equation when integrated does not satisfy the respective equations. We may treat each equation independently from the others. Let the solutions for S of the equations (5.13)-(5.15) be represented by the curves $S = f_1(t)$, $S = f_2(t)$ and $S = f_3(t)$, respectively. The equation (5.15) contains first and second derivatives of S. Hence the equations (5.13)-(5.15) are compatible at the points where

$$f_1(t) = f_3(t) \ , \quad \frac{df_1(t)}{dt} = \frac{df_3(t)}{dt} \ , \qquad (5.17)$$

and

$$f_2(t) = f_3(t) \ , \quad \frac{df_2(t)}{dt} = \frac{df_3(t)}{dt} \ . \qquad (5.18)$$

These are algebraic equations for t, m and m_1. If the spectra of t, m and m_1 obtained from solving the set (5.17) are the same as those obtained from the set (5.18) then the field equations (5.13)-(5.15) are compatible for these t, m and m_1 alone. Thus, in principle, the compatibility of the field equations for the interior regions of zero magnetic charge density should lead to discrete spectra of t and the corresponding "eigenvalues" m and m_1. The knowledge of m and m_1 together with their definitions by (5.16) would yield the allowed spectrum of partial magnetic charges g_n, n = 0,1,2,.., whose signs, because of the $(-1)^s$-dependence of $f_j(t)$, j=1,2,3, should alternate, even though $f_3(t)$ does not depend on the sign of magnetic charge.

The integrated forms of the equations (5.13)-(5.15) are given by

$$\frac{1}{m_1} f_1(t)=\frac{t^2}{2(1+t^2)}[1-\sqrt{(1+m^2t^2)}]+\frac{(-1)^s mt^2}{2(1+t)(1+t^2)}(2t^2-t+3)+$$

$$\frac{1}{4}\frac{t^{\frac{3}{2}}}{\sqrt{(1+t^2)}}[(3(-1)^s m-1)K(\alpha,\tfrac{1}{\sqrt{2}})-6(-1)^s mE(\alpha,\tfrac{1}{\sqrt{2}})]+$$

$$\frac{3}{2}\frac{t^{\frac{3}{2}}}{\sqrt{(1+t^2)}}L_1(t)+\frac{1}{4\sqrt{(1+m)}}\frac{t^{\frac{3}{2}}}{\sqrt{(1+t^2)}}[K(\omega,\tau_2)-K(\gamma,\tau_1)], \quad (5.19)$$

$$\frac{1}{m_1}f_2(t)=-\frac{2mG_o}{c^2\ell_o m_1}\frac{t^{\frac{1}{2}}}{\sqrt{(1+t^2)}}+\frac{t^2}{2(1+t^2)}[\sqrt{(1+m^2t^2)}-1]$$

$$+t[\tfrac{3}{2}(\tfrac{1}{1+t})-\frac{(-1)^s m}{2(1+t^2)}+\tfrac{1}{3}m]+\tfrac{1}{3}m]+\frac{1}{4}\frac{t^{\frac{1}{2}}}{\sqrt{(1+t^2)}}\times$$

$$[((3+(-1)^s m)-\tfrac{2}{3}m)K(\alpha,\tfrac{1}{\sqrt{2}})-6E(\alpha,\tfrac{1}{\sqrt{2}})]$$

$$-\frac{5}{2}\frac{t^{\frac{1}{2}}}{\sqrt{(1+t^2)}}L_3(t)+\frac{3}{8}\frac{t^{\frac{1}{2}}}{\sqrt{m}\sqrt{(1+m)}\sqrt{(1+t^2)}}\times$$

$$[K(\omega,\tau_2)+K(\gamma,\tau_1)], \quad (5.20)$$

$$\frac{1}{m_1}\frac{df_3(t)}{dt}=\frac{mG_o}{c^2\ell_o m_1}\frac{t^{\frac{3}{2}}}{(1+t^2)^{\frac{3}{2}}}+\frac{m}{3}\frac{t^2}{1+t^2}-\frac{t^{\frac{3}{2}}}{(1+t^2)^{\frac{3}{2}}}\times$$

$$[L_3(t)+\tfrac{1}{6}mK(\alpha,\tfrac{1}{\sqrt{2}})]+\frac{3}{8}\frac{1}{\sqrt{m}\sqrt{(1+m)}}\frac{t^{\frac{3}{2}}}{(1+t^2)^{\frac{3}{2}}}\times$$

$$[K(\gamma,\tau_1)+K(\omega,\tau_2)], \quad (5.21)$$

where various elliptic integrals are defined in Appendix 1. The presence of a branch point at $t = m^{-\frac{1}{2}}$ for the integral $K(\gamma,\tau_1)$, as shown in the appendix 1, leads to the noncommutation of the limit $m=1$ (or e=0) and the integration over γ. This fact implies that transitions like proton or electron, by setting e=0, to transform into neutrinos are not allowed.

The derivatives of $f_1(t)$ and $f_2(t)$, as defined by (5.13), (5.14), can be written as

$$\frac{df_1}{dt} = \frac{1}{2}\,\frac{t^2+3}{t(t^2+1)}\,f_1 + \frac{1}{2}\,\frac{m_1\ t^3}{(t^2+1)^2}\,[\sqrt{(m^2t^2+1)}-1 + (-1)^2 mt], \tag{5.22}$$

$$\frac{df_2}{dt} = -\,\frac{1}{2}\,\frac{t^2-1}{t(t^2+1)}\,f_2 + \frac{1}{2}\,\frac{m_1\ t^2}{(t^2+1)^2}\ \times$$

$$[t - t\sqrt{(m^2t^2+1)}+m(t^2+1)+(-1)^s m]\ , \tag{5.23}$$

where f_1 and f_2 are given by (5.19) and (5.20), respectively. The magnetic spectral content of the equations (5.17) and (5.18) in the limit of large t can easily be obtained. We shall look for asymptotic solutions where $t = \frac{r^2}{\ell_o^2}$ and $mt = \frac{r^2}{r_o^2}$ are large compared to 1 so that terms proportional to t^{-1} and higher inverse powers of t can be neglected. In this case, using Appendix 1 and the equations (5.19)-(5.23), in the asymptotic limit of large t we obtain

$$K(\gamma,\tau_1) = \sqrt{2}\sqrt{(1+m)}\,m^{-\frac{1}{4}}\tan^{-1}[\sqrt{(\tfrac{t}{2})}\,m^{\frac{1}{4}}]\ ,\ K(\omega,\tau_2) \to 0\ ,$$

$$L_1(t) \to mt^{\frac{1}{2}}\ ,\ L_3(t) \to \frac{1}{3}\,mt^{\frac{3}{2}}\ , \tag{5.24}$$

$$K(\alpha, \frac{1}{\sqrt{2}}) \to -2t^{-\frac{1}{2}} \ , \ E(\alpha, \frac{1}{\sqrt{2}}) \to -2t^{-\frac{1}{2}} \ .$$

Hence it follows that

$$\frac{1}{m_1} f_1(t) \to 1+(-1)^s m+[1+(-1)^s]mt-\frac{t^{\frac{1}{2}}}{2\sqrt{2}} m^{-\frac{1}{4}} \tan^{-1}[\sqrt{(\tfrac{t}{2})} m^{\frac{1}{4}}] \ , \tag{5.25}$$

$$\frac{1}{m_1} f_2(t) \to 1 - \frac{2mG_0}{c^2 \ell_o m_1} t^{-\frac{1}{2}} + \frac{3t^{-\frac{1}{2}}}{4\sqrt{2}} m^{-\frac{3}{4}} \tan^{-1} [\sqrt{(\tfrac{t}{2})} m^{\frac{1}{4}}] \ , \tag{5.26}$$

$$\frac{1}{m_1} f_3(t) \to 1 - \frac{2mG_0}{c^2 \ell_o m_1} t^{-\frac{1}{2}} \ , \tag{5.27}$$

$$\frac{1}{m_1} \frac{df_1}{dt} \to m \ [1+(-1)^s] - \frac{t^{-\frac{1}{2}}}{4\sqrt{2}} m^{-\frac{1}{4}} \tan^{-1}[\sqrt{(\tfrac{t}{2})} m^{\frac{1}{4}}] \ , \tag{5.28}$$

$$\frac{df_2}{dt} \to 0 \ , \ \frac{df_3}{dt} \to 0 \ . \tag{5.29}$$

The equations (5.17) and (5.18) are now solved by

$$4\sqrt{(2t)} m^{\frac{5}{4}} [1+(-1)^s] = \tan^{-1}[\sqrt{(\tfrac{t}{2})} m^{\frac{1}{4}}] \ , \tag{5.30}$$

$$(-1)^s m+[1+(-1)^s]mt+\frac{2mG_0}{c^2 \ell_o m_1} t^{-\frac{1}{2}} = \frac{1}{2}\sqrt{(\tfrac{t}{2})} m^{-\frac{1}{4}} \tan^{-1}[\sqrt{(\tfrac{t}{2})} m^{\frac{1}{4}}] , \tag{5.31}$$

which can be replaced by the equations

$$(t^{\frac{1}{2}})^3 \ [1+(-1)^s] - (-1)^s (t^{\frac{1}{2}}) - \frac{2mG_0}{c^2 \ell_o m_1 m} = 0 \ , \tag{5.32}$$

$$\tan \ [4\sqrt{2} \ t^{-\frac{1}{2}} (-1)^s m^{\frac{5}{4}} m_1] = \frac{t^{\frac{1}{2}} m^{\frac{1}{4}}}{\sqrt{2}} \ . \tag{5.33}$$

From (5.32) for negative (s=1) and positive (s=0) magnetic charges we obtain

$$t_-^{\frac{1}{2}} = \frac{2mG_o}{c^2 \ell_o m_1 m} \, , \quad \text{or} \quad r_- = \frac{2mG_o}{c^2 m_1 m} \tag{5.34}$$

and

$$(t_+^{\frac{1}{2}})^3 - \frac{1}{2}(t_+^{\frac{1}{2}}) - \frac{mG_o}{c^2 \ell_o m_1 m} = 0 \, , \tag{5.35}$$

respectively. The discriminant of the cubic equation (5.35) is given by

$$\Delta = \frac{1}{4}\left[\frac{mG_o}{c^2 \ell_o m_1 m}\right]^2 - \frac{1}{8\text{x}27} \quad , \tag{5.36}$$

which for

$$\frac{g^3 m_1}{\sqrt{(e^2+g^2)}} < 27m^2 G_o \, , \tag{5.37}$$

is positive and the cubic equation (5.35) has only one root

$$t_+^{\frac{1}{2}} = \left[\frac{mG_o}{2c^2 \ell_o m_1 m}\right]^{\frac{1}{3}} [(1+b)^{\frac{1}{3}} + (1-b)^{\frac{1}{3}}] = \frac{r_+}{\ell_o} \tag{5.38}$$

where

$$b = \sqrt{\left[1 - \frac{1}{6}\left[\frac{c^2 \ell_o m_1 m}{3mG_o}\right]^2\right]} \, .$$

Thus, if, for example, r_- or r_+ are of the order of Compton wave length $\frac{\hbar}{Mc}$, where M is the observed mass of the particle, then ℓ_o is of the order of 10^{-50}cm and the condition (5.37) is satisfied.

Now, the infinite spectra of magnetic charges are obtained, using (5.33), as solutions of

$$\tan\left[4\sqrt{2}\; t_{+}^{-\frac{1}{2}}\, m^{\frac{5}{4}} m_1\right] = \frac{t_{+}^{\frac{1}{2}}\, m^{\frac{1}{4}}}{\sqrt{2}} \quad , \qquad (5.39)$$

for positive partial magnetic charges, (s=0), and

$$\tan\left[4\sqrt{2}\; t_{-}^{-\frac{1}{2}}\, m^{\frac{5}{4}} m_1\right] = -\frac{t_{-}^{\frac{1}{2}}\, m^{\frac{1}{4}}}{\sqrt{2}} \quad , \qquad (5.40)$$

for negative partial magnetic charges (s=1), where $t_{\pm}^{\frac{1}{2}}$ are given by (5.34) and (5.38), respectively. Hence we see that to each solution of (5.39) and (5.40) corresponds through the definitions (5.34) and (5.38) a unique $r_{\pm}$. Furthermore, it is clear from (5.34) and (5.38) that with increasing distance r from the origin the "eigenvalue" m must decrease and eventually (i.e. as $n \to \infty$) tend to zero where $g=0$ and $r_{\pm} \sim \frac{\hbar}{Mc}$. Thus the solutions (5.39) and (5.40) establishe a distribution of magnetic charges of alternating signs and decreasing (with distance from the origin) magnitude. The figure 1 for the structure of elementary particle is an illustration of the solutions (5.39) and (5.40).

6. CONSERVATION OF ELECTRIC CHARGE MULTIPLICITY

The symmetry of the electric charge multiplicity (2.39) together with the symmetry (2.26) (i.e $q \to -q$) imply the existence of particles with +1, -1, 0 units of electric charges and their corresponding antiparticles with -1, +1, 0 units of electric charges, respectively. In order to find other distinguishing characteristics of the particles with different electric charges we may further classify the solutions according to the <u>invariance properties</u> of the function Φ as it appears in the field equations (2.22)-(2.25). Thus from (2.25) we may, formally, deduce the <u>invariant statement</u>

$$\Phi^b = A + (-1)^{b+1} \int \surd[-(2\ddot{\rho}+\dot{\rho}^2)]d\beta \ , \tag{6.1}$$

where A is a constant of integration and where b is the charge multiplicity number which assumes the values b=0 and b=1.

From (6.1) it is clear that we may, in principle, substitute for Φ in the field equations (2.22)-(2.24) to determine two sets of solutions ρ_b, S_b and the corresponding relations between r_o, ℓ_o, λ_o, where b=0,1. For the solutions where $\exp(\dot{\rho})=0$ and $\exp(\rho)=\beta^2$ ($\beta \neq$ constant) the constant A assumes the values 0 and $\frac{\pi}{2}$, respectively. For these special solutions and also for the solutions where $\beta = \pm \ell_o$, $A = (s+1)\pi$, the invariant function Φ is independent of b. However, all other solutions for which the integral in (6.1) does not vanish will depend on b. Thus the solutions corresponding to b=0 and b=1 must, in general, represent different systems or particles. All the currents, being determined by the solutions of the field equations, will acquire a new degree of freedom defined by b=0,1. The same applies to all other derived quantities like the electric, magnetic fields and also the corresponding energy and momenta. Based on the prediction by the theory for the existence of particles with electric charges ± e, 0, the most reasonable interpretation for the <u>electric charge multiplicity number</u> b is to regard it as a conservation law for electrons (b=0) and protons (b=1). For the case e=0, in view of the symmetries discussed in sections 2 and 3, the natural conclusion would be to assign b=0,1 to electron-neutrino and muon-neutrino, respectively.

The number b should also play an important role in the study of the maxima and minima of the fields. For example, the extrema of the electric field (2.27) occur at the points where

$$E'_e = -\frac{q}{f}\sinh\Gamma\left(\frac{f'}{f} + \rho'\right) = 0 \quad , \qquad (6.2)$$

where we used the relation

$$\Gamma' = -\rho' \tanh\Gamma \quad .$$

For the minima of E_e we have:

(i) $$\frac{1}{f} = \frac{\exp(\rho)}{\upsilon(R^2+r_o^2)} = 0 \quad \text{or} \quad \exp(\rho) = 0 \quad ,$$

which satisfies the field equations (2.1')-(2.4') at the point r=0;

(ii) $$\sinh\Gamma = \pm\lambda_o^2 \exp(-\rho) = 0, \quad \exp(-\rho) = 0 \quad ,$$

which satisfies the field equations at $r=\infty$. Thus the field equations have solutions for which the electric field assumes its minimum value $E_e=0$. Both r=0 and $r=\infty$ are points of inflexion.

The remaining extrema of E_e satisfy the equation

$$\frac{f'}{f} + \rho' = 0 \quad , \qquad (6.3)$$

which together with the four field equations (2.1')-(2.4') constitute a system of five equations to determine the four functions ρ, Φ, f, S and also to yield a relation between the extremal values of E_e and the constants r_o, λ_o, ℓ_o. By substituting for the function f from (2.27)

$$f = \frac{\pm\lambda_o^2 q}{E_e} \exp(-\rho) \quad ,$$

where E_e is now a constant referring to extremal values of the electric field, and using (6.3) in the field equations (2.1')-(2.4') we obtain

$$\frac{1}{2}\,(S\,\Phi'_{ns\tau})' = B^2 \exp(\rho_{n\tau})[R^2\cos\Phi_{ns\tau}+\ell_o^2\sin\Phi_{ns\tau}] \;, \qquad (6.4)$$

$$\frac{1}{2}(S\,\rho'_{n\tau})'=B^2 \exp(\rho_{n\tau})[-R^2\sin\Phi_{ns\tau}+\ell_o^2\cos\Phi_{ns\tau}+\exp(\rho_{n\tau})] \;, \qquad (6.5)$$

$$\frac{1}{2}\,S'' = B^2 \exp(2\rho_{n\tau})[1 - \frac{\exp(\rho_{n\tau})\sin\Phi_{ns\tau}}{R^2+r_o^2}] \;, \qquad (6.6)$$

$$\Phi'^2 = \rho'^2 - 2\rho'' \;, \qquad (6.7)$$

where the constant B is given by

$$E_e = \pm\, B\, r_o\, q\, \lambda_o^2 = \pm\, \frac{c^2\lambda_o^2}{\sqrt{(2G_b)}}\, B \;. \qquad (6.8)$$

The solutions of (6.4)-(6.7) should determine Φ, ρ, S and the constant B. As implied by (6.7) one should obtain two sets of extremal values corresponding to b=0,1.

For the magnetic field $\mathcal{B} = q\exp(\rho)\cos\Phi\sin\theta$ the extremal values occur at the points where magnetic charge density vanishes. Thus we easily obtain

$$\mathcal{B}_n = (-1)^s\, g_n \sin\theta \;, \quad n=0,1,2,\ldots \;, \qquad (6.9)$$

and at the origin as well as at $r=r_c$ (i.e $\Phi=\frac{\pi}{2}$)

$$\mathcal{B}(o) = 0 \quad , \quad \mathcal{B}(r_c) = 0 \quad .$$

In the case of the vacuum magnetic field H_o the extrema satisfy

$$H'_o = - \frac{q}{f} \left(\frac{f'}{f} \cos\Phi + \Phi'\sin\Phi\right) \; .$$

Hence as in the case of $\mathcal{B}$ we find its minima at r=0 (i.e Φ=0) and $r=r_c$ (i.e $\Phi=\frac{\pi}{2}$). The remaining extrema occur at the points where

$$\frac{f'}{f} = - \Phi'\tan\Phi \quad ,$$

where $\Phi \neq n\pi$, $\Phi=\frac{\pi}{2}$. The rest of the discussion for H_o proceed as in E_e.

7. ENERGY-MOMENTUM TENSOR

The field equations, derived in (I), are given by

$$R_{\{\mu\nu\}} = \frac{1}{2} \kappa^2 (b_{\mu\nu} - g_{\mu\nu}) \quad , \tag{7.1}$$

$$R_{[\mu\nu\rho]} = - \frac{1}{2} \kappa^2 \; I_{\mu\nu\rho} \quad , \tag{7.2}$$

$$\hat{g}^{[\mu\nu]}{}_{,\nu} = 0 \quad , \tag{7.3}$$

where the triple subscript $[\mu\nu\rho]$ implies cyclic derivatives of $R_{[\mu\nu]}$, and where the nonsymmetric curvature tensor $R_{\mu\nu}$ is given by

$$R_{\mu\nu} = - \Gamma^{\rho}_{\mu\nu,\rho} + \Gamma^{\rho}_{\mu\rho,\nu} + \Gamma^{\rho}_{\mu\sigma}\Gamma^{\sigma}_{\rho\nu} - \Gamma^{\rho}_{\mu\nu}\Gamma^{\sigma}_{\rho\sigma} \quad . \tag{7.4}$$

The affine connections are obtained as the algebraic solutions of the "transposition invariant" equations

$$\hat{g}_{\mu\nu;\rho} = \hat{g}_{\mu\nu,\rho} - \hat{g}_{\sigma\nu}\Gamma^{\sigma}_{\mu\rho} - \hat{g}_{\mu\sigma}\Gamma^{\sigma}_{\rho\nu} = 0 \quad . \tag{7.5}$$

The equations (7.5), because of (7.3), yield the results

$$\hat{\mathfrak{g}}^{\mu\nu}{}_{;\rho} = 0 \quad , \quad \Gamma^{\rho}_{[\mu\rho]} = 0 \quad , \tag{7.6}$$

where

$$\hat{\mathfrak{g}}^{\mu\nu} = \surd(-\hat{g})\hat{g}^{\mu\nu} \quad , \quad \hat{g}^{\mu\rho}\hat{g}_{\nu\rho} = \delta^{\mu}_{\nu} \, , \ \hat{g} = \mathrm{Det}(\hat{g}_{\mu\nu}) \quad , \tag{7.7}$$

$$\hat{g}_{\mu\nu} = g_{\mu\nu} + q^{-1}\,\Phi_{\mu\nu} \quad .$$

By separating out symmetric and antisymmetric parts of the equations (7.5) and by solving formally the resulting equations we obtain the results

$$\Gamma^{\rho}_{\mu\nu} = \{^{\rho}_{\mu\nu}\} + S^{\rho}_{\mu\nu} + \Gamma^{\rho}_{[\mu\nu]} \quad , \tag{7.8}$$

$$g_{\mu\sigma}\Gamma^{\sigma}_{[\nu\rho]} + g_{\nu\sigma}\Gamma^{\sigma}_{[\rho\mu]} + g_{\rho\sigma}\Gamma^{\sigma}_{[\mu\nu]} = -\frac{1}{2}\, I_{\mu\nu\rho} \quad , \tag{7.9}$$

$$g_{\mu\sigma}S^{\sigma}_{\nu\rho} + g_{\nu\sigma}S^{\sigma}_{\rho\mu} + g_{\rho\sigma}S^{\sigma}_{\mu\nu} = 0 \quad , \tag{7.10}$$

where the tensor $S^{\rho}_{\mu\nu}(= S^{\rho}_{\nu\mu})$ and the fully anti-symmetric tensor $I_{\mu\nu\rho}$ are given by

$$S^{\rho}_{\mu\nu} = g^{\rho\sigma}[\Phi_{\mu\alpha}\Gamma^{\alpha}_{[\sigma\nu]} + \Phi_{\nu\alpha}\Gamma^{\alpha}_{[\sigma\mu]}] \quad , \tag{7.11}$$

$$I_{\mu\nu\rho} = \Phi_{\mu\nu,\rho} + \Phi_{\nu\rho,\mu} + \Phi_{\rho\mu,\nu} \quad , \tag{7.12}$$

and where $I_{\mu\nu\rho} = -4\pi\ \varepsilon_{\mu\nu\rho\sigma}\mathfrak{s}^{\sigma}$ represents magnetic current density. The metric tensor $g_{\mu\nu}$ and its inverse $g^{\mu\nu}$ are related by

$$g^{\mu\rho}g_{\nu\rho} = \delta^{\mu}_{\nu} \quad .$$

Furthermore, the tensor $b^{\mu\nu}$, which will be shown to be part of the energy tensor, is defined by

$$b^{\mu\nu} = \frac{1}{\sqrt{(-g)}}\,\hat{g}^{\{\mu\nu\}} = \frac{g^{\mu\nu}(1+\frac{1}{2}\Omega)+T^{\mu\nu}}{\sqrt{(1+\Omega-\Lambda^2)}} \tag{7.13}$$

and its inverse by

$$b_{\mu\nu} = \frac{g_{\mu\nu}(1+\frac{1}{2}\Omega)-T_{\mu\nu}}{\sqrt{(1+\Omega-\Lambda^2)}} \quad , \tag{7.14}$$

where

$$T_{\mu\nu} = \frac{1}{2}\,g_{\mu\nu}\Omega-\Phi_{\mu\rho}\Phi_{\nu}{}^{\rho} \;,\; \Omega = \frac{1}{2}\Phi^{\mu\nu}\Phi_{\mu\nu} \;,\; \Lambda = \frac{1}{4}f^{\mu\nu}\Phi_{\mu\nu} \;, \tag{7.15}$$

$$f^{\mu\nu} = \frac{1}{2\sqrt{(-g)}}\,\varepsilon^{\mu\nu\rho\sigma}\,\Phi_{\rho\sigma} \quad ,$$

and

$$b^{\mu\rho}b_{\nu\rho} = \delta^{\mu}_{\nu} \quad .$$

If we set $\Gamma^{\rho}_{[\mu\nu]} = 0$ we obtain, following from (7.9), $I_{\mu\nu\rho} = 0$. Therefore the antisymmetric part $\Gamma^{\rho}_{[\mu\nu]}$ of the affine connection $\Gamma^{\rho}_{\mu\nu}$ can be interpreted as the structure tensor giving rise to magnetic charge content of an elementary particle and hence to its spin. In terms of $\Gamma^{\rho}_{[\mu\nu]}$ the magnetic current s^{μ} can be expressed as

$$s^{\mu} = \varepsilon^{\mu\nu\rho\sigma}\,\Gamma_{[\nu\rho]\sigma} \quad , \tag{7.16}$$

where

$$\Gamma_{[\nu\rho]\sigma} = g_{\sigma\alpha} \Gamma^{\alpha}_{[\nu\rho]} \quad .$$

Furthermore, as shown in (2.53) and (8.5) of (I), the field equations (7.2) have the linearized asymptotic form

$$(\nabla^2 - \frac{\partial^2}{c^2 \partial t^2} + \kappa^2) s^{\mu} = 0 \quad , \tag{7.17}$$

implying the confinement of the magnetic current density s^{μ} to distances of the order of r_c $(\sim \frac{\hbar}{Mc})$ since beyond r_c

$$\kappa = \frac{2}{r_o^2} \sim 10^{33} \text{ cm}^{-1} \quad , \tag{7.18}$$

tends to infinity and s^{μ} vanishes. It must be noted, as pointed out earlier, that r_o is not the range of the magnetic charge distribution but, as a function of the partial magnetic charges g_n, it measures the degree of deviation of this theory from general relativity and classical electrodynamics.

Now, the gravitational field is described in terms of the curvature of space and, therefore it is necessary to know all the interaction energy densities which act as the source of the gravitational field. The generalized theory provides a complete microscopic basis for calculating, uniquely, the energy momentum tensor as the source of the gravitational field. Thus by using the definition (7.4) and the result (7.8), the field equations (7.1) can be written as

$$G^{\nu}_{\mu} - \frac{1}{2} \delta^{\nu}_{\mu} G = \frac{8\pi G_o}{c^4} T^{\nu}_{\mu} \quad , \tag{7.19}$$

where the subscript (o) of G_o for the gravitational constant on the right hand side differentiates it from the scalar curvature G. In the derivation of (7.19) we used the relations

$$R_{\{\mu\nu\}} = - G_{\mu\nu} + S_{\mu\nu} + \Gamma^{\rho}_{[\mu\sigma]}\Gamma^{\sigma}_{[\rho\nu]} = \frac{1}{2}\kappa^2(b_{\mu\nu}-g_{\mu\nu}) \quad ,(7.20)$$

where

$$S_{\mu\nu} = - S^{\rho}_{\mu\nu|\rho} + S^{\rho}_{\mu\rho|\nu} + S^{\rho}_{\mu\sigma}S^{\sigma}_{\rho\nu} - S^{\rho}_{\mu\nu}S^{\sigma}_{\rho\sigma} \quad , \qquad (7.21)$$

$$S^{\rho}_{\mu\rho} = \frac{\partial}{\partial x^{\mu}}[Ln\surd(1+\Omega-\Lambda^2)] = -\frac{1}{2} g_{\mu\nu}g^{\rho\sigma}S^{\nu}_{\rho\sigma} \quad , \qquad (7.22)$$

$$G_{\mu\nu} = \{^{\rho}_{\mu\nu}\}_{,\rho} - \{^{\rho}_{\mu\rho}\}_{,\nu} + \{^{\rho}_{\mu\nu}\}\{^{\sigma}_{\rho\sigma}\} - \{^{\rho}_{\mu\sigma}\}\{^{\sigma}_{\rho\nu}\} \quad , \qquad (7.23)$$

and where the symbol (|) signifies the covariant derivative with respect to the metric tensor $g_{\mu\nu}$.

Because of the known properties of the left hand side of the equations (7.19), the symmetric energy momentum tensor $T_{\mu\nu}$ satisfies the covariant conservation laws

$$[\surd(-g)T^{\nu}_{\mu}]_{|\nu} = 0 \quad , \qquad (7.24)$$

where the energy tensor T^{ν}_{μ} is given by

$$4\pi\, T_{\mu\nu} = q^2[(\tfrac{1}{2}b^{\rho}_{\rho}-1)g_{\mu\nu}-b_{\mu\nu}] + r_o^2 \text{ trace}[S_{\mu|\nu} - \tfrac{3}{2}g_{\mu\nu}S^{\rho}_{|\rho}$$
$$+ S_{\mu}S_{\nu} - \tfrac{1}{2}g_{\mu\nu}S^{\rho}S_{\rho} - \Gamma_{\mu}\Gamma_{\nu} + \tfrac{1}{2}g_{\mu\nu}\Gamma^{\rho}\Gamma_{\rho}] - r_o^2 \quad \times$$
$$[S_{\mu})^{\rho}{}_{\nu|\rho} + (S_{\mu})^{\rho}_{\nu} \text{ trace } S_{\rho} + g_{\mu\nu} \text{ trace } S^{\rho} \text{ trace } S_{\rho}] \quad ,$$
$$(7.25)$$

and where the 4×4 matrices S_μ and Γ_μ are defined by

$$(S_\mu)^\rho_\nu = S^\rho_{\mu\nu} = (S_\nu)^\rho_\mu \ , \ (\Gamma_\mu)^\rho_\nu = \Gamma^\rho_{[\mu\nu]} = -(\Gamma_\nu)^\rho_\mu \ . \tag{7.26}$$

These transform according to

$$S'_\mu = \frac{\partial x^\rho}{\partial x'^\mu} U S_\rho U^{-1} \ , \ \Gamma'_\mu = \frac{\partial x^\rho}{\partial x'^\mu} U \Gamma_\rho U^{-1} \ ,$$

where U is the 4×4 matrix

$$U = [\frac{\partial x'}{\partial x}] \ , \ U^{-1} = [\frac{\partial x}{\partial x'}] \ .$$

Thus

$$\text{trace } S_\mu = (S_\mu)^\rho_\rho = S^\rho_{\mu\rho} = \partial_\mu[\text{Ln}\sqrt{(1+\Omega-\Lambda^2)}] \ ,$$

$$\text{trace } \Gamma_\mu = (\Gamma_\mu)^\rho_\rho = \Gamma^\rho_{[\mu\rho]} = 0 \ .$$

In terms of the matrices S_μ and Γ_μ the neutral field $R_{[\mu\nu]}$ can be written as

$$R_{[\mu\nu]} = -(\Gamma_\mu)^\rho_{\nu|\rho} - (\Gamma_\mu)^\rho_\nu \text{ trace } S_\rho + \text{trace } (\Gamma_\mu S_\nu - \Gamma_\nu S_\mu) . \tag{7.27}$$

The trace of the energy tensor is given by

$$4\pi\, T = 4\pi\, T^\rho_\rho = q^2(b^\rho_\rho - 4) + r_o^2 \text{ trace } [\Gamma^\rho\Gamma_\rho - 3\, S^\rho_{|\rho} - S^\rho S_\rho - 2S^\rho \text{ trace } S_\rho] \ . \tag{7.28}$$

In the energy tensor $T_{\mu\nu}$ in the limit $r_o = 0$ the first term tends to the energy tensor

$$T_{\mu\nu} = \frac{1}{2} g_{\mu\nu} \Omega - \Phi_{\mu\rho} \Phi_{\nu}{}^{\rho} , \qquad (7.29)$$

of the pure radiation field since in this case $\Phi_{\mu\nu} = \partial_{\mu}A_{\nu} - \partial_{\nu}A_{\mu}$. In the limit $r_o = 0$ the remaining terms vanish. Thus, they evidently refer to the short range interaction energy density of the field and the associated particles. We note that for the limit $r_o = 0$ we have used the relation

$$r_o^2 q^2 = \frac{c^4}{2G_o} . \qquad (7.30)$$

At this point it is interesting to note that the presence of a short range interaction energy density in a theory obtained from the unification of two long range fields like electromagnetic and gravitational fields may seem quite surprising. There are at least two reasons for this result: (i) The antisymmetric tensor $\Phi_{\mu\nu}$, not being derivable from a potential, led to the existence of magnetic charges of the field; (ii) If in primeval times the field consisted of free magnetic charge units g_n ($n = 0,1,2,...$) then through the gravitational attraction of this "primordial field" a gravitational condensation, occurring over the enormous times available to nature, could have placed all these units g_n into their most stable states (in order to conserve magnetic currents), that of stratified layers and thus give rise to the birth of elementary particles and hence to their short range interactions.

The fundamental constant r_o, which measures the degree of deviation from the conventional theory, in the energy tensor $T_{\mu\nu}$ also couples the field and its particles at short distances, where the strength of the coupling is measured by $\frac{Q_n^2}{\hbar c}$. We can now rewrite the tensor $T_{\mu\nu}$ in the form

$$4\pi T_{\mu\nu} = \Lambda_{\mu\nu} + W_{\mu\nu} \tag{7.31}$$

where the first term

$$\Lambda_{\mu\nu} = q^2 [(\frac{1}{2}b^\rho_\rho - 1)g_{\mu\nu} - b_{\mu\nu}]$$

represents total electromagnetic energy density as well as magnetic energy density throughout space time which is divided into the three regions $0 \leqq r < r_c$, $r \geqq r_c$ and $r >> r_c$. The indeterminate surface $r=r_c$ contains, in the rest frame, an amount of energy Mc^2 and therefore we may quantify r_c by assuming that $Mc\, r_c \sim \hbar$ or

$$r_c \sim \frac{\hbar}{Mc} \quad . \tag{7.32}$$

Thus at distances

$$r \geqq \frac{\hbar}{Mc}$$

we have only an electromagnetic interaction between the particle and the radiation field and in this region of space time the fundamental constant r_o has the small value

$$r_o^2 = \frac{2G_o e^2}{c^4} \quad .$$

Therefore at these distances the deviation from the conventional theory is still smaller. For distances

$$r >> r_c$$

the constant $r_o = 0$ and we have only the radiation energy density.

We see that the larger r_o (approaching inner regions of magnetic structure) the greater is the deviation of this theory from electrodynamics and general relativity. In the first region ($0 \leqq r \leqq r_c$) of space time the electromagnetic energy density is a function of the magnetic charge and hence of mass also. Therefore this portion of the electromagnetic energy density exists also in the rest frame of the particles.

The $W_{\mu\nu}$ part of the energy tensor, which refers to terms of $T_{\mu\nu}$ containing r_o^2 explicitly, contributes only in the regions $0 \leqq r \leqq r_c$ and $r_c \leqq r < r$ (radiation) since for $r >> r_c$, we can set $r_o = 0$ and the short range energy density vanishes. The contribution of $W_{\mu\nu}$ to the energy density in $r_c \leqq r$ (radiation) is purely electromagnetic because $g_n = 0$. Thus the fundamental constant r_o, from the point of view of the energy density, partitions space time into three regions specified by the values

$$r_{on}^2 = \frac{2G_o}{c^4}(e^2+g_n^2) \ , \ r_{o\infty}^2 = \frac{2G_o}{c^4}e^2 \ , \ r_o^2 = 0 \ . \qquad (7.33)$$

The maximum value of r_o^2, which occurs at $\Phi = \pm(s+1)\pi$, is given by

$$r_{oo}^2 = \frac{2G_o}{c^4}(e^2+g_o^2) \ , \qquad (7.34)$$

where g_o is the maximum value of the magnetic charge. Hence in addition to the deviations from classical behavior due to the existence of Planck's constant h (i.e. quantum mechanics) we have here an additional change measured by the discrete values of the constant r_o. The new modification of the classical theory arises from finite description of the structure of elementary particles.

It will be shown that the energy tensor $T_{\mu\nu}$, as discussed here briefly, and its conservation laws (7.24) contains the laws of motion of particles in the field, including that of the chargeless particles or neutrino like particles.

The inertial properties of the energy momentum tensor (7.25) can be studied in terms of the time independent spherically symmetric field. By using the results of (I), it is easily shown that

$$\frac{c^4}{4\pi G_o} \sqrt{(-g)}G_4^4=\sqrt{(-g)}(T_4^4-T_1^1-T_2^2-T_3^3) = \frac{c^4}{8\pi G_o} \times$$

$$[\upsilon \exp(u+\rho)u'\sin\Phi]'\sin\theta \quad . \qquad (7.35)$$

The volume integral of G_4^4 yields the interesting result

$$\int \sqrt{(-g)}(T_4^4-T_1^1-T_2^2-T_3^3)\,dr d\theta\, d\phi = \pm\, mc^2 \quad ,(7.36)$$

where ± signs signify the presence of positive and negative energy states. The mass m appearing in (7.36) is the "bare gravitational mass" which is, because of principle of equivalence, equal to the "bare inertial mass". The gravitational force does not distinguish between different masses. Therefore any mass splitting must have a nongravitational origin. In the present theory the electron and proton mass ratio will have to arise from the extremum values of the <u>intrinsic fields</u> (i.e. the field strength inside the region $0 \leqq r \leqq r_c$).

In order to complete the discussion of the energy-momentum tensor we must also consider the use of the conservation laws

$$\mathfrak{T}^{\nu}_{\mu,\nu} = 0 \qquad (7.37)$$

where the nonsymmetrized pseudo energy tensor $\mathfrak{T}^{\nu}_{\mu}$ is given by

$$-4\pi q^{-2}\kappa^{2}\mathfrak{T}^{\nu}_{\mu} = \hat{g}^{\nu\rho}\, R_{\mu\rho}+\hat{g}^{\rho\nu}\, R_{\rho\mu}-\delta^{\nu}_{\mu}\hat{g}^{\rho\sigma}\, R_{\rho\sigma}+\hat{g}^{\rho\sigma}{}_{,\mu}\, \mathcal{B}^{\nu}_{\rho\sigma}-\delta^{\nu}_{\mu}\mathcal{B} \;, \qquad (7.38)$$

and where, because of the field equations (7.3), we have

$$\mathcal{B}^{\rho}_{\mu\nu}=\tfrac{1}{2}(\delta^{\rho}_{\mu}\, \Gamma^{\sigma}_{\nu\sigma}+\delta^{\rho}_{\nu}\, \Gamma^{\sigma}_{\mu\sigma})-\Gamma^{\rho}_{\mu\nu} \;, \quad \mathcal{B}=\hat{g}^{\mu\nu}(\Gamma^{\rho}_{\mu\sigma}\, \Gamma^{\sigma}_{\rho\nu}-\Gamma^{\rho}_{\mu\nu}\, \Gamma^{\sigma}_{\rho\sigma}) \;. \qquad (7.39)$$

The quantities $\mathfrak{T}^{\nu}_{\mu}$, except in flat space time, do not, under general coordinate transformations, behave as components of a tensor. Therefore the somewhat undecided status of $\mathfrak{T}^{\nu}_{\mu}$ as an energy tensor is carried over from general relativity. However the conservation laws (7.37), with the application of Gauss' theorem, enable us, formally, to define a "4-momentum vector" by

$$P_{\mu} = \int \mathfrak{T}^{\nu}_{\mu}\, d\sigma_{\nu} \;. \qquad (7.40)$$

We shall treat $\mathfrak{T}^{\nu}_{\mu}$, at least in the rest frame of a spherically symmetric system, as a respectable physical quantity.

We may calculate the corresponding quantity (7.36) for the tensor $\mathfrak{T}^{\nu}_{\mu}$. Thus by multiplying the field equation (2.23) by $\cosh^2\Gamma$ and rewriting it in the form

$$\frac{d}{d\beta}[\dot{S}\exp(\rho)\cosh^2\Gamma] =$$

$$\kappa^2 \cosh\Gamma\left(1-\frac{\sin\Phi}{\cosh\Gamma}\right)\exp(\rho)-2R_{[41]}\sinh\Gamma\exp(\rho)f$$

where

$$R_{[41]} = \frac{1}{f}\dot{\rho}\dot{S}\sinh\Gamma \quad .$$

From the definition (7.38) and the field equations (7.1)-(7.3) we get

$$\mathfrak{F}^4_4-\mathfrak{F}^1_1-\mathfrak{F}^2_2-\mathfrak{F}^3_3 = \frac{q^2}{2\pi}[\sqrt{(-\hat{g})}-\sqrt{(-g)}] - \frac{q^2}{2\pi}\hat{g}^{[41]} F_{41} \quad , \quad (7.42)$$

where

$$F_{41}=\Phi_{41}+r_o^2 R_{[41]}, \quad \Phi_{41}=-\frac{\sinh\Gamma}{f}, \quad \hat{g}^{[41]} = \exp(\rho)\sinh\Gamma\sin\theta \quad ,$$

we obtain

$$\mathfrak{F}^4_4-\mathfrak{F}^1_1-\mathfrak{F}^2_2-\mathfrak{F}^3_3=\frac{q^2}{2\pi}\left[\frac{\cosh^2\Gamma}{f}\left(1-\frac{\sin\Phi}{\cosh\Gamma}\right)\exp(\rho)-r_o^2\exp(\rho)\sinh\Gamma\, R_{[41]}\right]\sin\theta$$

$$= \frac{r_o^2 q^2}{4\pi}\frac{1}{f}\frac{d}{d\beta}[\exp(\rho)\dot{S}\cosh^2\Gamma]\sin\theta \quad .$$

Hence

$$\int(\mathfrak{F}^4_4-\mathfrak{F}^1_1-\mathfrak{F}^2_2-\mathfrak{F}^3_3)\,dr\,d\theta\,d\phi =$$

$$\pm r_o^2 q^2 \frac{2mG_o}{c^4} - r_o^2 q^2 [\exp(\rho)\dot{S}\cosh^2\Gamma]_{r=0}$$

$$= \pm mc^2 \qquad (7.43)$$

Now, in order to compare the energy tensors T_μ^ν and $\mathfrak{F}_\mu^\nu$ let us consider the pseudo tensor $\mathfrak{F}_\mu^\nu$ of the gravitational field alone i.e. we construct $\mathfrak{F}_\mu^\nu$ in terms of the $g_{\mu\nu}$ of this theory. The tensor $\mathfrak{F}_\mu^\nu$ can be obtained in terms of $g_{\mu\nu}$ by setting $\Phi_{\mu\nu} = 0$ in the $\mathfrak{F}_\mu^\nu$ of the nonsymmetric theory. Thus using the relations

$$\mathfrak{F}_\mu^\nu = \frac{c^4}{16\pi G_0}\frac{\partial}{\partial x^\rho}(S_\mu^{\nu\rho}) \;, \quad \mathfrak{F} = \mathfrak{F}_\rho^\rho = \frac{\partial U^\rho}{\partial x^\rho} \;, \tag{7.44}$$

with

$$U^\rho = \frac{c^4}{8\pi G_0}\frac{1}{\sqrt{(-\hat{g})}}\frac{\partial}{\partial x^\sigma}[\sqrt{(-\hat{g})}\hat{g}^{\rho\sigma}] \;, \tag{7.45}$$

$$S_\mu^{\nu\rho} = \hat{g}^{\nu\sigma}\, B_{\mu\sigma}^\rho + \hat{g}^{\sigma\nu}\, B_{\sigma\mu}^\rho - \delta_\mu^\nu\, \hat{g}^{\alpha\beta}\, B_{\alpha\beta}^\rho \;, \tag{7.46}$$

for the nonsymmetric theory and setting $\Phi_{\mu\nu} = 0$ we can obtain the corresponding results for the pure gravitational field. Hence for the gravitational part alone we have

$$\overset{o}{\mathfrak{F}}{}_4^4 = \frac{c^4}{8\pi G_0}\frac{\sin\theta}{\upsilon} - \frac{c^4\sin\theta}{8\pi G_0}[\upsilon\ \exp(u+\rho)(\rho'+\Phi'\cot\Phi)\sin\Phi]' \tag{7.47}$$

$$\overset{o}{\mathfrak{F}} = \frac{c^4}{4\pi G_0}\frac{\sin\theta}{\upsilon} - \frac{c^4}{4\pi G_0}[\upsilon\ \exp(u+\rho)(\rho'+\Phi'\cot\Phi)\sin\Phi + \tfrac{1}{2}\upsilon u'\exp(u+\rho)\sin\Phi]' \tag{7.48}$$

Hence we obtain

$$\overset{o}{\mathfrak{F}}{}_4^4 - \overset{o}{\mathfrak{F}}{}_1^1 - \overset{o}{\mathfrak{F}}{}_2^2 - \overset{o}{\mathfrak{F}}{}_3^3 = \frac{c^4}{8\pi G_0}[\upsilon\ u'\ \exp(u+\rho)\sin\Phi]' \;, \tag{7.49}$$

and

$$\int(\overset{\circ}{\mathcal{T}}{}^4_4-\overset{\circ}{\mathcal{T}}{}^1_1-\overset{\circ}{\mathcal{T}}{}^2_2-\overset{\circ}{\mathcal{T}}{}^3_3)\,dr\,d\theta\,d\phi = \pm\, mc^2 \quad , \tag{7.50}$$

yielding the same result obtained in (7.36) and (7.43) for the energy tensors T^{ν}_{μ} and $\mathcal{T}^{\nu}_{\mu}$, respectively.

8. THE FOUR FUNDAMENTAL CURRENTS

Some of the properties of elementary particles and their interactions can be expressed and analysed in terms of their currents. The currents of the generalized theory of gravitation can be defined in a unique manner. One of the basic premises of the theory lies in the fact that the antisymmetric part $\Phi_{\mu\nu}$ of the fundamental field variables $\hat{g}_{\mu\nu} = g_{\mu\nu} + q^{-1}\Phi_{\mu\nu}$, is not derivable from a potential viz.,

$$\Phi_{\mu\nu} \neq \partial_{\mu}A_{\nu}-\partial_{\nu}A_{\mu} \quad ,$$

and therefore, as discussed in (I), its cyclic derivatives

$$\Phi_{\mu\nu,\rho} + \Phi_{\nu\rho,\mu} + \Phi_{\rho\mu,\nu} \; (\equiv I_{\mu\nu\rho})$$

can be used to define a conserved magnetic current density by

$$\mathfrak{s}^{\mu} = \frac{1}{4\pi}\frac{\partial}{\partial x^{\nu}}[\sqrt{(-g)}f^{\mu\nu}] \quad , \tag{8.1}$$

where

$$I_{\mu\nu\rho} = -4\pi\,\varepsilon_{\mu\nu\rho\sigma}\,\mathfrak{s}^{\sigma} \; , \quad \mathfrak{s}^{\mu} = \sqrt{(-g)}s^{\mu} \quad , \tag{8.2}$$

and where $f^{\mu\nu}$ was defined by (7.15). It is clear that if $\Phi_{\mu\nu}$ were derivable from a potential then the current

s^μ would vanish. The integral of (8.1) which is conveniently evaluated in its spherically symmetric rest frame, is given by

$$\int s^\mu \, d\sigma_\mu = 0 \quad . \tag{8.3}$$

Now from the field equations

$$\Psi_{\mu\nu,\rho} + \Psi_{\nu\rho,\mu} + \Psi_{\rho\mu,\nu} = 0 \quad , \tag{8.4}$$

we can derive another (axial) conserved magnetic current, where

$$\Psi_{\mu\nu} = \frac{1}{2} \varepsilon_{\mu\nu\rho\sigma} \hat{g}^{[\rho\sigma]} \quad , \quad g^{[\mu\nu]}{}_{,\nu} = 0 \quad , \tag{8.5}$$

$$\Psi_{\mu\nu} = \partial_\mu B_\nu - \partial_\nu B_\mu \quad , \quad \hat{g}^{[\mu\nu]} = \hat{g}^{[\mu\nu\rho]}{}_{,\rho} \quad , \tag{8.6}$$

$$\hat{g}^{[\mu\nu\rho]} = \varepsilon^{\mu\nu\rho\sigma} B_\sigma \quad .$$

A neutral magnetic vacuum current can now be defined by

$$\zeta^\mu = \frac{1}{4\pi} \frac{\partial}{\partial x^\nu} [\surd(-g) \Psi^{\mu\nu}] \quad , \tag{8.7}$$

where, as before we use its spherically symmetric rest frame for convenience and obtain

$$\int \zeta^\mu \, d\sigma_\mu = 0 \quad . \tag{8.8}$$

The conserved current ζ^μ will be interpreted as the magnetic "vacuum current" where the vacuum contains distributions of equal amounts of net positive and negative magnetic charges (e.g. particles and anti-particles).

The two electric currents can be derived from the field equations

$$F_{\mu\nu,\rho} + F_{\nu\rho,\mu} + F_{\rho\mu,\nu} = 0 \quad . \tag{8.9}$$

Thus the generalized conserved electric current can be defined by

$$J^{\mu} = \frac{1}{4\pi} \frac{\partial}{\partial x^{\nu}} [\sqrt{(-g)} F^{\mu\nu}] \quad , \tag{8.10}$$

where in accordance with the field equations (7.2) we have

$$F_{\mu\nu} = \Phi_{\mu\nu} + r_o^2 R_{[\mu\nu]} = \partial_\mu A_\nu - \partial_\nu A_\mu \quad . \tag{8.11}$$

Hence

$$J^{\mu} = j_e^{\mu} + j_o^{\mu} \quad , \tag{8.12}$$

where

$$j_e^{\mu} = \frac{1}{4\pi} \frac{\partial}{\partial x^{\nu}} [\sqrt{(-g)} \Phi^{\mu\nu}] \quad , \tag{8.13}$$

represents charged electric current, which, because of the field equation (7.3), for linearized fields, vanishes, and

$$j_o^{\mu} = \frac{r_o^2}{4\pi} \frac{\partial}{\partial x^{\nu}} [\sqrt{(-g)} R^{[\mu\nu]}] \quad , \tag{8.14}$$

represents the neutral electric current. Their integrals are given by

$$\int j_e^\mu \, d\sigma_\mu = \pm e \quad , \quad \int j_o^\mu \, d\sigma_\mu = 0 \quad . \tag{8.15}$$

The neutral currents ζ^μ, j_o^μ together with the corresponding fields $\Psi_{\mu\nu}$, $R_{[\mu\nu]}$, respectively, describe the properties of the vacuum where there are equal amounts of positive and negative electric charges residing on particles and antiparticles. It is clear from the above definitions that the four currents s^μ, ζ^μ, j_e^μ, j_o^μ are obtained from the solutions of the field equations, contrary to classical electrodynamics where one assumes distributions of currents and then calculates the fields they generate.

In (I) it was found that the neutral charge density j_o^4 and the corresponding field E_o fall off as $1/r^5$ and they, therefore, have short range character. But both of these quantities, like their charged counterparts j_e^4, E_e, vanish at the origin. Furthermore, both j_o^4 and E_o, through their dependence in magnetic charge, have mass dependence. Therefore all four currents because of their dependence on ℓ_o^2 have an indirect functional relationship to mass. The four currents j_e^μ, j_o^μ, s^μ, ζ^μ represent either the proton (anti-proton) or the electron (positron) currents depending on the electric charge multiplicity number discussed in section 6. Furthermore, currents in classical theories are prescribed to generate fields and not vice versa. For zero electric charge the only surviving currents are s_o^μ and ζ_o^μ, where subscript o is used to distinguish them from those belonging to charged particles. In this case the two currents s_o^μ and ζ_o^μ together represent either ν_e-like or ν_μ-like particle currents. Thus the four currents of the electron and proton can be represented by $s^\mu(b)$, $\zeta^\mu(b)$, $j_e^\mu(b)$, $j_o^\mu(b)$ where b=0 for electron and b=1 for proton. For neutrinos we have e=0 and the surviving currents are $s_{0}^{\mu}(b)$, $\zeta_{0}^{\mu}(b)$ where b=0 for electron-neutrino and b=1 for muon neutrino.

An important observation is the fact that, in this theory, the absence of currents in the field equations (8.4) and (8.9) is a fundamental symmetry. The field equations (8.4) and (8.9) can also be written in the form (2.55) and (2.56) of I, which contain the sources of the fields in the nonlinearity of the equations. Thus instead of introducing a monopole current as stipulated by Dirac[4], the complete symmetry is restored by having no currents at all in the theory. The latter are to be derived from the solutions of the field equations alone, as defined above.

9. EQUATIONS OF MOTION

A) Action Principle Derivation of the Equations of Motion

One of the most important tests of the unified field theories has always been the fundamental requirement that the field equations must yield the laws of motion of the particles. In this section we shall demonstrate that the generalized theory of gravitation does, in a straightforward way, yield the equations of motion of a particle (anti-particle) in an external field as well as the action of the self-field on the particle (anti-particle) itself. We shall give two equivalent methods for the derivations of the equations of motion, one of which is the action principle derivation and the other is based, as in general relativity, on the existence of Bianchi identities. We may begin by obtaining, from the field equations (7.1)-(7.3), the most general decomposition of the field $\Phi_{\mu\nu}$. The field equations (8.4) and (8.9), which follow from

(7.3) and (7.2), respectively, are satisfied not only by $\Psi_{\mu\nu}$, $F_{\mu\nu}$, but also by $\Psi_{\mu\nu} + f_{1\mu\nu}$ and $F_{\mu\nu} + \Phi_{1\mu\nu}$, where $\Phi_{1\mu\nu}$ and $f_{1\mu\nu}$ obey the field equations

$$\Phi_{1\mu\nu,\rho} + \Phi_{1\nu\rho,\mu} + \Phi_{1\rho\mu,\nu} = 0 \ , \quad \frac{\partial}{\partial x^\nu} [\sqrt{(-g)}\Phi_1^{\mu\nu}] = 0 \ , \tag{9.1}$$

$$f_{1\mu\nu,\rho} + f_{1\nu\rho,\mu} + f_{1\rho\mu,\nu} = 0 \ , \quad \frac{\partial}{\partial x^\nu} [\sqrt{(-g)}f_1^{\mu\nu}] = 0 \ , \tag{9.2}$$

and where

$$f_1^{\mu\nu} = \frac{1}{2\sqrt{(-g)}} \varepsilon^{\mu\nu\rho\sigma} \Phi_{1\rho\sigma} \ . \tag{9.3}$$

Thus $\Phi_{1\mu\nu}$ represents radiative solutions of the field equations (7.2) and (7.3). Hence the most general decomposition of the field variables can be represented by

$$\Phi_{\mu\nu} = \Phi_{o\mu\nu} + \Phi_{1\mu\nu} \ , \tag{9.4}$$

$$f_{\mu\nu} = f_{o\mu\nu} + f_{1\mu\nu} \ , \tag{9.5}$$

where

$$\frac{1}{4\pi} \frac{\partial}{\partial x^\nu} [\sqrt{(-g)}\Phi_o^{\mu\nu}] = j_e^\mu \ , \quad \frac{1}{4\pi} \frac{\partial}{\partial x^\nu} [\sqrt{(-g)}f_o^{\mu\nu}] = \mathscr{s}^\mu \ . \tag{9.6}$$

By using (9.4) and (9.5) we obtain decomposition of Ω and Λ, defined by (7.15), in the form

$$\Omega = \Omega_o + \Omega_1 + \Omega_I \ , \quad \Lambda = \Lambda_o + \Lambda_1 + \Lambda_I \ , \tag{9.7}$$

where

$$\Omega_o=\frac{1}{2}\Phi_o^{\mu\nu}\ \Phi_{o\mu\nu}\ ,\ \Omega_1=\frac{1}{2}\Phi_1^{\mu\nu}\ \Phi_{1\mu\nu}\ ,\ \Omega_I=\Phi_o^{\mu\nu}\ \Phi_{1\mu\nu}\ , \tag{9.8}$$

and

$$\Lambda_o=\frac{1}{4}f_o^{\mu\nu}\ \Phi_{o\mu\nu},\ \Lambda_1=\frac{1}{4}f_1^{\mu\nu}\ \Phi_{1\mu\nu}\ ,\ \Lambda_I=\frac{1}{4}(f_o^{\mu\nu}\Phi_{1\mu\nu}+f_1^{\mu\nu}\Phi_{o\mu\nu})\ . \tag{9.9}$$

The Maxwell energy tensor T_μ^ν , defined by (7.15), becomes

$$T_\mu^\nu = T_{o\mu}^\nu + T_{1\mu}^\nu + T_{I\mu}^\nu\ , \tag{9.10}$$

where

$$T_{o\mu}^\nu = \frac{1}{2}\ \delta_\mu^\nu\ \Omega_o - \Phi_{o\mu\rho}\Phi_o^{\nu\rho}\ ,\quad T_{1\mu}^\nu = \frac{1}{2}\ \delta_\mu^\nu\ \Omega_1 - \Phi_{1\mu\rho}\Phi_1^{\nu\rho}\ , \tag{9.11}$$

$$T_{I\mu}^\nu = \frac{1}{2}\ \delta_\mu^\nu\ \Omega_I - (\Phi_{o\mu\rho}\Phi_1^{\nu\rho} + \Phi_{1\mu\rho}\Phi_o^{\nu\rho})\ .$$

Hence, we obtain the results

$$[\sqrt{(-g)}T_{1\mu}^\nu]_{|\nu} = 0\ ,\ \frac{1}{4\pi}\ [\sqrt{(-g)}T_{o\mu}^\nu]_{|\nu} = \Phi_{o\mu\rho}j_e^\rho + f_{o\mu\rho}\ s^\rho\ ,$$

$$\frac{1}{4\pi}\ [\sqrt{(-g)}T_{I\mu}^\nu]_{|\nu} = \Phi_{1\mu\rho}j_e^\rho + f_{1\mu\rho}\ s^\rho\ , \tag{9.12}$$

which, as will be seen, represent force densities corresponding to the interactions between the radiation and current densities j_e^μ, s^μ and also between the self-fields and their corresponding currents.

The square root term which appears in the action integral of the theory can now be written as

$$\sqrt{(1+q^{-2}\Omega-q^{-4}\Lambda^2)}=\sqrt{(1+q^{-2}\Omega_o-q^{-4}\Lambda_o^2)}\Big[1 + \frac{q^{-2}(\Omega_1+\Omega_I)}{1+q^{-2}\Omega_o-q^{-4}\Lambda_o^2}$$

$$- q^{-4}\,\frac{\Lambda_1^2+\Lambda_I^2+2(\Lambda_o\Lambda_1+\Lambda_o\Lambda_I+\Lambda_1\Lambda_I)}{1+q^{-2}\Omega_o-q^{-4}\Lambda_o^2}\Big]^{\frac{1}{2}} \quad .(9.13)$$

In order to expand (9.13) in powers of q^{-2} we shall use the dimensionless constants

$$\alpha_n = \frac{e^2}{g_n^2} = \frac{1}{\gamma_n}\frac{e^2}{\hbar c} \quad , \quad n=0,1,2,... \qquad .,(9.14)$$

in the limits of small and large n. Thus, with $r_c \sim \frac{\hbar}{Mc}$ we have the following four possibilities:

(i) In the region $r < r_c$ for large n (i.e. small g_n) from

$$q_n = \frac{\sqrt{(e^2+g_n^2)}}{r_{on}^2} = \frac{c^4}{2G_o}(e^2+g_n^2)^{-\frac{1}{2}} \quad ,$$

we obtain the result

$$q_n^{-1} \sim \frac{2G_o}{c^4}e(1 + \frac{1}{2}\alpha_n^{-1}) \quad , \qquad (9.15)$$

which is of the order of 10^{-58} $(esu)^{-1}$. The effect of the small q^{-2} is, of course, offset by the factor q^2 appearing in the energy tensor (7.25) as well as in the action function itself.

(ii) In the region $r < r_c$ for $\alpha_n <<1$ (i.e. large g_n) we obtain

$$q_n^{-1} \sim \frac{2G_o}{c^4}|g_n|(1 + \frac{1}{2}\alpha_n) \quad , \qquad (9.16)$$

which, for reasonable values of g_n with smaller n, would

still be a small quantity but somewhat larger than in the case of (i).

(iii) In the region $r > r_c$ we have $g_n = 0$ and

$$q^{-1} = \frac{2G}{c^4} e^2 \; , \tag{9.17}$$

a value that is smaller than the values obtained for q^{-1} in the cases of (i) and (ii).

(iv) In the region where $r >> r_c$ we have $q \to \infty$ (or $r_o \to o$) and in this case the theory reduces to general relativity plus classical electromagnetic field in the absence of electric charges.

Hence to order q^{-2}, in the regions (i) and (ii) where $r < r_c$, we obtain

$$\sqrt{(1+q^{-2}\Omega - q^{-4}\Lambda^2)} \sim \sqrt{(1+q^{-2}\Omega_o - q^{-4}\Lambda_o^2)}$$
$$+\frac{1}{2}q^{-2}(\Omega_1+\Omega_I)(1+q^{-2}\Omega_o - q^{-4}\Lambda_o^2)^{-\frac{1}{2}}, (1+q^{-2}\Omega - q^{-4}\Lambda^2)^{-\frac{1}{2}}$$
$$\sim (1+q^{-2}\Omega_o - q^{-4}\Lambda_o^2)^{-\frac{1}{2}} - \frac{1}{2}q^{-2}(\Omega_1+\Omega_I)(1+q^{-2}\Omega_o - q^{-4}\Lambda_o^2)^{-\frac{3}{2}} \; .$$

For the region (iii) the corresponding expansions are given by

$$\sqrt{(1+q^{-2}\Omega - q^{-4}\Lambda^2)} \sim \sqrt{(1+q^{-2}\Omega_o - q^{-4}\Lambda_o^2)} + \frac{1}{2} q^{-2}(\Omega_1+\Omega_I) \; , \tag{9.18}$$

$$(1+q^{-2}\Omega - q^{-4}\Lambda^2)^{-\frac{1}{2}} \sim (1+q^{-2}\Omega_o - q^{-4}\Lambda_o^2)^{-\frac{1}{2}} - \frac{1}{2} q^{-2}(\Omega_1+\Omega_I) \; . \tag{9.19}$$

The increasing values of q (i.e. $n\to\infty$) correspond to increasing distances from the origin of the particle where the self-fields are small compared to q, and therefore in region (iv) we have the approximate values

$$\surd(1+q^{-2}\Omega-q^{-4}\Lambda^{2}) \sim 1 + \frac{1}{2} q^{-2}(\Omega_{o}+\Omega_{1}+\Omega_{I}) , \quad (9.20)$$

$$(1+q^{-2}\Omega-q^{-4}\Lambda^{2})^{-\frac{1}{2}} \sim 1 - \frac{1}{2} q^{-2}(\Omega_{o}+\Omega_{1}+\Omega_{I}) . \quad (9.21)$$

In this theory particles (anti-particles) are represented by the magnetic regions (generated by the magnetic charge) of the field. The action principle of the theory yields the field equations which are valid everywhere, including the magnetic regions of the field. In order to study the time dependent behavior of the magnetic regions under the influence of an external field (i.e. the space-time trajectories of magnetic regions), as well as under the action of their self-fields, we must minimize (by displacing each point on the worl-line of the particle's center by an infinitesimal amount) the action function which was obtained by substituting the field equations back into the original action function, which yielded the field equations as a result of the variation of the field variables $\hat{g}_{\mu\nu} = g_{\mu\nu} + q^{-1} \Phi_{\mu\nu}$. The action principle for the field equations (7.1)-(7.3) is based on the variation of the action function

$$S = \frac{q^{2}r_{o}^{2}}{8\pi c} \int \{\hat{g}^{\mu\nu}R_{\mu\nu} + \kappa^{2}[\surd(-\hat{g})-\surd(-g)]\}d^{4}x , \quad (9.22)$$

where

$$\hat{g}^{[\mu\nu]} = g^{[\mu\nu\rho]}{}_{,\rho} \quad (\text{i.e. } \hat{g}^{[\mu\nu]}{}_{,\nu} = 0) ,$$

and where, as shown in (I), the variation of S with

respect to $\hat{g}^{\{\mu\nu\}}$ and $g^{[\mu\nu\rho]}$ yields the field equations (7.1)-(7.3).

On substituting the field equations (7.1)-(7.3) in (9.22) and dropping the divergence term, we obtain the extremum value of the action function S in the form

$$S_o = -\frac{q^2}{4\pi}\int[\sqrt{(-\hat{g})}-\sqrt{(-g)}]d^4x \quad , \tag{9.23}$$

where now the integrand is a function of the solutions $\hat{g}_{\mu\nu}$ of the field equations (7.1)-(7.3). Thus whenever the field equations (7.1)-(7.3) are solved, of which particles are prescribed as a special spectrum of values of $\hat{g}_{\mu\nu}$, the value of the action function S_o for these solutions can be minimized to obtain the equations of motion. Such equations of motion would then describe all interactions between particles and fields (external as well as self-fields). In this way the evolution of the particle aspects of the field are separated out, leading, under the variation of the appropriate variables, to the actual trajectories of these particles. The simplest application of this method is to consider the motion of a chargeless bare particle in the gravitational field. In this case the action function (9.23) in the rest frame of a spherically symmetric field becomes

$$S_o = -\frac{q^2}{c}\int\frac{\exp(\rho)}{\upsilon}(1-\sin\Phi)dr\ dx^4 \quad . \tag{9.24}$$

From the exact solution (4.21), setting $\lambda_o^2 = 0$, it is seen that for the chargeless particle metric the function $\exp(u)$ at $r = r_c$ (or $\Phi = \frac{\pi}{2}$) is given by

$$\exp(u) = 1 - \frac{2mG_o}{c^2 r_c} \quad ,$$

and also

$$\upsilon = \pm 1 \ , \ \exp(\rho) = r^2 \ .$$

Hence using the solutions (4.14) and the field equation (3.22) we obtain the rest frame value of (9.24) in the form

$$S_o = -\frac{1}{2}(\pm mc)\int dx^4 \qquad (9.25)$$

where m refers to "bare gravitational mass". The flat space-time generalization of (9.25) to a moving frame of reference is of the form

$$S_o = -\frac{1}{2}(\pm mc)\int v_\mu dx^\mu \ , \qquad (9.26)$$

where

$$v^\mu = \frac{dx^\mu}{ds} \ , \quad v^\mu v_\mu = 1 \ , \quad dx^\mu = v^\mu ds \ , \qquad (9.27)$$

is the velocity vector of the moving frame of reference. In the rest frame of a flat space-time we have $v^4 = 1$, $v_j = 0$, $(j=1,2,3)$ and we regain (9.25). Therefore (because of the principle of equivalence) in a moving frame of reference of a curved space-time we obtain the action function (9.25) in the form

$$S_o = -\frac{1}{2}(\pm mc)\int ds \ , \qquad (9.28)$$

where now

$$ds^2 = g_{\mu\nu}\, dx^\mu\, dx^\nu \ , \qquad (9.29)$$

and where x^{μ} represents the coordinates of the "particle's center". Thus, in the gravitational field the bare mass m moves along a geodesic world-line (obtained from variation of S_o) spanning the space-time region of a tube comprised of a bundle of world-lines of an extended bare particle (with spatial dimension $r_c \sim \frac{\hbar}{mc}$).

For the more general case of a charged particle we may decompose the action function S_o of (9.24) in the form

$$S_o \cong S_{oo} + S_{1I} + S_{2I} \quad , \tag{9.30}$$

where

$$S_{oo} = -\frac{q^2}{4\pi c}\int [\sqrt{(1+q^{-2}\Omega_o - q^{-4}\Lambda_o^2)}-1]\sqrt{(-g)}d^4x \quad , \tag{9.31}$$

$$S_{1I} = -\frac{1}{8\pi c}\int_{r<r_c} \frac{(\Omega_1+\Omega_I)\sqrt{(-g)}d^4x}{\sqrt{(1+q^{-2}\Omega_o - q^{-4}\Lambda_o^2)}} \quad , \tag{9.32}$$

$$S_{2I} = -\frac{1}{8\pi c}\int_{r\geqq r_c} (\Omega_1+\Omega_I)\sqrt{(-g)}d^4x \quad . \tag{9.33}$$

In the rest frame of the particle the action function S_{oo} for a spherically symmetric field becomes

$$S_{oo} = -\frac{q^2}{c}\int \frac{\exp(\rho)}{\upsilon}\left(\frac{1}{\cosh\Gamma} - \sin\Phi\right) dr\, dx^4 \quad , \tag{9.34}$$

where we used the results (3.16) and (3.18) of (I), and where the integration over r extends from 0 to ∞. In a moving coordinate system S_{oo} can be written as

$$S_{oo} = -(\pm Mc)\int ds \quad , \tag{9.35}$$

where now the observed mass M is related to the "bare gravitational mass" m according to

$$M = \frac{1}{2} m + \frac{2E_s}{c^2} \quad , \qquad (9.36)$$

where E_s is the finite electromagnetic self-energy. In order to see this we write the integrand in (9.34) in the form

$$\frac{\exp(\rho)}{\upsilon}\left(\frac{1}{\cosh\Gamma} - \sin\Phi\right) = \frac{\exp(\rho)}{f}\left(1 - \frac{\sin\Phi}{\cosh\Gamma}\right) - \upsilon^2\left(\frac{\sinh^2\Gamma}{f^2}\right)\left(\frac{\exp(\rho)}{\upsilon}\sin\Phi\right). \qquad (9.37)$$

Hence, by using the potentials A_μ $(= A^o_\mu + A^1_\mu)$ introduced by the equation (8.11), definition of the electric field by (2.27) and the relation

$$\upsilon^2 = - g^{11} g^{44} \quad ,$$

between the coefficients of the spherically symmetric metric tensor $g_{\mu\nu}$, the action function S_{oo} can be written as

$$S_{oo} = -\frac{q^2}{c}\int \exp(\rho)\left(1 - \frac{\sin\Phi}{\cosh\Gamma}\right) d\beta \; dx^4 - \frac{1}{4\pi c}\int \Phi_o^{41} \; \Phi_{o41}\sqrt{(-g)}\, d^3x$$

$$= -\left(\pm\frac{1}{2}mc\right)\int dx^4 + \frac{1}{4\pi c}\int \sqrt{(-g)}\,\Phi_o^{41}\left(\frac{dA^o_4}{dr} + r_o^2 \, R_{[41]}\right) d^3x \quad , \qquad (9.38)$$

where $d^3x = dr d\theta d\phi$ and where the first term was obtained by using the field equation (2.24) and by integrating the result. Now by integrating by parts over r and noting the vanishing of the integrated terms and using the definition of the electric current by (9.6) the last integral in (9.38) results in

$$\int j_e^4 A_4 \, d^3x - \frac{1}{4\pi}\int r_o^2 \sqrt{(-g)} \, \Phi_o^{41} R_{[41]} \, d^3x = 2E_s \ .$$

The <u>finite self-energy</u> E_s in a moving frame of reference can be written in a manifastly invariant form

$$E_s = \frac{1}{2}\int j^\mu A_\mu^o \, d\sigma - \frac{r_o^2}{8\pi}\int \Phi_o^{\rho\nu} R_{[\rho\mu]} v_\nu v^\mu \, d\sigma \quad , \qquad (9.39)$$

where

$$\sqrt{(-g)} d^3x \to d\sigma \ , \ j_e^\mu = \sqrt{(-g)} j^\mu \quad .$$

The selfenergy E_s, in the correspondence limit $r_o \to 0$, reduces to the classical infinite selfenergy. The second term in (9.39) describes the contribution to selfenergy from the interaction between the proper field $\Phi_o^{\mu\nu}$ and the vacuum field $R_{[\mu\nu]}$. It is clear from (9.37) that for +m (i.e $\upsilon > 0$) the selfenergy E_s is negative, while for -m (i.e $\upsilon < 0$) we obtain a positive E_s. Thus the observed mass (9.36) (i.e the mass that moves according to the laws of motion) is the difference of two unobserved but large masses.

The action function S_{2I} can be brought to the conventional form by writing

$$\Omega_I \sqrt{(-g)} = \Phi_o^{\mu\nu}\Phi_{1\mu\nu}\sqrt{(-g)} = 2 \ \frac{\partial}{\partial x^\mu} \ [\Phi_o^{\mu\nu} \ A_\nu \sqrt{(-g)}] - 2A_\nu \ \frac{\partial}{\partial x^\mu}[\sqrt{(-g)}\Phi_o^{\mu\nu}]$$

$$= 8\pi \ A_\mu \ j_e^\mu \quad ,$$

where the divergence term has been dropped. Hence the total action function (9.30) can be written as

$$S_o = S_{1I} - (\pm Mc)\int ds - \frac{1}{c}\int\left[\int_\sigma A_\mu j^\mu d\sigma\right]ds - \frac{1}{16\pi c}\int \Phi^{\mu\nu}\Phi_{\mu\nu}\sqrt{(-g)}\,d^4x \ , \tag{9.40}$$

where

$$d\sigma = v^\mu d\sigma_\mu \ , \quad v^\mu = \frac{dz^\mu}{ds} \ , \quad v^\mu v_\mu = 1 \ ,$$

and

$$\sqrt{(-g)}\,d^4x \rightarrow d\sigma\; ds \ , \quad j^\mu = \frac{1}{\sqrt{(-g)}}\, j^\mu_e \ .$$

In the result (9.40) the last three terms are the same as the classical action function for a charged particle moving according to Lorentz's equations of motion in an external electromagnetic field. The last term in (9.40), because of the field equations (9.1), does not contribute to the action principle. The action S_{1I} defined by (9.32) contains the effects of the particle's own fields on its motion in the external field $\Phi_{\mu\nu}$. The term S_{1I} will not be discussed in this paper. It must be observed that the approximation adopted here, in view of not including the motions of g_n in the structure of the particle, describes only the average or classical motion of the particle. The exact treatment, where, the effects of the particles magnetic structure are included, should yield results implying the spin degree of freedom as well as other possible interaction.

It is interesting to observe that in Einstein's[2] as well as in Schrödinger's[3] versions of the generalized theory of gravitation the action function S_o defined by (9.23) vanishes, and that therefore these theories cannot yield equations of motion.

B) Derivation of the Equations of Motion From The Bianchi Identities

The field equations (7.19), via Bianchi identities, imply the covariant conservation laws

$$[\sqrt{(-g)}\,T^{\nu}_{\mu}]_{|\nu} = 0 \quad , \qquad (9.41)$$

where the energy momentum tensor T^{ν}_{μ} can be decomposed in the form

$$T^{\nu}_{\mu} = T^{\nu}_{o\mu} + T^{\nu}_{1\mu} + T^{\nu}_{I\mu} \quad , \qquad (9.42)$$

and where $T^{\nu}_{o\mu}$ is constructed in terms of $\Phi_{o\mu\nu}$ of (9.4) in the form given by (7.25) and has the linearized form (see appendix 2)

$$T^{\mu\nu}_{o} = \frac{1}{4\pi}\, \mathrm{T}^{\mu\nu}_{o} + 4\pi\, r_o^2[\tfrac{1}{2}(s^{\mu}s^{\nu} + \tfrac{1}{2}g^{\mu\nu}s^{\rho}s_{\rho}) + j^{\mu}j^{\nu} + \tfrac{1}{2}\, g^{\mu\nu}j_{\rho}j^{\rho}] +$$

$$\frac{1}{2}\, r_o^2\, (f^{\mu}_{o\rho}\, s^{\rho\nu} + f^{\nu}_{o\rho}\, s^{\rho\mu}) \quad . \qquad (9.43)$$

The tensor $T^{\mu\nu}_{o}$ vanishes on the tubular surface (dimensions $r_c \sim \frac{\hbar}{Mc}$) generated during the motion of the particle. The remaining two terms in (9.42) are given by (9.11) and

$$T^{\mu\nu}_{I} = \frac{1}{4\pi}\, \mathrm{T}^{\mu\nu}_{I} + \frac{1}{2}\, r_o^2\, (f^{\mu}_{1\rho}\, s^{\rho\nu} + f^{\nu}_{1\rho}\, s^{\rho\mu}) . \qquad (9.44)$$

By using the relations

$$g^{\rho\sigma}(s^{\mu}{}_{|\rho\sigma}) = \kappa^2 s^{\mu} \;,\; (s^{\nu|\rho})_{|\nu} \cong 0 \quad , \qquad (9.45)$$

we obtain

$$T^{\mu\nu}_{I|\nu} = \Phi^{\mu}_{1\rho}\ j^{\rho} - \frac{1}{4}\,(g^{\mu\sigma}\ f_1^{\rho\alpha}\ s_{\rho\alpha})_{|\sigma} \quad ,$$

where

$$\frac{1}{2}\,(g^{\mu\sigma} f_1^{\rho\alpha}\ s_{\rho\alpha})_{|\sigma} = (g^{\mu\sigma}\ f_1^{\rho\alpha}\ s_{\alpha})_{|\rho\sigma} \quad ,$$

and where the first equation of (9.45) follows from linearization of the field equations (7.2). Hence

$$T^{\mu\nu}_{I}{}_{|\nu} = \Phi^{\mu}_{1\rho}\ j^{\rho} \quad , \qquad (9.46)$$

where the last term has been dropped since its contribution to the energy integral vanishes (see Appendix 2). For the divergence of $T^{\mu\nu}_{o}$ we obtain

$$T^{\mu\nu}_{o}{}_{|\nu} = \Phi^{\mu}_{o\rho}\ j^{\rho} + 4\ \pi\ r_o^2\ (j^{\rho} j^{\mu}{}_{|\rho} + j_{\rho} j^{\rho|\mu}) \quad .(9.47)$$

In terms of a hydrodynamic picture we can introduce a velocity vector field to describe a continuous flow of matter. The density and flow of electric and magnetic charges can be described in terms of the electric and magnetic current densities by defining them in the form

$$j^{\mu} = c\ v^{\mu} \quad , \quad s^{\mu} = a\ v^{\mu} \quad , \qquad (9.48)$$

so that c, a and v^{μ} are not independent variables. The scalar c and pseudo-scalar a determine the electric and magnetic charge densities. The conservation laws,

$$j^{\mu}{}_{|\mu} = 0 \quad , \quad s^{\mu}{}_{|\mu} = 0 \quad , \qquad (9.49)$$

lead to the relation

$$\frac{1}{a}\frac{da}{ds} = \frac{1}{c}\frac{dc}{ds} \quad ,$$

or

$$c = \sqrt{(\alpha_n)}\, a_n \quad , \qquad (9.50)$$

where the α_n were defined by (9.14). By using the above definitions the internal part of the energy density can be written as

$$T_0^{\mu\nu} = (\varepsilon_n + P_n) v^\mu v^\nu + P_n\, g^{\mu\nu} + \frac{1}{2} r_0^2 (f^\mu_{o\rho} s^{\rho\nu} + f^\nu_{o\rho} s^{\mu\rho}) - \frac{1}{4\pi}\, \Phi^\mu_{o\rho}\, \Phi^{\nu\rho}_o \quad , \qquad (9.51)$$

where the energy density ε_n and the isotropic pressure P_n are defined by

$$\varepsilon_n = 2\pi\, r_{on}^2\, a_n^2 \left(\frac{1}{2} + \frac{e^2}{\hbar c}\frac{1}{n}\right) - \frac{1}{8\pi}\,\Omega_o \quad , \qquad (9.52)$$

$$P_n = 2\pi\, r_{on}^2\, a_n^2 \left(\frac{1}{2} + \frac{e^2}{\hbar c}\frac{1}{\gamma_n}\right) + \frac{1}{8\pi}\,\Omega_o \quad . \qquad (9.53)$$

The remaining terms in (9.51) represent self-couplings of the field and the particle. In the result (9.51) the subscript n of $T_0^{\mu\nu}$ has been suppressed, but in fact it describes the energy densities in each magnetic layer specified by n=0,1,2,... .

Let us now rewrite the conservation law (9.41) in the form

$$[\sqrt{(-g)}\, T_0^{\mu\nu}]_{|\nu} + \Phi^\mu_{1\rho}\, j^\rho_e = 0 \quad , \qquad (9.54)$$

where the first term will be treated exactly and therefore the linearized form will not be used in (9.54).

The integration of (9.54) over a four-dimensional tubular region spanned by the motion of the extended particle can be written as

$$\int_T \sqrt{(-g)}\, T_0^{\mu\nu}\, d\sigma_\nu + \int \sqrt{(-g)}\, T_0^{\rho\sigma} \left\{ {\mu \atop \rho\sigma} \right\} d^4x + \int \Phi^\mu_{1\rho}\, j_e^\rho\, d^4x = 0 \quad , \tag{9.55}$$

where in the first integral we used Gauss' theorem and where T represents a region bounded by the hypersurfaces at the instants $x^4 = \sigma_1$ and $x^4 = \sigma_2$ (where $\sigma_1 < \sigma_2$) and by the tubular region between these hypersurfaces. The total energy in the tubular region consists of an influx of energy at the instant σ_1 and an efflux of energy at the instant σ_2 plus the energy on the surface of the tubular region spanned during the motion of the particle from σ_1 to σ_2. Hence the first integral in (9.55) can be written as

$$\int \sqrt{(-g)}\; T_0^{\mu\nu}\; d\sigma_\nu = P^\mu(\sigma_1) - P^\mu(\sigma_2) + P^\mu(\text{tubular}), \tag{9.56}$$

where

$$P^\mu(\sigma_1) = \int_{\sigma_1} \sqrt{(-g)}\; T_0^{\mu\nu}\; d\sigma_\nu \quad , \quad P^\mu(\sigma_2) = \int_{\sigma_2} \sqrt{(-g)}\; T_0^{\mu\nu}\; d\sigma_\nu \quad , \tag{9.57}$$

and where the term P^μ (tubular), as seen from the linearized form (9.51), vanishes.

Now the integral (7.36) of the energy tensor, because of its time independence, can be taken to refer to either of the instants σ_1 and σ_2. We can write (7.36) as

$$\int \sqrt{(-g)}(T^4_{o\nu} - \frac{1}{2}\delta^4_\nu T_o)d\sigma^\nu = \pm\frac{1}{2}mc^2 ,$$

or as

$$\int \sqrt{(-g)}T^4_{o\nu}\, d\sigma^\nu = \pm\frac{1}{2}mc^2 + \frac{1}{2}\int \sqrt{(-g)}T_o\,\delta^4_\nu\, d\sigma^\nu = P^4(\sigma) , \tag{9.58}$$

where

$$T_o = T^\mu_{o\mu} .$$

In a moving frame of reference the equation (9.58) can be replaced by the equation (9.57), where, as in the action principle derivation, the observed particle mass is defined by

$$Mc^2 = \frac{1}{2}mc^2 + 2E_s ,$$

$$E_s = \frac{1}{2}\int \sqrt{(-g)}\, T_o\, d\sigma . \tag{9.59}$$

In the limit $\sigma_2 \to \sigma_1$ the integral (9.55) can be replaced by

$$\int_{\sigma_1}^{\sigma_2} Mc^2 \frac{dv^\mu}{ds}\, ds + \int_{\sigma_1}^{\sigma_2}[\, T^{\rho\sigma}_o \{^{\mu}_{\rho\sigma}\} d\sigma] ds + \int_\sigma^\sigma [\int \Phi^\mu_{1\rho} j^\rho d\sigma] ds = 0 , \tag{9.60}$$

which, in view of the arbitrariness of σ_1 and σ_2, implies the equations motion

$$Mc^2 \frac{dv^\mu}{ds} + \int T^{\rho\sigma}_o \{^{\mu}_{\rho\sigma}\} d\sigma + \int \Phi^\mu_{1\rho} j^\rho d\sigma = 0 . \tag{9.61}$$

The velocity vector v^μ of the particle's center is the same vector which defines the current vector in the form $j^\mu = cv^\mu$. It was shown in section 4 that at the origin the electric and magnetic field (of the magnetic charge) vanish. If we wish we can use the linearized energy tensor (9.51) in the second term of (9.61) and calculate it in the approximation where in the magnetic core of the particle self-couplings and pressure vanish. In this case the equations of motion (9.61) are replaced by Lorentz's equations of motion in a gravitational field,

$$Mc^2 \left[\frac{dv^\mu}{ds} + \{^{\mu}_{\rho\sigma}\} v^\rho v^\sigma\right] + \int \Phi^\mu{}_\rho \, j^\rho \, d\sigma = 0 \quad . \qquad (9.62)$$

The derivation of the equations of motion as a consequence of the Bianchi identities, as can be seen from the result (9.61), is less general than the derivation based on action principle. This can, for example, be seen from the self-interaction effects contained in (9.40) which are absent in the present derivation. However, Bianchi identities are of great physical significance for the generalized theory of gravitation. In fact the original derivation of the field equations (7.1)-(7.3) was based on the Bianchi identities of the nonsymmetric theory (see Appendix 3).

10. CONCLUSIONS

Unification of gravitational and electromagnetic forces by the nonsymmetric generalization of the general theory of relativity has led to a new understanding of the nature of short range forces in the elementary particle interactions. The short range interactions arise from the presence of screening caused by the magnetic charge distribution with alternating signs. The magnetic structure of the elementary particle, where $\sum_{n=o}^{\infty} g_n = 0$ and $\lim_{n\to\infty} g_n = 0$, provides a field theoretical basis whose solutions are regular everywhere and in which spin is a consequence of such neutral distribution of magnetic charges. The exact solutions of the nonlinear field equations revealed the remarkable result that the magnetic charge spectrum g_n, n=0,1,2,. ... ., depends on mass m. The symmetries like electric, magnetic charge and mass conjugations, parity, time reversal, are, for $e\neq 0$, conserved. However for e=0 it was found that charge conjugation and parity are not conserved. The latter solutions, despite their masses, are interpreted as referring to neutrino-like particles.

It appears that the interpretation of the constant of integration m as a "bare mass" is the only way to obtain reasonable values for g_n where m is of the order of Planck's mass $\sqrt{(\frac{\hbar c}{G_0})}$. The observed mass

$$Mc^2 = \frac{1}{2}\, mc^2 + 2E_s \quad ,$$

which moves in accordance with the laws of motion, is also a function of m and the selfenergy E_s. The mass m and E_s have opposite signs and hence the observed mass M emerges as the difference between $\frac{1}{2}$m and the self-energy. In the correspondence limit $r_0 = 0$ the self-energy E_s becomes infinite and in this case, as in quantum electrodynamics, the observed mass M is also infinite. The same conclusions apply to the observed electric charge e defined in terms of the constant of integration $\lambda_0^2 = eq^{-1}$. Thus if λ_0 is fixed then in the limit r_0=0 we have $q=\infty$ and therefore e must also tend to infinity. From these results we see that the finiteness of M and e in this theory are consequences of a very small but nonzero bare length r_0 ($\sim 10^{-34}$cm). The existence of the fundamental "bare length" r_0 as a result of the role of Bianchi identities (see Appendix 3) in the nonsymmetric theory, laid the foundations of a correspondence principle and, furthermore, induced an extended structure ($r_c \sim \frac{\hbar}{Mc}$) for an elementary particle. Thus the length r_o which is zero in the conventional theory, measures the degree of deviation of the generalized theory of gravitation from general relativity plus classical electrodynamics, just as Planck's constant $\hbar$ can be thought as measuring the degree of deviation of quantum theory from the classical theory. The effort in the paper (I) was in trying to bring the length r_o in line with the "magic" quantity 10^{-13}cm, by choosing g_o (largest member of the spectrum g_n, n=0,1,2,...) to be of the order of 10^{18}e, and was mainly due to personal prejudices accumulated over many years. It is now clear that g_o is, at most, one or two orders of magnitude of the electric charge.

Furthermore, as shown in (I) and in this paper, there exist no monopoles in this theory and the fields produced by g_n (with $\sum_{n=o}^{\infty} g_n = 0$) are short range fields. In the limit $n \to \infty$ ($g_n \to 0$) the coupling, between the field and the particle, of strength $\frac{Q_n^2}{\hbar c}$ ($= \frac{e^2}{\hbar c} + \frac{g_n^2}{\hbar c}$) tends to $\frac{e^2}{\hbar c}$ and the interaction takes place at and beyond the distance $r_c (\sim \frac{\hbar}{Mc})$. The self-energy is the total energy of interaction between particle's own field and its electric current augmented by the total energy of interaction (at high frequencies and short distances) between particles own field and the new vacuum field generated by equal number of particles and antiparticles. The vacuum consists of <u>vacuum pairs</u> whose total spin, total energy, total electric and magnetic charges vanish. The definition of the observed mass as given in this paper is a consequence of the equations of motion whose derivation is based on the use of the solutions of the field equations in the original action function and the variation of the resulting action function with respect to the displacement of the particle coordinates. This is a new method for deriving the laws of motion from the field equations. Its validity and usefulness comes from the regularity of the solutions of the field equations everywhere. It is hoped that the extremum action function (9.23) obtained in this manner may also be used to study the quantum mechanical behavior of the field. In fact the indeterminacy with regard to the size of an elementary particle, as a consequence of the invariance principle of the theory under general coordinate transformations, due to its stratified magnetic structure has already emerged without the use of quantum mechanics. Moreover many of the results from the theory, as described in this paper, like particle, antiparticle, self-energy, vacuum, finite renormalization of mass and electric charges, various descrete symmetries, have their counterparts in quantum electrodynamics except the fact that all of the observed quantities are, in this case, finite. A further important result refers to the derivation of the spin $\frac{1}{2}$ and its new status where its direction and the signs of partial magnetic charges g_n are correlated.

A formal result refers to the fact that for $n\to\infty$ (i.e $g_n=0$) and for distances large compared to $r_c(\sim \frac{\hbar}{Mc})$ the spherically symmetric time independent solutions reduce to Nordström solution of general relativity with the surprising result that the bare gravitational mass m appears along with an electric charge that does not represent the usual elementary charge carried by an elementary particle. The latter is a consequence of the limiting process where in the equation $e_o = q\lambda_o^2$, for fixed constant of integration λ_o^2, the electric charge e_o for very large q ($\sim \infty$) does not correspond to the elementary unit of charge e. The Nordström solution is obtained from this theory only in the limit of very large q or very small $r_o (\to o)$. The e_o/m ratio in the solution

$$\exp(u) = 1 - \frac{2mG_o}{c^2 r} + \frac{G e_o^2}{c^4 r^2} ,$$

must be related to the corresponding ratio e/M for the observed quantities according to

$$\frac{e_o}{m} = \frac{e}{M} .$$

Thus Nordström solution, as obtained here, contains the bare mass m ($\sim 10^{-6}$ gr.) and bare charge e_o ($\sim 10^9$ esu) carried by m.

Furthermore, the classification of the solutions in terms of the <u>invariant</u> function $\Phi(r)$ led to the existence of two classes of solutions characterized by the <u>electric charge multiplicity number</u> b. The conservation of electrons (b=0) and protons (b=1) and, as well as, conservation of ν_e (b=0), ν_μ (b=1) together with their antiparticles correspond to conservation of electric charge multiplicity. The possibility of constructing all other elementary particles as bound or resonance states of the fundamental "quartet" p, e, ν_e, ν_μ and the "antiquartet" $\bar{p}, e^+, \bar{\nu}_e, \bar{\nu}_\mu$ will be derived in the next paper.

Finally, as a possible conjecture, I would like to state that the two-body, three-body and, more generally, many-body problem (as in the construction of elementary particles other than proton, electron, neutrinos and their antiparticles, as well as nuclei, atoms, molecules, etc.) can be based on the space-time geometrical symmetries. The general linear group of coordinate transformations were, in general relativity, assumed to pertain to the description of the gravitational field alone. In the generalized theory of gravitation, in view of the nonlinearity of its electrodynamics, the consequences of the general covariance can be extended to all the fields and their interactions. Thus a spherically symmetric space-time describing a one-particle system (e.g. electron, proton, ν_e, ν_μ) and its interaction with fields can be modified, say, into an ellipsoidal or a spheroidal symmetry to describe a three-particle bound system like, for example, a neutron having for its constituents the proton plus electron plus $\bar{\nu}_e$ or its antiparticle $\bar{n}(=\bar{p}+e^{+}+\nu_e)$. Another example of a three-body system is a μ^-, consisting of a bound state of $e+\bar{\nu}_e+\nu_\mu$ and its antiparticle $\mu^+ = e^{+}+\nu_e+\bar{\nu}_\mu$. A further modification of the symmetry could correspond to other three-body, five-body etc. interactions. An "open space-time symmetry" such as a cylindrically symmetric system could describe systems like $e^{-}+\nu_e(=\pi^-)$, $e^{+}+\nu_e$ $(=\pi^+)$, $e^{+}+e^{-}(=\pi^o)$. In this case one could envisage an infinite number of similar particles of integral spin with even numbers of constituant particles. It thus appears to be possible to relate the geometry and spin-statistics properties of elementary particles.

APPENDIX 1

In this appendix we shall carry out the integration of the various integrals used in section 5 for the magnetic charge spectrum. We begin with

$$I_1 = \frac{1}{2}\int \frac{t^{3/2}\sqrt{(1+m^2t^2)}}{(1+t^2)^{3/2}}\,dt \tag{A.1}$$

$$= -\frac{1}{2}\frac{x\sqrt{(1+m^2x^4)}}{\sqrt{(1+x^4)}} + \frac{3}{2}L_1 - L_- \ ,$$

where

$$x^2 = t \ , \quad \frac{d}{dm}\left(\frac{L_1}{m}\right) = -\frac{L_-}{m^2} \ , \tag{A.2}$$

$$L_1 = \int \frac{\sqrt{(1+m^2x^4)}}{\sqrt{(1+x^4)}}\,dx \ , \quad L_- = \int \frac{dx}{\sqrt{(1+x^4)}\sqrt{(1+m^2x^4)}} \ . \tag{A.3}$$

For the integral L_- we use the substitution

$$x = m^{-\frac{1}{4}}\,[z+(z^2-1)]^{\frac{1}{2}} = m^{-\frac{1}{4}}\,U^{\frac{1}{2}}, \ U^2 = 2Uz-1 \ , \tag{A.4}$$

$$dx = \frac{m^{-\frac{1}{4}}}{2\sqrt{2}}\left[\frac{1}{\sqrt{(z-1)}} + \frac{1}{\sqrt{(z+1)}}\right]dz, \ U^{\frac{1}{2}} = \frac{1}{\sqrt{2}}[\sqrt{(z-1)}+\sqrt{(z+1)}],$$

$$U^{-\frac{1}{2}} = \frac{1}{\sqrt{2}}\,[\sqrt{(z+1)}-\sqrt{(z-1)}] \ ,$$

and obtain

$$L_- = \frac{1}{4\sqrt{(1+m)}}\,[K(\gamma,\tau_1) - K(\omega,\tau_2)] \ , \tag{A.5}$$

where the elliptic integrals of the first kind are defined as

$$K(\gamma,\tau_1)=\int^{\gamma}\frac{d\gamma}{\sqrt{(1-\tau_1^2\sin^2\gamma)}}\ ,\quad \gamma=\cos^{-1}\left[\frac{x^2(1+\ \ )-(x^2\sqrt{m}-1)^2}{x^2(1+\ \ )+(x^2\sqrt{m}-1)^2}\right], \tag{A.6}$$

$$K(\omega,\tau_2)=\int^{\omega}\frac{d\omega}{\sqrt{(1-\tau_2^2\sin^2\omega)}}\ ,\quad \omega=\cos^{-1}\left[\frac{x^2(1+m)-(x^2\sqrt{m}+1)^2}{x^2(1+m)+(x^2\sqrt{m}+1)^2}\right], \tag{A.7}$$

$$\tau_1^2=\frac{(1-\sqrt{m})^2}{2(1+m)}\ ,\quad \tau_2^2=\frac{(1+\sqrt{m})^2}{2(1+m)}\ , \tag{A.8}$$

$$\tau_1^2+\tau_2^2=1\ . \tag{A.9}$$

The behavior of the elliptic integral $K(\gamma,\tau_1)$ at $x=m^{-\frac{1}{4}}$ can be seen more clearly by writing

$$K(\gamma,\tau_1)=\sqrt{2}\ m^{-\frac{1}{4}}\sqrt{(1+m)}\int\frac{dz}{\sqrt{(z-1)}\sqrt{[4z^2+\frac{(1-m)^2}{m}]}}\ , \tag{A.10}$$

$$K(\omega,\tau_2)=\sqrt{2}\ m^{-\frac{1}{4}}\sqrt{(1+m)}\int\frac{dz}{\sqrt{(z+1)}\sqrt{[4z^2+\frac{(1+m)^2}{m}]}}\ , \tag{A.11}$$

where

$$z-1=\frac{1}{2}\,\frac{1+m}{\sqrt{m}}\,\frac{1-\cos\gamma}{1+\cos\gamma}=\frac{1}{2}\,\frac{1+m}{\sqrt{m}}\,\frac{\sin^2\gamma}{(1+\cos\gamma)^2}\ , \tag{A.12}$$

$$z+1=\frac{1}{2}\,\frac{1+m}{\sqrt{m}}\,\frac{1-\cos\omega}{1+\cos\omega}=\frac{1}{2}\,\frac{1+m}{\sqrt{m}}\,\frac{\sin^2\omega}{(1+\cos\omega)^2}\ . \tag{A.13}$$

From the relations

$$\sqrt{(z-1)}=\frac{x^2\sqrt{m}-1}{xm^{\frac{1}{4}}\sqrt{2}}\ ,\quad \sqrt{(z+1)}=\frac{x^2\sqrt{m}+1}{xm^{\frac{1}{4}}\sqrt{2}}\ , \tag{A.14}$$

and from (A.10) we see that there is a branch point at $z=1$ or at $x = m^{-\frac{1}{4}}$ (where $x > 0$). The point $z = -1$ corresponds to an imaginary x and therefore the integral (A.11) does not have a branch point. One of the consequences of the branch point in (A.10) is the fact that the limiting process of $m \to 1$ and integration do not commute. Thus in the forms (A.10) and (A.11) we have for $m = 1$ the results

$$\int \frac{dz}{z\sqrt{(z-1)}} = 2\varepsilon \tan^{-1}(z-1)^{\frac{1}{2}\varepsilon} , \qquad (A.15)$$

$$\int \frac{dz}{z\sqrt{(z+1)}} = 2 \tanh^{-1}(z+1)^{\frac{1}{2}\varepsilon} , \qquad (A.16)$$

where $\varepsilon = \pm 1$. However from (A.6) we see that for $m = 1$ we have $\tau_1 = 0$, and $K(\gamma,\tau_1) \to \gamma$ which is not equal to (A.15). However for $m = 1$ we have $\tau_2 = 1$ and the integral (A.7) is the same as (A.16) with $\varepsilon = 1$. The $\varepsilon = -1$ value in (A.16) corresponds to a complex number.

Now, if we apply the above substitutions we can integrate

$$I_2 = \frac{1}{2}\int \frac{t^{5/2}\sqrt{(1+m^2t^2)}}{(1+t^2)^{3/2}}\, dt = \qquad (A.17)$$

$$-\frac{1}{2}\ \frac{x^3\sqrt{(1+m^2x^4)}}{\sqrt{(1+x^4)}} + \frac{5}{2} L_3 - \frac{3}{2} L_+ ,$$

where

$$L_3 = \int \frac{x^2\sqrt{(1+m^2x^4)}}{\sqrt{(1+x^4)}}\, dx , \qquad (A.18)$$

$$L_+ = \int \frac{x^2}{\sqrt{(1+m^2x^4)}\sqrt{(1+x^4)}} dx = \frac{1}{4\sqrt{m}\sqrt{(1+m)}} [K(\gamma,\tau_1)+K(\omega,\tau_2)], \qquad (A.19)$$

$$\frac{d}{dm}\left(\frac{L_3}{m}\right) = -\frac{1}{m^2} L_+ \quad . \tag{A.20}$$

In an analogous way for the integral

$$I_3 = \frac{1}{2}\int \frac{t^{5/2}\, dt}{\sqrt{(1+t^2)}\sqrt{(1+m^2t^2)}} \quad , \tag{A.21}$$

we obtain

$$I_3 = \frac{1}{m^2}(L_3 - L_+) \quad . \tag{A.22}$$

Other relevant integrals are given by

$$\text{(i)} \int \frac{t^{3/2}\, dt}{1+t^2} = 2t^{\frac{1}{2}}\left[1 - \frac{1}{2\sqrt{(2t)}}\left|\tan^{-1}\left(\frac{\sqrt{(2t)}}{1-t}\right) + \tanh^{-1}\left(\frac{\sqrt{(2t)}}{1+t}\right)\right|\right] \tag{A.23}$$

$$\text{(ii)} \int \frac{t^{5/2}\, dt}{1+t^2} = 2t^{\frac{1}{2}}\left[\frac{t}{3} - \frac{1}{2\sqrt{(2t)}}\left|\tan^{-1}\left(\frac{\sqrt{(2t)}}{1-t}\right) + \tanh^{-1}\left(\frac{\sqrt{(2t)}}{1+t}\right)\right|\right] \tag{A.24}$$

$$\text{(iii)} \int \frac{t^{\frac{1}{2}} dt}{\sqrt{(1+t^2)}} = K\left(\alpha, \frac{1}{\sqrt{2}}\right) - 2E\left(\alpha, \frac{1}{\sqrt{2}}\right) + 2\,\frac{\sqrt{t}\sqrt{(1+t^2)}}{1+t} \quad , \tag{A.25}$$

$$\text{(iv)} \int \frac{t^{3/2}\, dt}{\sqrt{(1+t^2)}} = \frac{2}{3}\left[\sqrt{t}\sqrt{(1+t^2)} - \frac{1}{2} K\left(\alpha, \frac{1}{\sqrt{2}}\right)\right] \quad , \tag{A.26}$$

$$\text{(v)} \int \frac{t^{3/2}\, dt}{(1+t^2)^{3/2}} = -\frac{t^{\frac{1}{2}}}{\sqrt{(1+t^2)}} + \frac{1}{2} K\left(\alpha, \frac{1}{\sqrt{2}}\right) \quad , \tag{A.27}$$

$$\text{(vi)} \int \frac{t^{5/2}\, dt}{(1+t^2)^{3/2}} = -\frac{t^{3/2}}{\sqrt{(1+t^2)}} + \frac{3}{2} \frac{t^{\frac{1}{2}} dt}{\sqrt{(1+t^2)}} \quad , \tag{A.28}$$

$$\text{(vii)} \int \frac{t^{7/2}\, dt}{(1+t^2)^{3/2}} = -\frac{t^{5/2}}{\sqrt{(1+t^2)}} + \frac{5}{3}\left[\sqrt{t}\ \sqrt{(1+t^2)} - \frac{1}{2} K\left(\alpha, \frac{1}{\sqrt{2}}\right)\right], \tag{A.29}$$

where

$$K(\alpha, \tfrac{1}{\sqrt{2}}) = \int^{\alpha} \frac{d\alpha'}{\sqrt{(1-\frac{1}{2}\sin^2\alpha')}} \quad , \quad E(\alpha, \tfrac{1}{\sqrt{2}}) = \int^{\alpha} \sqrt{(1-\tfrac{1}{2}\sin^2\alpha')}\,d\alpha',$$

$$\alpha = \cos^{-1}\left(\frac{1-t}{1+t}\right) \quad , \quad \frac{dK}{dt} = \frac{1}{\sqrt{t}\ \sqrt{(1+t^2)}} \ .$$

APPENDIX 2

Linearization of Energy Tensor

From the equations

$$\hat{g}_{\mu\nu;\rho} = \hat{g}_{\mu\nu,\rho} - \hat{g}_{\sigma\nu}\Gamma^{\sigma}_{\mu\rho} - \hat{g}_{\mu\sigma}\Gamma^{\sigma}_{\rho\nu} = 0 \quad , \tag{B.1}$$

by separating out the anti-symmetric part of the affine connection $\Gamma^{\rho}_{\mu\nu}$ we can easily show that

$$\Gamma^{\rho}_{[\mu\nu]} = g^{\rho\sigma}(-\tfrac{1}{2}\, I_{\mu\nu\sigma} + \Phi_{\mu\nu o\sigma}) \quad , \tag{B.2}$$

where the symbol (o) implies covariant differentiation with respect to the symmetric part $\Gamma^{\rho}_{\{\mu\nu\}}$ of $\Gamma^{\rho}_{\mu\nu}$, viz.,

$$\Phi_{\mu\nu o\sigma} = \Phi_{\mu\nu,\sigma} - \Phi_{\alpha\nu}\,\Gamma^{\alpha}_{\{\mu\nu\}} - \Phi_{\mu\alpha}\,\Gamma^{\alpha}_{\{\sigma\nu\}} \tag{B.3}$$

$$= \Phi_{\mu\nu|\sigma} - \Phi_{\alpha\nu}\,S^{\alpha}_{\mu\sigma} - \Phi_{\mu\alpha}\,S^{\alpha}_{\sigma\nu} \quad ,$$

where we used the relations

$$\Gamma^{\rho}_{\{\mu\nu\}} = \left\{{\rho \atop \mu\nu}\right\} + S^{\rho}_{\mu\nu} \quad , \quad S^{\rho}_{\mu\nu} = S^{\rho}_{\nu\mu} \quad . \tag{B.4}$$

The tensor $S^{\rho}_{\mu\nu}$, as follows from B.1, is given by

$$S^{\rho}_{\mu\nu} = g^{\rho\sigma}[\Phi_{\mu}{}^{\alpha}\,\Gamma_{[\sigma\nu]\alpha} + \Phi_{\nu}{}^{\alpha}\,\Gamma_{[\sigma\mu]\alpha}] \,, \qquad (B.5)$$

where

$$\Gamma_{[\mu\nu]\rho} = g_{\sigma\rho}\,\Gamma^{\sigma}_{[\mu\nu]} \quad .$$

From the equations B.2 and B.3 it is clear that the tensor $\Gamma^{\rho}_{[\mu\nu]}$ can be expanded in powers of q^{-1}. Because of very large size of q ($\sim 10^{58}$ esu) it will not be necessary to retain terms beyond q^{-2} orders. Thus to q^{-2} order the equation B.2, dropping (o)-covariant derivatives in favor of (|)-covariant derivatives, becomes

$$\Gamma^{\rho}_{[\mu\nu]} = g^{\rho\sigma}\,(-\tfrac{1}{2}\,I_{\mu\nu\sigma} + \Phi_{\mu\nu|\sigma}) \quad . \qquad (B.6)$$

Hence for the tensor $S^{\rho}_{\mu\nu}$, as defined by B.5, we can write

$$S^{\rho}_{\mu\nu} = (g_{\mu\beta}\Phi_{\nu\alpha}+g_{\nu\beta}\Phi_{\mu\alpha})(\tfrac{1}{2}\,I^{\alpha\beta\rho}+\Phi^{\rho\beta|\alpha}) \,, \qquad (B.7)$$

where

$$\Phi^{\alpha\beta|\rho} = g^{\rho\sigma}\Phi^{\alpha\beta}{}_{|\sigma} \quad . \qquad (B.8)$$

In the linearization of the energy tensor $T_{\mu\nu}$ defined by (7.25) we shall use the following relations and definitions: From

$$f^{\mu\nu} = \frac{1}{2\sqrt{(-g)}}\,\varepsilon^{\mu\nu\rho\sigma}\Phi_{\rho\sigma} \quad ,$$

we obtain

$$\Phi_{\mu\nu} = \frac{1}{2}\sqrt{(-g)}\varepsilon_{\mu\nu\rho\sigma}f^{\rho\sigma} , \quad \Phi^{\mu\nu} = -\frac{1}{2\sqrt{(-g)}}\varepsilon^{\mu\nu\rho\sigma}f_{\rho\sigma} , \tag{B.9}$$

where we used the well-known relations

$$\varepsilon_{\alpha\beta\gamma\delta}\, g^{\alpha\mu}g^{\beta\nu}g^{\gamma\rho}g^{\delta\sigma} = \frac{1}{g}\varepsilon^{\mu\nu\rho\sigma} ,$$

$$\varepsilon^{\mu\nu\rho\gamma}\varepsilon_{\alpha\beta\sigma\gamma} = \delta^{\mu\nu\rho}_{\alpha\beta\sigma} , \quad \delta^{\mu\nu\rho}_{\alpha\beta\rho} = 2\delta^{\mu\nu}_{\alpha\beta} = 2(\delta^{\mu}_{\alpha}\delta^{\nu}_{\beta}-\delta^{\mu}_{\beta}\delta^{\nu}_{\alpha}). \tag{B.10}$$

From the definitions

$$I_{\mu\nu\rho} = -4\pi\,\varepsilon_{\mu\nu\rho\sigma}\,\mathfrak{s}^{\sigma} , \quad \mathfrak{s}^{\rho} = \sqrt{(-g)}s^{\rho} , \tag{B.11}$$

we obtain

$$\mathfrak{s}^{\mu} = \frac{1}{4\pi}\frac{\partial}{\partial x^{\nu}}[\sqrt{(-g)}f^{\mu\nu}] = \frac{1}{4\pi}[\sqrt{(-g)}f^{\mu\nu}]_{|\nu}, \tag{B.12}$$

$$I^{\mu\nu\rho} = \frac{4\pi}{\sqrt{(-g)}}\varepsilon^{\mu\nu\rho\sigma}s_{\sigma} ,$$

where

$$s^{\mu} = \frac{1}{4\pi}f^{\mu\nu}{}_{|\nu} . \tag{B.13}$$

We may now express the linearized tensor $S^{\rho}_{\mu\nu}$ in the form

$$S^{\rho}_{\mu\nu} = 4\pi[-g_{\mu\nu}f^{\rho\sigma}s_{\sigma}+\tfrac{1}{2}(f_{\mu\sigma}\delta^{\rho}_{\nu}+f_{\nu\sigma}\delta^{\rho}_{\mu})s^{\sigma}-\tfrac{1}{2}(f_{\mu}{}^{\rho}s_{\nu}+f_{\nu}{}^{\rho}s_{\mu})] +$$
$$4\pi(\Phi_{\mu}{}^{\rho}j_{\nu}+\Phi_{\nu}{}^{\rho}j_{\mu})-(\Phi_{\mu}{}^{\rho}\Phi_{\nu}{}^{\sigma}+\Phi_{\nu}{}^{\rho}\Phi_{\mu}{}^{\sigma})_{|\sigma} , \tag{B.14}$$

where the electric density vector j_{μ} is given by

$$j^{\mu} = \frac{1}{4\pi}\,\Phi^{\mu\nu}{}_{|\nu} \quad , \tag{B.15}$$

which is related to the electric current vector density j^{μ}_{e} by

$$\sqrt{(-g)}\,j^{\mu} = j^{\mu}_{e} \quad . \tag{B.16}$$

Hence the covariant divergence of $S^{\rho}_{\mu\nu}$ can be written as

$$\begin{aligned} S^{\rho}_{\mu\nu|\rho} = {} & 4\pi[4\pi(g_{\mu\nu}s_{\rho}s^{\rho}-s_{\mu}s_{\nu})+\tfrac{1}{2}(f_{\mu}{}^{\sigma}s_{\nu\sigma}+f_{\nu}{}^{\sigma}s_{\mu\sigma}-g_{\mu\nu}f^{\rho\sigma}s_{\rho\sigma}) + \\ & \tfrac{1}{2}(f_{\mu\sigma|\nu}+f_{\nu\sigma|\mu})s^{\sigma}]+4\pi(\Phi_{\mu}{}^{\rho}j_{\nu|\rho}+\Phi_{\nu}{}^{\rho}j_{\mu|\rho}) \\ & -(\Phi_{\mu}{}^{\rho}\Phi_{\nu}{}^{\sigma}+\Phi_{\nu}{}^{\rho}\Phi_{\mu}{}^{\sigma})_{|\sigma\rho} \quad , \end{aligned} \tag{B.17}$$

where

$$s_{\rho\sigma} = s_{\sigma|\rho}-s_{\rho|\sigma} = \partial_{\rho}s_{\sigma}-\partial_{\sigma}s_{\rho} \quad . \tag{B.18}$$

For the bilinear term of the energy tensor we have

$$\begin{aligned} \Gamma^{\rho}_{[\mu\sigma]}\Gamma^{\sigma}_{[\rho\nu]} = {} & 4\pi[4\pi\tfrac{1}{2}(g_{\mu\nu}s_{\rho}s^{\rho}-s_{\mu}s_{\nu})+\tfrac{1}{2}(f_{\mu\rho|\nu}+f_{\nu\rho|\mu})s^{\rho}] + \\ & 4\pi[4\pi j_{\mu}j_{\nu}+\Phi_{\mu}{}^{\rho}j_{\nu|\rho}+\Phi_{\nu}{}^{\rho}j_{\mu|\rho}]-\tfrac{1}{2}[\Phi_{\mu}{}^{\rho}\Phi_{\nu}{}^{\sigma}+\Phi_{\nu}{}^{\rho}\Phi_{\mu}{}^{\sigma}]_{|} \end{aligned} \tag{B.19}$$

and

$$g^{\mu\nu}\Gamma^{\rho}_{[\mu\sigma]}\Gamma^{\sigma}_{[\rho\nu]}=4\pi[4\pi(\tfrac{1}{2}s_{\rho}s^{\rho}+j_{\rho}j^{\rho})-\Phi^{\rho\sigma}j_{\rho\sigma}]-(\Phi^{\nu\rho}\Phi_{\nu}{}^{\sigma})_{|\rho\sigma} \tag{B.20}$$

where

$$j_{\mu\nu} = j_{\nu|\mu} - j_{\mu|\nu} = \partial_\mu j_\nu - \partial_\nu j_\mu \quad . \qquad (B.21)$$

For the derivation of B.19 we have used the relations

$$\Phi_\mu{}^\rho{}_{|\sigma\rho} - \Phi_\mu{}^\rho{}_{|\rho\sigma} = G_{\rho\sigma}\Phi_\mu{}^\rho + G^\alpha{}_{\mu\rho\sigma}\Phi_\alpha{}^\rho \quad , \qquad (B.22)$$

where $G_{\mu\nu}$ was defined by (7.23) and where $G^\sigma{}_{\mu\nu\rho}$ is the corresponding curvature tensor. Hence, using the field equations (7.19) in B.22, we obtain

$$\Phi_\mu{}^\rho{}_{|\sigma\rho} \cong 4\pi j_{\mu|\sigma} \quad . \qquad (B.23)$$

The linearized energy tensor, as follows from (7.25), is given by

$$4\pi T_{\mu\nu} = T_{\mu\nu} + r_o^2[\Gamma^\rho_{[\mu\sigma]}\Gamma^\sigma_{[\rho\nu]}$$

$$-\tfrac{1}{2}g_{\mu\nu}g^{\alpha\beta}\Gamma^\rho_{[\alpha\sigma]}\Gamma^\sigma_{[\rho\beta]} + S^\rho_{\rho\mu|\nu} - \tfrac{3}{2}g_{\mu\nu}g^{\alpha\beta}S^\rho_{\rho\alpha|\beta} - S^\rho_{\mu\nu|\rho}] \quad .$$

$$(B.24)$$

On substituting from the above relations in B.24 we obtain the linearized energy tensor in the form

$$T_{\mu\nu} = \frac{1}{4\pi} T_{\mu\nu} + 4\pi r_o^2 \, [\tfrac{1}{2}(s_\mu s_\nu + \tfrac{1}{2}g_{\mu\nu}s_\rho s^\rho) + j_\mu j_\nu + \tfrac{1}{2}g_{\mu\nu}j_\rho j^\rho] +$$

$$\frac{1}{2}\, r_o^2 \, (f_\mu{}^\rho s_{\rho\nu} + f_\nu{}^\rho s_{\rho\mu}) + \frac{r_o^2}{4\pi} M^{\rho\sigma}_{\mu\nu|\rho\sigma}$$

where

$$T_{\mu\nu} = \frac{1}{2}\,\Omega g_{\mu\nu} - \Phi_{\mu\rho}\Phi_{\nu}{}^{\rho}$$

and where

$$M^{\rho\sigma}_{\mu\nu} = \frac{1}{2}\left[\left(\frac{1}{2}(\delta^{\rho}_{\mu}\delta^{\sigma}_{\nu}+\delta^{\rho}_{\nu}\delta^{\sigma}_{\mu}) - g_{\mu\nu}g^{\rho\sigma}\right)\Omega + g_{\mu\nu}T^{\rho\sigma} + \Phi_{\mu}{}^{\rho}\Phi_{\nu}{}^{\sigma} + \Phi_{\nu}{}^{\rho}\Phi_{\mu}{}^{\sigma}\right] \tag{B.26}$$

is a symmetric tensor in both upper and lower indices. Now, by an appropriate choice of coordinates, we can bring the quantities $g_{\mu\nu}$, which describe gravitational field, into their Galilean form at any individual point of the curved space-time. In fact we can regard a small neighborhood of such a point as a flat space-time and replace the covariant derivatives in the last term of B.25 by ordinary partial derivatives. The integration over the four-volume enclosed by this small hypersurface of the last term in B.25 is given by

$$\int M^{\mu\nu\rho\sigma}{}_{,\rho\sigma}\, d^4x = \int M^{\mu\nu\rho\sigma}{}_{,\sigma}\, dS_{\rho} = \frac{1}{2}\int M^{\mu\nu\rho\sigma}\, dS_{\rho\sigma} = 0 \ , \tag{B.27}$$

where the infinitesimal surface elements $dS_{\rho\sigma}$ are given by

$$dS_{\rho\sigma} = \frac{1}{2}\,\varepsilon_{\rho\sigma\mu\nu}(dx^{\mu}dx'^{\nu} - dx^{\nu}dx'^{\mu})$$

and where the hypersurface elements dS_{ρ} are defined in the usual way. In (B.27) we used Gauss' theorem and the relations $M^{\mu\nu\rho\sigma} = M^{\mu\nu\sigma\rho}$, $dS_{\rho\sigma} = -\,dS_{\sigma\rho}$. We may now go back to a curved space-time and extend the last integral to the entire space-time without reference to a Galilean point. Hence we see that the contribution of the last term in (B.25) to the total energy flow in space-time vanishes. Thus the last term in (B.25) may be interpreted as the energy density of the "new vacuum".

APPENDIX 3

Here we shall derive the fundamental indentities of the nonsymmetric theory. One of the ways to obtain these identities can be based on the use of infinitesimal coordinate transformations. First we note that under an arbitrary transformation of the coordinates by

$$x'^{\mu} = f^{\mu}(x) \tag{C.1}$$

we have the transformation rules

$$\hat{g}'_{\mu\nu} = \frac{\partial x^{\rho}}{\partial x'^{\mu}} \frac{\partial x^{\sigma}}{\partial x'^{\nu}} \hat{g}_{\rho\sigma} \quad , \tag{C.2}$$

and, as follows from the infinitesimal parallel displacement law

$$\delta A^{\mu} = -\Gamma^{\mu}_{\rho\sigma} A^{\rho} \delta x^{\sigma} \quad ,$$

the rules

$$\Gamma'^{\rho}_{\mu\nu} = \frac{\partial x^{\alpha}}{\partial x'^{\mu}} \frac{\partial x^{\beta}}{\partial x'^{\nu}} \frac{\partial x'^{\rho}}{\partial x^{\sigma}} \Gamma^{\sigma}_{\alpha\beta} - \frac{\partial^2 x'^{\rho}}{\partial x^{\alpha} \partial x^{\beta}} \frac{\partial x^{\alpha}}{\partial x'^{\mu}} \frac{\partial x^{\beta}}{\partial x'^{\nu}} . \tag{C.3}$$

From (C.2) we obtain

$$\sqrt{(-\hat{g}')} = \left|\frac{\partial x}{\partial x'}\right| \sqrt{(-\hat{g})} \quad , \tag{C.4}$$

where

$$\left|\frac{\partial x}{\partial x'}\right| = \mathrm{Det}\left[\frac{\partial x^{\mu}}{\partial x'^{\nu}}\right] \quad .$$

We also have

$$\hat{g}'^{\mu\nu} = \left|\frac{\partial x}{\partial x'}\right| \frac{\partial x'^{\mu}}{\partial x^{\rho}} \frac{\partial x'^{\nu}}{\partial x^{\sigma}} \hat{g}^{\rho\sigma} \quad . \tag{C.5}$$

We may now apply an infinitesimal transformation of the coordinates

$$x'^{\mu} = x^{\mu} + \xi^{\mu}(x) \quad , \tag{C.6}$$

and obtain

$$\left|\frac{\partial x'}{\partial x}\right| \cong 1 + \xi^{\rho}_{,\rho} \quad , \quad \left|\frac{\partial x}{\partial x'}\right| \cong 1 - \xi^{\rho}_{,\rho} \quad , \tag{C.7}$$

where ξ^{μ} are small compared to 1. Hence

$$\delta\hat{g}^{\mu\nu} = \hat{g}'^{\mu\nu}(x) - \hat{g}^{\mu\nu}(x) = \hat{g}^{\mu\rho}\,\xi^{\nu}_{,\rho} + \hat{g}^{\rho\nu}\,\xi^{\mu}_{,\rho} - \hat{g}^{\mu\nu}\,\xi^{\rho}_{,\rho} - \hat{g}^{\mu\nu}_{,\rho}\,\xi^{\rho} \ , \tag{C.8}$$

$$\delta\Gamma^{\rho}_{\mu\nu} = \Gamma^{\alpha}_{\mu\nu}\,\xi^{\rho}_{,\alpha} - \Gamma^{\rho}_{\mu\alpha}\,\xi^{\alpha}_{,\nu} - \Gamma^{\rho}_{\alpha\nu}\,\xi^{\alpha}_{,\mu} - \Gamma^{\rho}_{\mu\nu,\alpha}\,\xi^{\alpha} - \xi^{\rho}_{,\mu\nu} \ . \tag{C.9}$$

From the definition of the nonsymmetric curvature tensor by

$$R_{\mu\nu} = -\Gamma^{\rho}_{\mu\nu,\rho} + \Gamma^{\rho}_{\mu\rho,\nu} + \Gamma^{\rho}_{\mu\sigma}\Gamma^{\sigma}_{\rho\nu} - \Gamma^{\rho}_{\mu\nu}\Gamma^{\sigma}_{\rho\sigma} \ , \tag{C.10}$$

and from using (C.9) we obtain the change

$$\delta R_{\mu\nu} = -R_{\mu\nu,\rho}\,\xi^{\rho} - R_{\rho\nu}\,\xi^{\rho}_{,\mu} - R_{\mu\rho}\,\xi^{\rho}_{,\nu} \quad , \tag{C.11}$$

which follows also from

$$R'_{\mu\nu}(x+\xi) = \frac{\partial x^{\rho}}{\partial x'^{\mu}} \frac{\partial x^{\sigma}}{\partial x'^{\nu}} R_{\rho\sigma} = (\delta^{\rho}_{\mu} - \xi^{\rho}_{,\mu})(\delta^{\sigma}_{\nu} - \xi^{\sigma}_{,\nu}) R_{\rho\sigma} \quad ,$$

as

$$R'_{\mu\nu}(x) + R'_{\mu\nu,\rho}\ \xi^{\rho} \cong R_{\mu\nu}(x) - R_{\mu\sigma}\ \xi^{\sigma}_{,\nu} - R_{\rho\nu}\ \xi^{\rho}_{,\mu} \quad .$$

Hence

$$\delta R_{\mu\nu} = R'_{\mu\nu}(x) - R_{\mu\nu}(x) \cong - R_{\mu\nu,\rho}\ \xi^{\rho} - R_{\mu\sigma}\ \xi^{\sigma}_{,\nu} - R_{\rho\nu}\ \xi^{\rho}_{,\mu} \quad .$$

The above results lead to

$$R_{\mu\nu}\ \delta\hat{g}^{\mu\nu} = \hat{g}^{\mu\nu}\ (R_{\underset{+-}{\mu\nu};\rho} - R_{\underset{++}{\mu\rho};\nu} - R_{\underset{--}{\rho\nu};\mu})\xi^{\rho} \quad , \tag{C.12}$$

$$\hat{g}^{\mu\nu}\ \delta R_{\mu\nu} = \hat{g}^{\underset{+-}{\mu\nu}}{}_{;\rho}\ \delta\Gamma^{\rho}_{\mu\nu} - \hat{g}^{\underset{+-}{\mu\nu}}{}_{;\nu}\ \delta\Gamma^{\rho}_{\mu\rho} \quad , \tag{C.13}$$

where the divergence terms have been dropped and where we assumed that the variations $\delta\hat{g}^{\mu\nu}$, $\delta\Gamma^{\rho}_{\mu\nu}$ arise from the infinitesimal change of the coordinates defined by (C.6). We also assumed that

$$\hat{g}^{\underset{+-}{\mu\nu}}{}_{;\rho} = 0 \quad . \tag{C.14}$$

The + and - signs under subscripts serve to rank the respective indices to be placed as first and second index, respectively, in the process of covariant differentiation with respect to the affine connection $\Gamma^{\rho}_{\mu\nu}$. Furthermore from the relations

$$\delta\surd(-\hat{g}) = \frac{1}{2}\ \hat{g}_{\mu\nu}\ \delta\hat{g}^{\mu\nu} \quad , \quad \delta\surd(-g) = \frac{1}{2}\ b_{\mu\nu}\ \delta\hat{g}^{\mu\nu}$$

we obtain

$$\delta\surd(-\hat{g}) = \frac{1}{2}\ \hat{g}_{\mu\nu}\ (\hat{g}^{\mu\rho}\ \xi^{\nu}_{,\rho} + \hat{g}^{\rho\nu}\ \xi^{\mu}_{,\rho} - \hat{g}^{\mu\nu}\ \xi^{\rho}_{,\rho} - \hat{g}^{\mu\nu}_{,\rho}\ \xi^{\rho})$$

$$= \frac{1}{2}\ \hat{g}^{\mu\nu}\ (\hat{g}_{\underset{+-}{\mu\nu};\rho} - \hat{g}_{\underset{++}{\mu\rho};\nu} - \hat{g}_{\underset{--}{\rho\nu};\mu})\xi^{\rho} \quad . \tag{C.15}$$

Hence the variational principle $\delta S=0$ lead to the differential identities

$$\hat{g}^{\mu\nu}(R_{\underset{+-}{\mu\nu;\rho}} - R_{\underset{++}{\mu\rho;\nu}} - R_{\underset{--}{\rho\nu;\mu}}) = 0 \quad , \quad (C.16)$$

$$\hat{g}^{\mu\nu}(\hat{g}_{\underset{+-}{\mu\nu;\rho}} - \hat{g}_{\underset{++}{\mu\rho;\nu}} - \hat{g}_{\underset{--}{\rho\nu;\mu}}) = 0 \quad . \quad (C.17)$$

Actually, the identities (C.16) and (C.17) which correspond to the Bianchi identities of the nonsymmetric field were derived by Einstein without using variational principle but only by using various symmetries of the curvature tensor

$$R^{\sigma}{}_{\mu\nu\rho} = - \Gamma^{\sigma}_{\mu\nu,\rho}+\Gamma^{\sigma}_{\mu\rho,\nu}+\Gamma^{\sigma}_{\alpha\nu}\Gamma^{\alpha}_{\mu\rho}-\Gamma^{\alpha}_{\mu\nu}\Gamma^{\sigma}_{\alpha\rho} \quad .(C.18)$$

We note that in (C.17) we have

$$\hat{g}_{\underset{+-}{\mu\nu;\rho}} = 0 \quad , \quad \hat{g}_{\underset{++}{\mu\rho;\nu}} \neq 0 \quad , \quad \hat{g}_{\underset{--}{\rho\nu;\mu}} \neq 0 \quad .$$

Now from (C.14) we have

$$\hat{g}^{\{\mu\nu\}}{}_{,\nu} = - \hat{g}^{\rho\sigma} \Gamma^{\mu}_{\rho\sigma} \quad . \quad (C.19)$$

Hence on carrying out the indicated covariant differentiations in (C.16), (C.17) and using (C.19) we may rewrite them in a more symmetrical form as

$$\left[\sqrt{(-g)}[b^{\mu\nu}R_{\{\mu\rho\}}-\tfrac{1}{2}\delta^{\nu}_{\rho}b^{\mu\sigma}R_{\{\mu\sigma\}}]\right]_{||\nu}$$

$$= \tfrac{1}{2}\hat{g}^{[\mu\nu]}\left[R_{[\mu\nu],\rho}+R_{[\nu\rho],\mu}+R_{[\rho\mu],\nu}\right] \quad , \quad (C.20)$$

$$\left[\sqrt{(-g)}[b^{\mu\nu}g_{\mu\rho}-\tfrac{1}{2}\delta^{\nu}_{\rho}b^{\mu\sigma}g_{\mu\sigma}]\right]_{||\nu}=\tfrac{1}{2}\,\hat{g}^{[\mu\nu]}\left[\Phi_{\mu\nu,\rho}+\Phi_{\nu\rho,\mu}+\Phi_{\rho\mu,\nu}\right], \tag{C.21}$$

where the sign(||) indicates covariant differentiation in terms of the Christoffel symbols constructed out of the tensor $b_{\mu\nu}$. If we set $\Phi_{\mu\nu} = 0$ the identities (C.20) reduce to

$$\left[\sqrt{(-g)}[g^{\mu\nu}G_{\mu\rho} - \tfrac{1}{2}\,\delta^{\nu}_{\rho}\,g^{\mu\sigma}G_{\mu\sigma}]\right]_{|\nu} = 0 \;, \tag{C.22}$$

which are the Bianchi identities of the symmetric theory, where $G_{\mu\nu}$ is the curvature tensor of general relativity. Hence the right hand side of (C.20), in analogy with general relativity, represents a force density. Einstein used the identities (C.16) alone in deriving his field equations even though the same identities were also satisfied by the field tensor $\hat{g}_{\mu\nu}$, as in (C.17), as well as by the nonsymmetric tensor $b_{\mu\nu} + F_{\mu\nu}$. In general these identities are satisfied by the tensor $A_{\mu\nu} = \hat{g}_{\mu\nu} + \lambda(b_{\mu\nu}+F_{\mu\nu})$ where λ is an arbitrary constant. By direct substitution of $A_{\mu\nu}$ in (C.17) we obtain

$$\hat{g}^{\mu\nu}(b_{\mu\nu,\rho}-b_{\mu\rho,\nu}-b_{\rho\nu,\mu}+2b_{\rho\sigma}\Gamma^{\sigma}_{\mu\nu}+F_{\mu\nu,\rho}+F_{\nu\rho,\mu}+F_{\rho\mu,\nu}) =$$

$$\sqrt{(-g)}b^{\mu\nu}(b_{\mu\nu,\rho}-2b_{\mu\rho,\nu}) + 2b_{\rho\sigma}\,\hat{g}^{\mu\nu}\,\Gamma^{\sigma}_{\mu\nu} = 0 \quad ,$$

where we used the relations

$$\tfrac{1}{2}\,b^{\mu\nu}\,b_{\mu\nu,\rho} = \frac{[\sqrt{(-g)}]_{,\rho}}{\sqrt{(-g)}} \qquad \hat{g}^{\mu\nu}\,\Gamma^{\rho}_{\mu\nu} = -\,[\sqrt{(-g)}b^{\rho\nu}]_{,\nu} \;.$$

Einstein assumed that the field equations should stipulate the vanishing of either $R_{\mu\nu}$ or that the vanishing of $R_{\{\mu\nu\}}$ and the way $R_{[\mu\nu]}$ enter (C.20) viz.

$$R_{[\mu\nu],\rho} + R_{[\nu\rho],\mu} + R_{[\rho\mu],\nu} = 0 \quad .$$

On the other hand Schrödinger assumed the existence of a cosmological constant and proposed the field equations

$$R_{\mu\nu} = \lambda \, \hat{g}_{\mu\nu} \quad .$$

In fact the simultaneous existence of (C.16) and (C.17) or (C.20) and (C.21) implies that the role of these differential identities in the derivation of the field equations is more general than the use of $R_{\mu\nu}$ alone. The identities (C.16) or (C.20) are, because of the identities (C.17) or (C.21), invariant under the substitution

$$R_{\mu\nu} \rightarrow R_{\mu\nu} + \lambda \, \hat{g}_{\mu\nu} + \gamma(b_{\mu\nu}+F_{\mu\nu}) \quad . \tag{C.23}$$

Thus the most general possible form of the field equations is contained in the statement

$$R_{\mu\nu} + \lambda \, \hat{g}_{\mu\nu} + \gamma(b_{\mu\nu}+F_{\mu\nu}) = 0 \quad .$$

Hence we obtain

$$R_{\{\mu\nu\}} = - \lambda \, g_{\mu\nu} - \gamma \, b_{\mu\nu} \quad ,$$

$$R_{[\mu\nu\rho]} = - \lambda \, I_{\mu\nu\rho} \quad .$$

If we assume that for $\Phi_{\mu\nu}=0$ the field equations must reduce to the field equations of a pure gravitational field without a cosmological constant we obtain

$$\gamma = -\lambda \ (= -\frac{1}{2}\kappa^2)$$

or

$$R_{\{\mu\nu\}} = \frac{1}{2}\kappa^2(b_{\mu\nu}-g_{\mu\nu})$$

$$R_{[\mu\nu\rho]} = -\frac{1}{2}\kappa^2 I_{\mu\nu\rho}, \quad \hat{g}^{[\mu\nu]}{}_{,\nu} = 0 \quad .(C.24)$$

For the sake of completeness we shall use the above formalism to drive the conservation law (7.37). From (C.13) we have

$$\hat{g}^{\mu\nu}\,\delta R_{\mu\nu} = [(\hat{g}^{\mu\nu}R_{\mu\rho}+\hat{g}^{\nu\mu}R_{\rho\mu}-\delta^{\nu}_{\rho}\hat{g}^{\alpha\beta}R_{\alpha\beta})_{,\nu} + \hat{g}^{\alpha\beta}{}_{,\rho}R_{\alpha\beta}]\xi^{\rho}, \tag{C.25}$$

where the divergence term has been dropped. In order to calculate the last term in (C.25) we shall calculate the variation of $\hat{g}^{\mu\nu}{}_{,\rho}$. From the definition

$$\hat{g}^{\mu\nu}{}_{,\rho} = -\hat{g}^{\alpha\nu}\Gamma^{\mu}_{\alpha\rho} - \hat{g}^{\mu\alpha}\Gamma^{\nu}_{\rho\alpha} + \hat{g}^{\mu\nu}\Gamma^{\alpha}_{\rho\alpha},$$

we obtain

$$\delta\hat{g}^{\mu\nu}{}_{,\rho} = \Gamma^{\sigma}_{\rho\sigma}\delta\hat{g}^{\mu\nu}-\Gamma^{\mu}_{\sigma\rho}\delta\hat{g}^{\sigma\nu}-\Gamma^{\nu}_{\rho\sigma}\delta\hat{g}^{\mu\sigma}+\hat{g}^{\mu\nu}\delta\Gamma^{\sigma}_{\rho\sigma}-\hat{g}^{\mu\sigma}\delta\Gamma^{\nu}_{\rho\sigma}-\hat{g}^{\sigma\nu}\delta\Gamma^{\mu}_{\sigma\rho}.$$

On multiplying through by

$$B^{\rho}_{\mu\nu} = \frac{1}{2}(\delta^{\rho}_{\mu}\Gamma^{\alpha}_{\nu\alpha} + \delta^{\rho}_{\nu}\Gamma^{\alpha}_{\mu\alpha}) - \Gamma^{\rho}_{\mu\nu}, \tag{C.26}$$

we obtain

$$B^{\rho}_{\mu\nu}\,\delta\hat{g}^{\mu\nu}{}_{,\rho} = \hat{g}^{\mu\nu}\,\delta B_{\mu\nu} + 2B_{\mu\nu}\,\delta\hat{g}^{\mu\nu}, \tag{C.27}$$

where

$$\mathcal{B}_{\mu\nu} = \Gamma^{\rho}_{\mu\sigma}\Gamma^{\sigma}_{\rho\nu} - \Gamma^{\rho}_{\mu\nu}\Gamma^{\sigma}_{\rho\sigma} \ , \ \mathcal{B} = \hat{g}^{\mu\nu}\mathcal{B}_{\mu\nu} = \frac{1}{2}\,\hat{g}^{\mu\nu}_{\ \ ,\rho}\ \mathcal{B}^{\rho}_{\mu\nu} \ . \quad (C.28)$$

By using the variation of $\mathcal{B}$ we get the result

$$\delta\mathcal{B} = \mathcal{B}_{\mu\nu}\ \delta\hat{g}^{\mu\nu} + \hat{g}^{\mu\nu}\ \delta\mathcal{B}_{\mu\nu} = -\ \mathcal{B}_{\mu\nu}\ \delta\hat{g}^{\mu\nu} + \mathcal{B}^{\rho}_{\mu\nu}\ \delta\hat{g}^{\mu\nu}_{\ \ ,\rho} \ .$$

Hence the variational derivatives of $\mathcal{B}$ yields the result

$$\frac{\partial\mathcal{B}}{\partial\hat{g}^{\mu\nu}} - \frac{\partial}{\partial x^{\rho}}\left\{\frac{\partial\mathcal{B}}{\partial\hat{g}^{\mu\nu}_{\ \ ,\rho}}\right\} = -\ R_{\mu\nu} \ . \quad (C.29)$$

Thus, using (C.29), we obtain

$$-\ g^{\mu\nu}_{\ \ ,\rho}\ R_{\mu\nu} = \frac{\partial}{\partial x^{\sigma}}\left[\delta^{\sigma}_{\rho}\mathcal{B} - \hat{g}^{\mu\nu}_{\ \ ,\rho}\ \mathcal{B}^{\sigma}_{\mu\nu}\right] \ . \quad (C.30)$$

On combining (C.30) with (C.25) we obtain the conservation laws

$$\mathcal{F}^{\nu}_{\mu,\nu} = 0 \ , \quad (C.31)$$

where the nonsymmetric tensor $\mathcal{F}^{\nu}_{\mu}$ is given by

$$-4\pi\kappa^{2}q^{-2}\ \mathcal{F}^{\nu}_{\mu} = \hat{g}^{\nu\rho}R_{\mu\rho}+\hat{g}^{\rho\nu}R_{\rho\mu}-\delta^{\nu}_{\mu}\hat{g}^{\rho\sigma}R_{\rho\sigma}+\hat{g}^{\rho\sigma}_{\ \ ,\mu}\mathcal{B}^{\nu}_{\rho\sigma}-\delta^{\nu}_{\mu}\mathcal{B} . \quad (C.32)$$

Finally we shall include, for future reference, a different form of the identities (C.21). The right hand side of (C.21) can be written as

$$\frac{1}{2}\,\hat{g}^{[\mu\nu]}(\Phi_{\mu\nu|\rho}+\Phi_{\nu\rho|\mu}+\Phi_{\rho\mu|\nu}) = \frac{1}{2}\,\hat{g}^{[\mu\nu]}\Phi_{\mu\nu|\rho}+\hat{g}^{[\mu\nu]}\Phi_{\nu\rho|\mu}$$

$$= \frac{1}{2}\sqrt{(-g)}\,\frac{(1+\Omega-\Lambda^2)_{|\rho}}{\sqrt{(1+\Omega-\Lambda^2)}} - \sqrt{(-g)}\left[\frac{\Phi^{\mu\nu}\Phi_{\rho\nu}-\Lambda^2\delta^\mu_\rho}{\sqrt{(1+\Omega-\Lambda^2)}}\right]_{|\mu}$$

$$= \sqrt{(-g)}\left[\frac{(1+\Omega)\delta^\mu_\rho-\Phi^{\mu\nu}\Phi_{\rho\nu}}{\sqrt{(1+\Omega-\Lambda^2)}}\right]_{|\mu} = [\sqrt{(-g)}b^{\mu\sigma}g_{\rho\sigma}]_{|\mu} \quad .(C.33)$$

Hence the covariant derivative in (C.21) with respect to $b_{\mu\nu}$ is replaced by the covariant derivative with respect to the metric tensor $g_{\mu\nu}$. From (C.33) we obtain

$$[(\tfrac{1}{2}\,b^\rho_\rho-1)\delta^\nu_\mu - g^{\nu\rho}b_{\mu\rho}]_{|\nu} = 4\pi\,\Psi_{\mu\nu}\,s^\nu \quad , \qquad (C.34)$$

where

$$\Psi_{\mu\nu} = -\tfrac{1}{2}\sqrt{(-g)}\varepsilon_{\mu\nu\rho\sigma}\,\Gamma^{\rho\sigma} \;,\; \sqrt{(-g)}\Gamma^{\rho\sigma} = \hat{g}^{[\rho\sigma]} \;, \qquad (C.35)$$

and

$$\frac{1}{4\pi}\,\frac{\partial}{\partial x^\nu}\,[\sqrt{(-g)}\Psi^{\mu\nu}] = \zeta^\mu \quad ,$$

or

$$\zeta_{\mu\nu\rho} = \Gamma_{\mu\nu,\rho} + \Gamma_{\nu\rho,\mu} + \Gamma_{\rho\mu,\nu} = -4\pi\varepsilon_{\mu\nu\rho\sigma}\,\zeta^\sigma \quad (C.36)$$

where $\zeta_{\mu\nu\rho}$ or ζ^μ is the vacuum magnetic current.

The derivation of the field equations and the corresponding conservation laws are thus entirely equivalent to those obtained from the action principle. However the above method demonstrates the uniqueness and generality of the proposed field equations on a geometrical basis. The use of the action principle alone could, formally, allow other alternatives which, in our opinion, might not have convincing physical bases.

REFERENCES

1. B. Kursunoglu, Phys. Rev. 9, No. 10, 2723 (1974).
2. A. Einstein, Can. J. Math. 2, 120 (1950),
   B. Kaufman, Helv. Phys. Acta Supp. 4, 227 (1956);
   A. Einstein and B. Kaufman, Ann. Math. 62, 128 (1955).
3. A. Schrödinger, Proc. R. Irish Acad. A LI, 213 (1948).
4. P.A.M. Dirac, Phys. Rev. 74, 817 (1948).
5. J. Schwinger, In Proceedings of the Third Coral Gables Conference on Symmetry Principles at High Energy, Univ. of Miami, 1966, edited by A. Perlmutter et al. (Freeman, San Francisco, 1966).
6. I.J. Aubert et al., Phys. Rev. Lett. 33, 1404 (1974);
   J.-E. Augustin et al., Phys. Rev. Lett. 33, 1406, 1453 (1974);
   C. Bacci et al., Phys. Rev. Lett. 33, 1408 (1974), 34, 43 (1975).
7. B. Kursunoglu, Phys. Rev. D, March 15 (1976).

ERRATA IN THE PAPER (I)

Phys. Rev. 9, No. 10, May 15, 1974

Equation	should read
(2.52)	$\int \mathcal{s}^{\mu} d\sigma_{\mu} = 0$
(2.50')	$\int \zeta^{\mu} d\sigma_{\mu} = 0$
In matrix (3.17) the elements (23) and (32)	$\frac{\cos\Phi}{\upsilon \cosh\Gamma}$, $-\frac{\cos\Phi}{\upsilon \cosh\Gamma}$
(4.8)	$\frac{q}{\upsilon} \frac{\cos\Phi}{\cosh\Gamma}$
(8.3)	$Q_o = 0$, $Q_g = 0$
(9.3)	$\kappa \ell_o = \sqrt{2}$
(9.6),(9.7),(9.8)	the value assigned to g is incorrect

The words "vacuum" and "neutral" magnetic currents and corresponding fields should refer to ζ^{μ}, $\mathcal{s}^{\mu}$ and H, B, respectively, not as used in the paper I for $\mathcal{s}^{\mu}$, ζ^{μ} and B, H.

Professor B. Kursunoglu and the most effective Orbis Scientiae organizer,
Dr. Sydney Meshkov

ELECTRIC CHARGE IN COMPOSITE MAGNETIC MONOPOLE THEORIES*

Alfred S. Goldhaber

Institute for Theoretical Physics

State University of New York at Stony Brook

It is observed that the spin approach to the interaction of electric charges with magnetic monopoles leads naturally to theories with composite monopoles made from SU(2) gauge fields. The key steps are to express the "electromagnetic field angular momentum" associated with a charge-pole system as a quantum-mechanical spin operator (as was done some years ago) and then to identify this spin operator as the generator of SU(2) transformations on charged-particle wave functions. It is emphasized that composite monopoles may not occur unless electric charge is one of the operators of an SU(2) symmetry. The formalism for conservation of electric charge, and quantization of the electric charge of a monopole, is developed at a "first-quantized" level. Speculations are given on high energy charge-pole scattering. Some problems with second quantization are discussed, and the possibility is raised that there may be an infinite mass renormalization of a composite monopole which has finite mass in the classical app-

*Research support in part by NSF Grant #GP-32998X.

roximation.

While the motive for discussing magnetic monopoles at this meeting is clearly the recent experimental indication of a monopole,[1] this year also is a significant anniversary for theoretical work on the subject. It is eighty years since Poincaré published a simple classical calculation for the trajectory of an electrically charged particle passing a fixed magnetic pole,[2] and 45 years since Dirac showed how a monopole could be fit (just barely) into quantum mechanics.[3] From then until last summer, the theory made progress, despite the complete lack of experimental support for the enterprise. Nonetheless, while monopoles were reconciled with quantum mechanics in a way so beautiful that one might say they must be true, there has still been no reconciliation with relativity in the fundamental sense. We can make formal or intuitive discussions of monopole creation, but not systematic quantitative calculations.

Today I wish to argue that practically all that is now known, along with some questions still unresolved, can be illuminated by the considerations of Poincaré, suitably extended to quantum mechanics. In solving for the classical trajectory, he exploited the conservation of the total angular momentum

$$\underset{\sim}{J} = \underset{\sim}{L} + \underset{\sim}{s}, \tag{1}$$

where $\underset{\sim}{L}$ is the orbital angular momentum of the charge q with respect to the monopole g, and $\underset{\sim}{s}$ is the "angular momentum in the electromagnetic field" with magnitude qg/c in the direction from charge to pole. The conservation of $\underset{\sim}{J}$ follows directly from the equations of motion, except for the case of a straight path directly through the pole, for which $\underset{\sim}{L}$ is uniformly zero, while $\underset{\sim}{s}$ reverses at the moment of overlap. This explains the discovery of Rosenbaum[4] that a Lagrangean formulation of charge-pole interactions must be constrained by the rule that the two may not intersect: The action principle leads to conservation of $\underset{\sim}{J}$, but $\underset{\sim}{J}$ is not conserved for intersecting trajectories - hence the constraint.[5]

The classical charge-pole interaction may be expressed in a Hamiltonian form

$$H = (\underset{\sim}{p} + \underset{\sim}{s} \times \underset{\sim}{r}/r^2)^2/2m \equiv m\underset{\sim}{v}^2/2 , \qquad (2)$$

where $\underset{\sim}{s}$ is taken to have the Poisson brackets of an angular momentum. The quantity $\underset{\sim}{s}\cdot\hat{r}$ is conserved and has the value $-qg/c$.

The same H may be used in quantum mechanics, where $\hbar\underset{\sim}{s}$ acts on a (2s+1) component spinor. Since $\underset{\sim}{s}\cdot\hat{r}$ commutes with H, a rotation of the spinor from a fixed basis to one in which $\hat{r}$ is the axis of quantization diagonalizes H (as well as $\underset{\sim}{v}$). The resulting H, acting on the state with $-\hbar s_z\psi = qg\psi/c$, yields the Schrödinger equation for a charged particle in a monopole field first found by Dirac, with a singular "string" in the vector potential along the negative half of the axis of quantization in the fixed basis.[6]

It is interesting that the constraint required on classical Lagrangean theory is automatically satisfied in the quantum theory: The solutions of Dirac's equation all vanish at the origin because the monopole field creates a quasi centrifugal potential proportional to Planck's constant. This is important, since otherwise the commutation relations of the velocity components

$(\underset{\sim}{v})_i$ would fail to satisfy the Jacobi identity.[7] On the other hand, in the spin formulation the Jacobi identity is satisfied identically.

It reads

$$0 = \underset{\sim}{s}\cdot r\ \delta(\underset{\sim}{r}) \quad , \qquad (3)$$

which is true for all matrix elements of the right hand side between continuous wave functions.

If we wish to generalize in any way from the original problem of a static non-relativistic charge-pole interaction, we should try to interpret the spin $\underset{\sim}{s}$. In particular, if the aim is to construct a theory with local interactions, then $\underset{\sim}{s}$ must be associated with a point in space. The obvious choice is the position of q (or equally well, of g). Taking the monopole as a fixed entity, we note that $\underset{\sim}{s}\cdot\hat{r}$ is a measure of the electric charge q, so that the components of $\underset{\sim}{s}$ transverse to $\hat{r}$ act as charge raising and lowering operators. To emphasize this recognition, we rename $\underset{\sim}{s}$ as $\vec{t}$, the isospin operator on a charged-particle wave function obeying the formula

$$q = -e\vec{t}\cdot\hat{r} \quad , \qquad (4)$$

with $e \equiv \hbar c/g$, and the arrow referring to a vector in isospin space. Now we may rewrite the charge-pole interaction in a suggestive way as

$$H = (\underset{\sim}{p} - e\underset{\sim}{\vec{A}}\cdot\vec{t}/c)^2/2m \quad ,$$

$$A^i_j = \hbar c\varepsilon_{ijk} r_k/er^2 \quad . \qquad (5)$$

This is nothing but the minimal SU(2)-gauge-invariant interaction of the charged particle with a Yang-Mills field![8] One may speculate that, if Poincaré had known quantum mechanics, his approach to the charge-pole interaction would have impelled him to invent the notion of an isovector photon interacting with itself and with other fields in an isospin-gauge-invariant manner. Both monopoles[3] and non-Abelian gauge theories[8] have been used to " explain" quantization of charge. This discussion suggests these are but different aspects of a single explanation. To avoid the singularities associated with monopoles in standard electrodynamics, one is led to introduce Yang-Mills fields; but by the same token, a gauge theory in which the charge operator is a member of an SU(2) group automatically has a place for monopoles. That the vector field A^i_j of Eq. 5 obeys the Yang-Mills field equations may be demonstrated by direct computation,[9] but also follows immediately from the fact that Eq. 5 is simply a position-dependent rotation (≡SU(2) gauge transformation) of Dirac's monopole Hamiltonian. Dirac's vector potential obeys Maxwell's equations and _a fortiori_ the Yang-Mills equations, so the gauge invariance of these equations implies that A^i_j also obeys the Yang-Mills equations.[10]

Having found a formulation in which the only singularity of the monopole vector potential is located at the origin, we may entertain the notion of a potential which is identical at large radii, but non-singular at the origin, so that the pole density distribution is no longer confined to a point. An explicit realization of this notion was found in a theory of Yang-Mills fields interacting with Higgs scalar isovector fields, by 't Hooft[11] and independently by Polyakov.[12] They were the first to note the relation between monopoles and non-Abelian gauge fields. It is evident that the identification could only work if the photon were a member of a triplet of vector fields, possibly embedded in a larger multiplet. Further consideration of the constraints imposed by rotational invariance, gauge invariance and integrability of gauge transformations implies that electric charge must be a generator of an O(3) or SU(2) symmetry if the existance of a monopole should arise from a non-Abelian gauge theory with only one massless vector field. This is stronger than a previous suggestion,[11] that the non-Abelian group should be compact. Even if compact, the group must contain an SU(2) which contains electric charge. For example, SU(3), with the photon identified as a U-spin singlet, does not exhibit the monopole phenomenon: There is no SU(3) transformation which can reverse the sign of the electric charge.[13] For such a transformation to exist, the non-zero eigenvalues of the charge matrix must come in pairs, equal in magnitude but opposite in sign. Whenever this be so, it is easy to construct an SU(2) symmetry of which the charge matrix is a generator. Without such a construction, $\vec{t}\cdot\hat{r}$ = constant is unobtainable. For similar reasons, any charged particles which interact with the monopole must lie in SU(2) families, which may be contained in even larger families. These considerations may influence subjective estimates of the *a priori* probability that monopoles would occur in spontaneously broken gauge theories.

We come now to the question, what is the electric charge of a composite magnetic monopole? If one treats the pole as a static object, and allows a charge single-particle wave packet to strike this static potential, it is easy to see that the electric charge of the packet (i.e., $-e\vec{t}\cdot\hat{r}$) is not conserved in the interior of the pole. At this level of discussion, the monopole must bear indefinite charge if total charge is to be conserved.

Although the non-conservation of charge may be seen by direct calculation, it is instructive to show it by a general argument. Far outside the pole, the interaction is imperceptibly different from that with a Dirac pole. Consequently, as already noted by Dirac,[3] there must be a nodal line in any charged-particle wave function. For wave packets not impinging on the pole, one can always arrange to avoid the nodal line, but for a direct hit this is impossible. Therefore, such a wave packet must have a zero, no matter how high its energy. In a single channel problem such a zero can only be produced by an infinitely strong potential, but such a potential evidently cannot be exerted by a finite-mass composite monopole. Hence, the zero must arise from leakage of the wave function into other channels, but as long as the pole is treated as static, the only available channels are other states of charge in the same isospin family with the incident particle. Finite energy of the pole is purchased at the cost of charge conservation, which may only be restored by tackling head-on the problem of excitation of the Yang-Mills field when charges scatter, as we shall discuss a little later.

To progress further, we may pass to a "first-quantized" description of the pole, in which a monopole state is labeled by a variable θ when it is obtained from the vector potential in Eq. 5 by a rotation through the angle θ about the $\hat{r}$ direction:

$$A'^{i}_{j} = \hbar c(r_i r_j - r^2\delta_{ij}) / er^3 \quad ,$$

$$\vec{\underset{\sim}{A}}(\theta) = \underset{\sim}{A} + \delta(r)\,[\cos\theta\, \vec{\underset{\sim}{A}} + \sin\theta\, \underset{\sim}{A}']. \quad (6)$$

Here 1 + δ(r) is the factor multiplying the point monopole potential, with δ(0) = -1.

Now we may construct superpositions of the monopole states |θ>, and the eigenstates of charge are

$$|m\rangle = (1/2\pi) \int d\theta\, e^{-im\theta}\, |\theta\rangle \quad , \quad (7)$$

with charge eigenvalues Q = -em, and m integer. This follows, since the generator of rotations about the $\hat{r}$ direction in isospin space is to be identified with the charge by

$$Q = -\,e\vec{T}\cdot\hat{r} \quad , \quad (8)$$

where $\vec{T}$ is the total isospin matrix for all fields at $\underset{\sim}{r}$. To verify that charge lost or gained by a wave packet is gained or lost by the pole, observe that the charge lowering and raising operators for the packet are simple functions, of θ, which only appear in H for δ≠0,

$$\sqrt{\underset{\sim}{L}^2}\, t_{\pm}(\theta) = e^{\pm i\theta}(\vec{t}\cdot\underset{\sim}{L} \pm i\vec{t}\cdot\hat{r} \times \underset{\sim}{L}) \quad , \quad (9)$$

where $\underset{\sim}{L}$ is the orbital angular momentum of the packet. When the charge raising operator t_- (sic) acts on the wave packet, the state |m> is changed to |m + 1>, meaning that the monopole has lost one unit of charge e.[11]

From first quantization we have learned to quantize the electric charge of monopoles, which cannot be done in the context of classical gauge field theory.[15] Nevertheless, we still don't have a fully developed framework to describe charge-monopole interactions, because excitations of the charged vector field will undoubtedly play a big part when high energy particles pass through the pole. To say it another way, when $e^{i\theta}$ acts on the monopole state, we know the charge is raised, but we don't know where it is localized.[16] Obviously then, we are in no position to discuss the dynamics of the high energy inelastic reactions, but one may speculate about them anyway. There seem to be two main alternatives for the case in which the incoming particle charge changes. Either the lost charge sticks to the pole, which might be kinematically favorable, or the fast particle is "diffractively" excited, with the net charge of particle plus excitation cloud unchanged from the incident charge. This latter mode would not entail high charge velocity transfer and hence bremsstrahlung. One might suppose the former dominates for energies comparable to the heavy vector meson mass, and the latter for much higher energies. This would be in line with present high energy phenomenology. However, if the first quantized theory is any guide, and therefore a useful starting point for second quantization, then small momentum transfer charge exchange will hold up even at the highest energies. This can be seen by following a packet whose dimensions are small compared to the diameter of the pole interior. When such a packet plunges into the pole, its isospin will stop precessing. For a central collision then, $t\cdot\hat{r}$ will reverse, so that the particle will emerge with opposite charge, leaving the residue on the monopole, at a small cost in energy. The total high energy charge-exchange cross section will be of order πR^2, where R is the monopole radius.

Let us go on to second quantization. The term "quantization" applied to classically stable objects is often used to describe quantization of small oscillations about the classical ground state. This is a formidable task, since even the enumeration of these small oscillations has not been carried out for monopoles, so far as I know. Once that is accomplished, it seems to me possible that a hitherto unknown kind of renormalization problem might exist. Namely, the twisting of boundary conditions might mean that the energy of every single-particle state is raised sufficiently to give an infinite difference between the zero-point energies for the monopole and the total vacuum. Presumably that would require an infinite renormalization of the pole mass, which would make the classically finite mass meaningless.

The string singularities found by Dirac[3] have posed a challenge ever since. The strings cause no practical difficulty in the single-particle non-relativistic[17] or relativistic[18] theory. They may be removed by extending the notion of a wave function in a gauge field to that of a "section," with different gauges used in different parts of space.[19] However, all these approaches forbid exploration of the interior of a pole. We have seen here what a Pandora's box is that interior, but despite potential difficulties, it holds the promise of some new and perhaps relevant physics.

REFERENCES

1. P. B. Price, E. K. Shirk, W. Z. Osborne and L. S. Pinsky, Phys. Rev. Lett. 35, 487 (1975); P. B. Price, these proceedings.
2. H. Poincaré, Compt. Rend. 123, 530 (1896).
3. P. A. M. Dirac, Proc. Roy. Soc. (London) A133, 60 (1931); P. A. M. Dirac, these proceedings.
4. D. Rosenbaum, Phys. Rev. 147, 891 (1966).
5. A. S. Goldhaber and J. Smith, Rep. Prog. Phys. 38, 731 (1975).
6. A. S. Goldhaber, Phys. Rev. 140, B1407 (1965). In this paper the Hamiltonian in Eq. 2 was written down, and shown to be equivalent to that of Dirac, avoiding the singular string in return for the introduction of 2s redundant components of the wave function. However, H was not recognized as the square of a velocity operator, and the connection with gauge theory described below was not made.
7. H. J. Lipkin, W. I. Weisberger and M. Peshkin, Ann. Phys. (N. Y.) 53, 203 (1969).
8. C. N. Yang and R. L. Mills, Phys. Rev. 96, 191 (1954).
9. T. T. Wu and C. N. Yang, in *Properties of Matter Under Unusual Conditions*, (H. Mark and S. Fernbach, eds., Interscience, N. Y.), 349 (1969). In this work the point monopole gauge field (Eq. 5) was written down, but not recognized as a magnetic pole.
10. An elementary consequence of this derivation of the SU(2) gauge field formulation of monopole theory is that the isospin of a single-particle wave function contributes to the total angular momentum. After all, the starting point was $\underset{\sim}{J} = \underset{\sim}{L} + \underset{\sim}{s}$, but from s $\underset{\sim}{s} \equiv \vec{t}$ we get $\underset{\sim}{J} = \underset{\sim}{L} + \vec{t}$. This result has been deduced

from the gauge field formalism by R. Jackiw and C. Rebbi in an MIT preprint, "Spin from Isospin in a Gauge Theory," C. T. P. 524, February, 1976. They refer to a recent preprint by M. Prasad and C. Sommerfield, which I have not seen. A similar observation has been made by P. Hasenfratz and G. 't Hooft, Phys. Rev. Lett., in press.

11. G. 't Hooft, Nuc. Phys. B79, 276 (1974).
12. A. M. Polyakov, ZhETF Pis. Red. 20, 430 (1974), [JETP Lett. 20, 194 (1974)].
13. There has been some confusion about this because the word "monopole" has been associted with topological properties of constructs which do not produce the Lorentz force of a magnetic pole on an electrically charged particle. From a matter-of-fact point of view then, these SU(3) gauge field solutions have nothing to do with magnetic monopoles, and do not contradict the assertions made here. A. Chakrabarti, Nucl. Phys. B101, 159 (1975); W. J. Marciano and H. Pagels, Phys. Rev. D12, 1093 (1975).
14. Quite a number of people have learned independently how to quantize electric charge. While I have not seen any of their work, I am aware of the following: J. Goldstone and R. Jackiw (unpublished); E. Tombouli and G. Woo, Nuc. Phys, in press; N. H. Christ, A. H. Guth and E. Weinberg (unpublished).
15. B. Julia and A. Zee, Phys. Rev. D11, 2227 (1975), M. K. Prasad and C. M. Sommerfield, Phys. Rev. Lett. 35, 760 (1975).
16. A. H. Guth has informed me that the charge density operator can be obtained in a straight-forward manner. It is a nonlocal function of deviations from the static monopole gauge field.

17. I. Tamm, Z. Phys. 71, 141 (1932).

18. P. P. Banderet, Helv. Phys. Acta 19, 503 (1946).

19. T. T. Wu and C. N. Yang, Phys, Rev. D12, 3845 (1975) and to be published. The different regions overlap, and in the overlap of 2 regions both gauges are defined and non-singular. The transformation between the two gauges must be single-valued.

NONCOVARIANCE OF THE SCHWINGER MONOPOLE*

C. R. Hagen

Department of Physics and Astronomy

University of Rochester, Rochester, N. Y. 14627

ABSTRACT

It is shown that a canonical formulation of the magnetic monopole cannot be covariant if one allows only couplings which are manifestly invariant under three dimensional rotation. Although Schwinger's monopole theory allows a more general class of interaction terms, Lorentz invariance is found to be violated there as well. One is forced to conclude that there does not exist a consistent field theory of the monopole at the present time.

I. INTRODUCTION

In 1965 I published[1] an article in which I claimed to prove the noncovariance of the magnetic monopole within the framework of quantum field theory. Very shortly after the appearance of that work Julian Schwinger[2] presented a theory which he claimed to possess a completely

*Research supported in part by the U. S. Energy Research and Development Administration.

consistent quantization of magnetic charge. Looking back now at events of the past decade and the almost complete disregard of my 1965 paper in recent years, I must conclude with some reluctance that I was apparently much less eloquent and/or persuasive than was Schwinger in our various approaches to the monopole problem.

In fact the prevailing consensus within the physics community concerning the field theoretical status of the monopole was forcefully recognized by Physics Today when it recently proclaimed[3] that "Julian Schwinger... developed a consistent field theory of monopoles." It would thus appear to be more than appropriate to use the occasion of this monopole session either to acknowledge an error on my part or to attempt to persuade you that the theory of Schwinger is less than adequate. It is in fact the latter of these two possibilities which is the unabashed aim of this talk.

II. NONCOVARIANCE OF THE MANIFESTLY ROTATIONALLY INVARIANT MONOPOLE

The framework within which Schwinger and I develope[d] our approaches to the magnetic monopole are extremely similar and it is therefore useful to present the monopo[le] problem using much of the formalism of my 1965 paper. By doing this it will be possible to indicate rather briefly the modifications necessary to accommodate Schwinger's theory as well as the crucial objections whi[ch] can be raised to that work.

We may begin by noting that the essential ingredients of a monopole theory are the equations for the divergence of the electromagnetic field tensor as given

$$\partial_\nu F^{\mu\nu} = ej^\mu, \qquad \partial_\nu \bar{\bar{F}}^{\mu\nu} = gj^\mu{}_5$$

where

$$\bar{F}^{\mu\nu} \equiv \frac{1}{2} \varepsilon^{\mu\nu\alpha\beta} F_{\alpha\beta}$$

and $\varepsilon^{\mu\nu\alpha\beta}$ is the usual Levi-Civita tensor. Thus such things as charge quantization (if any) are to be considered as derived results and not as input to the desired theory. The current j^μ may be taken to be bilinear in a fermion field ψ and of the form

$$j^\mu = \frac{1}{2} \psi\beta\gamma^\mu q\psi \quad ,$$

where q is a charge matrix

$$q = \begin{pmatrix} 0 & -i \\ i & 0 \end{pmatrix} ,$$

introduced to allow us to use Hermitian ψ. Similarly the magnetic current is taken to be

$$j_5^\mu = \frac{1}{2} \psi'\beta\gamma_5\gamma^\mu q'\psi' \quad ,$$

where ψ' is also a fermion field and q' is a magnetic charge matrix

$$q' = \begin{pmatrix} 0 & 1 \\ 1 & 0 \end{pmatrix} ,$$

which allows for Hermitian ψ' in conjunction with the operation of magnetic charge conjugation. While we choose here to take the magnetic current to be pseudovector rather than vector, none of the conclusions obtained here are altered by adopting the opposite choice. Similarly deserving of mention is the remark that the spin one-half nature of the electrically and magnetically charged particles plays no significant role in the demonstration of noncovariance.

It is convenient to decompose F^{ok} and $\overline{F}^{ok}$ into transverse and longitudinal components, i.e.

$$F^{ok} = F_T^{ok}+F_L^{ok},$$
$$\overline{F}^{ok} = F_T^{ok}+F_L^{ok},$$

where

$$\partial_\kappa F^{ok} = \partial_\kappa \overline{F}_T^{ok} = 0.$$

One can then introduce the usual (transverse) vector potential A_k by

$$\overline{F}^{ok} = \varepsilon^{k\ell m}\partial_\ell A_m \ ,$$

as well as the scalar potential

$$A^o = -\nabla^{-2}e_o j^o .$$

It is then easy to show that the Lagrangian

$$L=\tfrac{i}{2}\psi\beta\gamma^\mu\partial_\mu\psi-\tfrac{m}{2}\psi\beta\psi+\tfrac{i}{2}\psi'\beta\gamma^\mu\partial_\mu\psi+\tfrac{1}{4}F^{\mu\nu}F_{\mu\nu}$$
$$-\tfrac{1}{2}F^{\mu\nu}(\partial_\mu A_\nu-\partial_\nu A_\mu)+ej^\mu A_\mu+gF^{ok}\varepsilon_{k\ell m}\partial_\ell\nabla^{-2}j_5^m$$
$$-\tfrac{1}{2}gF^{\ell m}\varepsilon^{k\ell m}\partial_\kappa\nabla^{-2}j_5^o$$

implies the equations

$$\partial_\nu F^{\mu\nu}=ej^\mu ,$$
$$\partial_\nu F^{\mu\nu}=gj_5^\mu ,$$
$$[\gamma^\mu(\tfrac{1}{i}\partial_\mu-eqA_\mu)+m]\psi=0,$$

$$\gamma^\mu(\tfrac{1}{i}\partial_\mu - gq'\gamma_5\beta_\mu)\psi'=0,$$

$$F^{ok}=\partial^o A^k-\partial^k A^o+g\varepsilon^{k\ell m}\partial_\ell \nabla^{-2} j_5^m,$$

$$F^{\ell m}=\partial_\ell A_m-\partial_m A_\ell+g\varepsilon^{k\ell m}\partial_\ell \nabla^{-2} j_5^o,$$

where we have <u>defined</u>

$$B_k=-\varepsilon^{k\ell m}\partial_\ell\nabla^{-2}F^{om},$$

$$B^o=g\nabla^{-2}j_5^o.$$

Thus the equations of the theory <u>appear</u> to be covariant with the possible exception of the equations relating $F^{\mu\nu}$ and A^μ. Since furthermore A^μ is not expected to transform as a four-vector in the radiation gauge used here, one cannot infer at this point a noncovariance in the theory. The latter question can only be unamiguously answered by examining the transformation properties of the operators of the theory. These in turn require a knowledge of the only nonvanishing commutators of the theory, i.e.

$$\left\{\psi(x),\psi(x')\right\}\delta(x^o-x^{o'})=\delta(x-x'),$$

$$\left\{\psi'(x),\psi'(x')\right\}\delta(x^o-x^{0'})=\delta(x-x'),$$

$$[F_T^{ok}(x),A^\ell(x')]\delta(x^o-x^{o'})=i\delta_{k\ell}^T(x-x'),$$

as implied by straightforward application of the action principle.

In order to check the possible covariance or noncovariance of the theory, one seeks a P^μ and $J^{\mu\nu}$ with

$$P^\mu=\int d^3x T^{o\mu}$$

and

$$J^{\mu\nu}=\int d^3x[x^\mu T^{0\nu}-x^\nu T^{0\mu}],$$

such that one has the structure relations of the Poincaré group

$$[P^\mu,P^\nu]=0,$$

$$-i[P_\lambda,J_{\mu\nu}]=g_{\lambda\nu}P_\mu-g_{\lambda\mu}P_\nu,$$

$$-i[J_{\eta\lambda},J_{\mu\nu}]=g_{\mu\lambda}J_{\nu\eta}-g_{\nu\lambda}J_{\mu\eta}-g_{\mu\eta}J_{\nu\lambda}+g_{\nu\eta}J_{\mu\lambda}.$$

Since the P^μ and $J^{\mu\nu}$ should also generate the (Minkowski) translations and rotations respectively of all the field operators of the theory, one has a highly stringent set of conditions placed on the $T^{0\mu}$ operators. In order to proceed as directly as possible, it is convenient to consider as suitable candidates for these operators

$$T^{0k}=F^{0k}F^{k\ell}+\frac{1}{2}\psi(\frac{1}{i}\partial_k-eqA_k)\psi+\frac{1}{2}\psi'(\frac{1}{i}\partial_k-gq'\gamma_5B_k)\psi'$$
$$+\frac{1}{2}\partial_\ell(\psi\frac{1}{2}\sigma_{k\ell}\psi)+\frac{1}{2}\partial_\ell(\psi'\frac{1}{2}\sigma_{k\ell}\psi')-F_L^{0\ell}\varepsilon^{k\ell m}\bar{F}_L^{0m},$$

where

$$\sigma_{k\ell}=\frac{i}{4}[\gamma^k,\gamma^\ell]$$

and

$$T^{00}=\frac{1}{2}\psi\beta\gamma^k(\frac{1}{i}\partial_k-eqA_k)\psi+\frac{1}{2}m\psi\beta\psi+\frac{1}{2}\psi'\beta\gamma^k(\frac{1}{i}\partial_\kappa-g\gamma_5q'B_k)\psi'$$
$$+\frac{1}{2}[(F^{0k})^2+(\bar{F}^{0k})^2].$$

With the above forms one can after routine calculati

verify that the P^μ and $J^{k\ell}$ operators have the appropriate communtators with the field operators of the theory, i.e.

$$[P^\mu, \chi(x)] = i\partial^\mu \chi(x),$$

$$[J^{k\ell}, \chi(x)] = i(x^k \partial^\ell - x^\ell \partial^k)\chi(x) + \text{appropriate spin terms},$$

for all $\chi(x)$. This leaves only the commutators of J^{ok} to check; that is, only the pure Lorentz transformations can cause difficulty. Although one finds[1] that $F^{\mu\nu}$ and the currents j^μ and j_5^μ transform appropriately, the ψ, ψ' and A^μ commutators with J^{ok} are such that the equations of motion of those fields cannot transform covariantly, thereby displaying the inconsistency in the theory.

Before going on to discuss the Schwinger modification of these results it is instructive to consider the Dirac-Schwinger condition

$$-i[T^{oo}(x), T^{oo}(x')] = -(T^{ok}(x) + T^{ok}(x'))\partial_k \delta(\vec{x} - \vec{x}'),$$

the satisfaction of which is sufficient for the Lorentz invariance of a rotationally invariant theory. In the present case the Dirac-Schwinger condition fails because:

i) The commutator fails to generate the last term of T^{ok}

ii) The commutator of $j^k A_k$ with $j_5^k B_k$ is not accommodated by the structure of the Dirac-Schwinger condition. The latter of these two points in particular will prove to be of crucial importance in our subsequent discussion of the Schwinger theory of the monopole.

Finally, at least passing mention should be made of an interesting soluble monopole theory which can be obtained by taking j^μ and j_5^μ to be simply related to scalar fields ϕ and ψ, respectively, Thus by considering the Lagrangian of these fields in the absence of coupling, i.e.

$$L=\phi^\mu\partial_\mu\phi+\frac{1}{2}\phi^\mu\phi_\mu+\psi^\mu\partial_\mu\psi+\frac{1}{2}\psi^\mu\psi_\mu,$$

one sees that ϕ^μ and ψ^μ are conserved vectors which can be coupled to the electromagnetic field to obtain

$$\partial_\nu F^{\mu\nu}=e\phi^\mu,$$

$$\partial_\nu \bar{F}^{\mu\nu}=g\psi^\mu.$$

Such a theory is seen to be obtained from the Lagrangian

$$L=\frac{1}{4}F^{\mu\nu}F_{\mu\nu}-\frac{1}{2}F^{\mu\nu}(\partial_\mu A_\nu-\partial_\nu A_\mu)+\phi^\mu\partial_\mu\phi+\frac{1}{2}\phi^\mu\phi_\mu+\psi^\mu\partial_\mu\psi$$
$$+\frac{1}{2}\psi^\mu\psi_\mu+e\phi^\mu A_\mu+gF^{ok}\varepsilon^{k\ell m}\partial_\ell\nabla^{-2}\psi^m-\frac{1}{2}gF^{\ell m}\varepsilon^{k\ell m}\partial_k\nabla^{-2}\psi^o.$$

On working out the equations of motion (which can be readily solved in view of the bilinear nature of the interaction) one obtains

$$(-\partial^2+e^2)\phi=0,$$
$$(-\partial^2+g^2)\psi=0,$$

$$(-\partial^2+e^2+g^2)F_T^{ok}=e^2g^2\nabla^{-2}F_T^{ok},$$

thereby showing explicitly the noncovariant form of the excitation spectrum of the "photon" field F_T^{ok}.

III. EXTENSION TO THE SCHWINGER THEORY[4]

In the derivation of moncovariance in the theory just discussed it may have been noticed that the conclusion on the grounds that additional terms should have been included in those operators to restore covariance. This point in fact is precisely the origin of the difference between the theory just described and that advanced by Schwinger. Thus in my 1965 paper I recognized the possibility of such terms but concluded that manifest rotational invariance, if it is imposed, allows for no acceptable candidates for inclusion in $T^{\mu\nu}$. It is just this assumption (i.e. of manifest rotational invariance) that Schwinger gives up in the hope that full covariance can be salvaged even in the case where neither rotational invariance nor Lorentz invariance is manifest.

Using the formalism just described the Schwinger modification is easily described as basically consisting of the replacements in T^{oo} of

$$A_k \rightarrow A_k(x) + \int d^3x' a_k(x-x') j^o(x'),$$

$$B_k \rightarrow B_k(x) + \int d^3x' b_k(x-x') j_5^o(x'),$$

where

$$\vec{b}(x) = -\vec{a}(-x).$$

By requiring that the Dirac-Schwinger covariance condition be satisfied, one obtains

$$-\nabla D = \nabla x a,$$

where

$$D(x)=\frac{1}{4\pi|x|} \quad .$$

As every graduate student knows, an equation of thi type (a gradient equal to a curl) cannot be solved without exception. Thus Schwinger proposes that

$$\vec{a}(x)=D(x)\frac{1}{2}\left[\frac{\vec{n}\times\vec{x}}{|x|+n\cdot x}-\frac{\vec{n}\times\vec{x}}{|x|-n\cdot x}\right],$$

where $n^2=1$ but is otherwise arbitrary. This form for $\vec{a}$ satisfies the above equation except for all points that lie on the infinite line drawn through the origin in the direction of the unit vector $\vec{n}$.

With this observation Schwinger can infer covarianc provided that the transformation from one singularity line $\vec{n}$ to another can be shown to be a gauge transformation, thereby depriving any particular choice of the vector $\vec{n}$ of special significance. In fact Schwinger gives a form for the allegedly unitary operator U which rotates $\vec{n}$ by an angle $\delta\omega$.
It is

$$U(\delta\omega)=1+i\delta\vec{\omega}\cdot\vec{j}_n,$$

where > $\vec{j}_n=\int d^3x d^3x' j^0(x)\vec{f}_n(x-x')j_5^o(x')$
and $\vec{f}_n(x)=-[\vec{x}\times\vec{a}(x)+\vec{x}D(x)]$.

While one could take some encouragement from the explicit demonstration that U formally accomplishes the desired gauge transformation, it counts for nothing if U cannot reasonably be argued to be a well defined operator in Hilbert space.

In order to save writing let us consider instead of $\vec{j}_n$ an operator

$$Q=\int J^{o}d^{3}x.$$

If such an operator is to exist one must have

$$<o|QQ|o> <\infty$$

or

$$\int d^{3}x<o|QJ^{o}(0)|o> <\infty \quad ,$$

thereby implying that Q must annihilate the vacuum. On applying this result to $\vec{j}_n$ one readily sees that $\vec{j}_n$ must annihilate the vacuum (as must all its time derivatives as well). However, unlike the case of the usual operators which have the property of annihilating the vacuum (e.g. the generators of the Poincaré group, charge, etc.) and are associated with local conservation laws, there is no symmetry operation associated with $\vec{j}_n$ and consequently no reason for it to annihilate the vacuum. In the absence of such an argument $\vec{j}_n$ must be presumed to have unbounded norm and Schwinger's alleged covariance proof reduced to the status of conjecture.

There is an alternative argument which is given by Schwinger to assert covariance with which we must also reckon. He notes that since the gauge invariance argument (allegedly) eliminates the difficult terms in the Dirac-Schwinger commutator, surely it must be possible to make the cancellation occur explicitly. To this end he attempts to write the fermion terms in $T^{oo}(x)$ as a limit. He thus writes for the charge field

$$-\bar{\psi}(x)\gamma\cdot(\nabla-ieA)\psi(x)=\ell im[\bar{\psi}(x+\frac{\varepsilon}{2})\frac{3\gamma\cdot\varepsilon}{\varepsilon^2}\psi(x-\frac{\varepsilon}{2})\exp(ie\int_{x-\varepsilon/2}^{x+\varepsilon/2}A\cdot dx)],$$

with a similar limiting process for the magnetically charged field. Using these limits he succeeds in showin that the troublesome $j\cdot A$ commutator with $j_5\cdot B$ noted earlier can be made to disappear for appropriate quantization of the product eg.

In critically examining the limit advanced above one must avoid falling into the trap of imagining that limit is merely an application of Schwinger's often quoted result that currents must be defined as limits. The great success of limiting definitions of current operators in avoiding contradictions in field theory is legendary. On the other hand the limiting procedure giv above goes far beyond the mere assertion that currents a to be defined as limits and one must critically examine the two sides of the above equation to determine whether it can indeed be considered an identity.

In carrying out such an examination it is important to note that as an (alleged) identity it must be true independent of e, in particular for e=0. Thus an elementary test is provided by considering whether Schwinger's limit is valid for the free field and in particular whether the vacuum expectation value of that equation is valid. Since for $\varepsilon \to 0$ the fermion two point function goes as

$$\vec{\gamma}\cdot\vec{\varepsilon}/\varepsilon^4,$$

one must have for consistency

$$Tr\frac{3\gamma\cdot\varepsilon\gamma\cdot\varepsilon}{\varepsilon^6}=Tr\gamma\cdot\nabla_\varepsilon\frac{\gamma\cdot\varepsilon}{\varepsilon^4}$$

or

$$\frac{3}{\varepsilon^4} = \frac{3}{\varepsilon^4} - \frac{4}{\varepsilon^4} \; .$$

The contradiction is thus apparent and one has a clear demonstration that Schwinger's limiting procedure and the asserted covariance of the monopole are without basis.

IV. CONCLUSION

We have seen that the rather general acceptance achieved by the Schwinger theory of the monopole has been the result of a less than adequately critical appraisal, and that in fact there is no field theory of the monopole at this present time. On the other hand there does not appear to exist any completely convincing way to rule out all possible ways in which rotationally noninvariant terms could be introduced into the theory. Thus while a general proof of non-covariance is not likely to be immediately forthcoming, one now has considerable insight into the rather significant theoretical problems which must hinder the development of a consistent theory. In any event can one view these enormous difficulties confronting the monopole without seriously questioning the symmetry argument which supposedly implies the monopole?

Nonetheless the theoretical physicist must recognize the fact that physics is basically an experimental science and that the logical possibility of a monopole can hardly be excluded. Yet if as argued here the monopole must be accompanied by a failure of covariance, it would be the rare physicist indeed who would expect the monopole to prevail over Lorentz invariance.

Should a clever experimentalist succeed despite all adversity in unambiquously identifying a magnetic monopole we shall certainly be faced with formidable theoretical problems in seeking a consistent framework for the description of such a particle.

REFERENCES

1. C. R. Hagen, Phys. Rev. 140, B804 (1965).

2. J. Schwinger, Phys. Rev. 144, B1087 (1966).

3. *Physics Today*; October, (1975).

4. A brief summary of the objections raised here to the Schwinger theory will shortly appear in *Physics Today*.

EXPERIMENTAL SEARCHES FOR MAGNETIC MONOPOLES*

Ronald R. Ross

Lawrence Berkeley Laboratory

University of California, Berkeley, Ca. 94720

ABSTRACT

Analysis of the sensitivity of previous negative searches for magnetic monopoles shows that they constitute prior evidence against the monopole interpretation of the event reported as "evidence for detection of a moving magnetic monopole." The strength of the evidence varies with the unknown mass of the monopole. For $M \lesssim 10^5$ GeV, odds are greater than $10^6 : 1$ against. For larger masses, the limits depend strongly on assumption about the range of monopoles and the threshold for detection of monopole tracks in obsidian. In no case are the odds less then 8 : 1 and they may be no less than 8000 : 1 against. Since the reported event may also be due to an electrically charged heavy particle, it is probably not due to a monopole.

I. INTRODUCTION

This afternoon we will review the sensitivity and

*Work supported by the U.S. Energy Research & Development Administration under Contract W - 7405 - ENG - 48.

assumptions of the most sensitive monopole searches[1,5] published prior to the report of "Evidence for Detection of a Moving Magnetic Monopole."[6] None of them detected a monopole. It is appropriate to consider this prior evidence when judging the monopole interpretation of the event in question since the searches reviewed are those that would be most capable of detecting monopoles with properties similar to those deduced from the reported evidence.

II. PROPERTIES OF THE REPORTED MONOPOLE

Figure 1 taken from ref. 1 is a schematic representation of the stack of detectors in which the monopole candidate was detected. The evidence, taken at face value as published, indicates that <u>if</u> the particle responsible for the emulsion track and Lexan etching carried a magnetic charge, then the magnitude of its charge would be 137e, or 2 times the minimum charge allowed in Dirac's theory,[7] its velocity would be $\beta = 0.5^{+.1}_{-.05}$ and its mass would be greater than 200 m_p. Furthermore, its large mass implies that it could not have been produced in the atmosphere and slowed to $\beta = 0.5$,[8] and, therefore, must have been a component of the primary cosmic rays. The searches reviewed were sensitive to a flux of this type of monopole.

III. SENSITIVITY OF PREVIOUS SEARCHES FOR MONOPOLES IN COSMIC RAYS

The reported event was taken from a set of exposures in balloon flights and one satellite exposure which had a combined event collecting power of 3 m^2 - yr. Assuming a solid angle factor of π steradians, the one

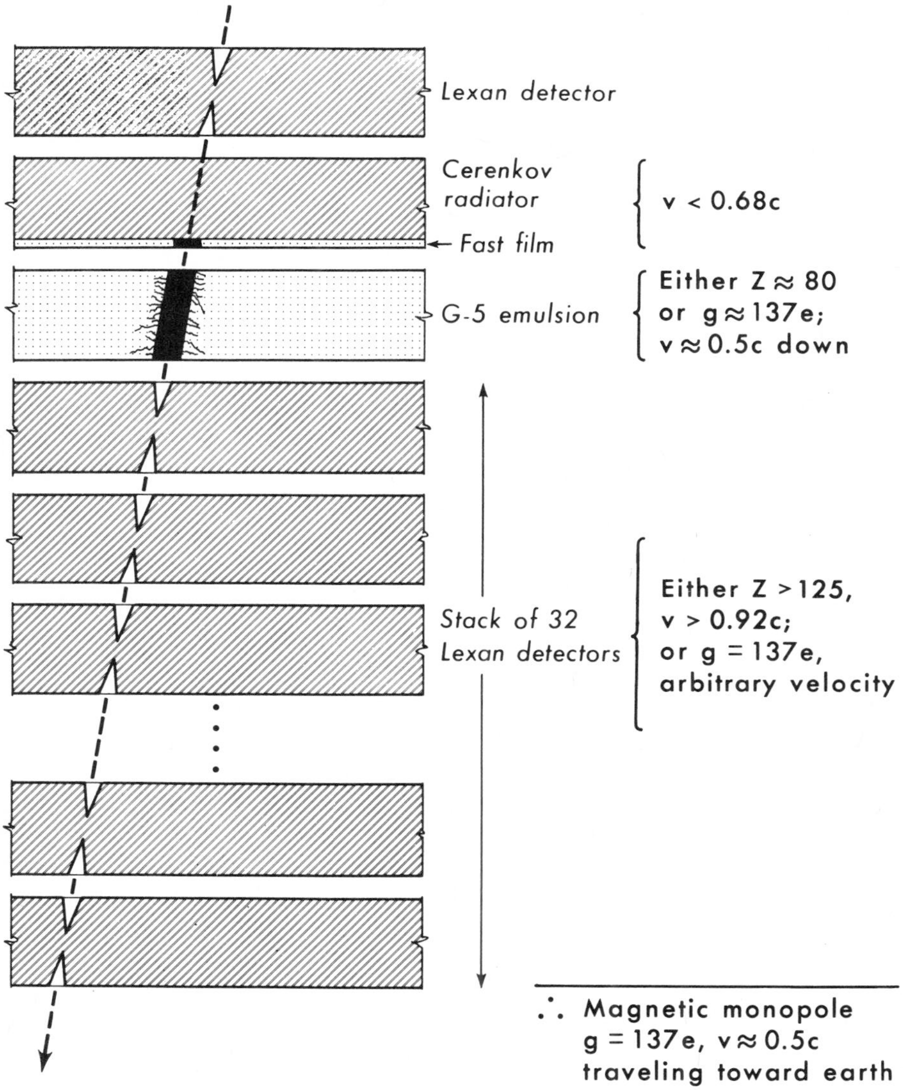

Figure 1 Schematic of a stack of balloon borne detectors in which an event was detected and reported as evidence for a magnetic monopole. Figure taken from ref. 6.

event corresponds to a measured flux of

$$3.4 \times 10^{-13}\ cm^{-2}\ sec^{-1}\ sr^{-1}$$

Using this flux as an estimate of monopole flux in the cosmic rays we predict for those previous searches that detected no monopoles the number they should have seen. This is a means of indicating their sensitivity, the larger the number predicted, the more sensitive the experiment. The results are shown as a function of the kinetic energy of the incident monopole in Fig. 2.

The sensitivity of the various experiments depend on the unknown range of monopoles of a given energy. We parameterized the range in terms of an effective charge N,[3] such that, if the only energy loss was due to ionization, the effective charge would be qual to the actual charge, ν, in units of the minimum charge allowed in the Dirac theory.[7] Of course, we expect that other energy loss mechanisms such as nuclear interactions or bremsstrahlung will be present so that $N \geqslant \nu$. The range R, in g/cm^2, of a monopole of kinetic energy E is then given by $R = E\ (GeV)\ /\ 8N^2$. To show the variability of the sensitivity of the experiments in case energy loss processes other than ionization are taken into account, the predictions have been evaluated for both N = 2 and N = 20. The curves on Fig. 2 are drawn dotted for N = 2 and dashed for N = 20. Where dashed and dotted curves overlap they are shown as solid lines.

It is clear from Fig. 2 that the previous experiments ABCD,[1-4] taken at face value rule out the possibility that the reported event was due to a particle to which they were sensitive. We must consider the assumptions and any limitations of these experiments.

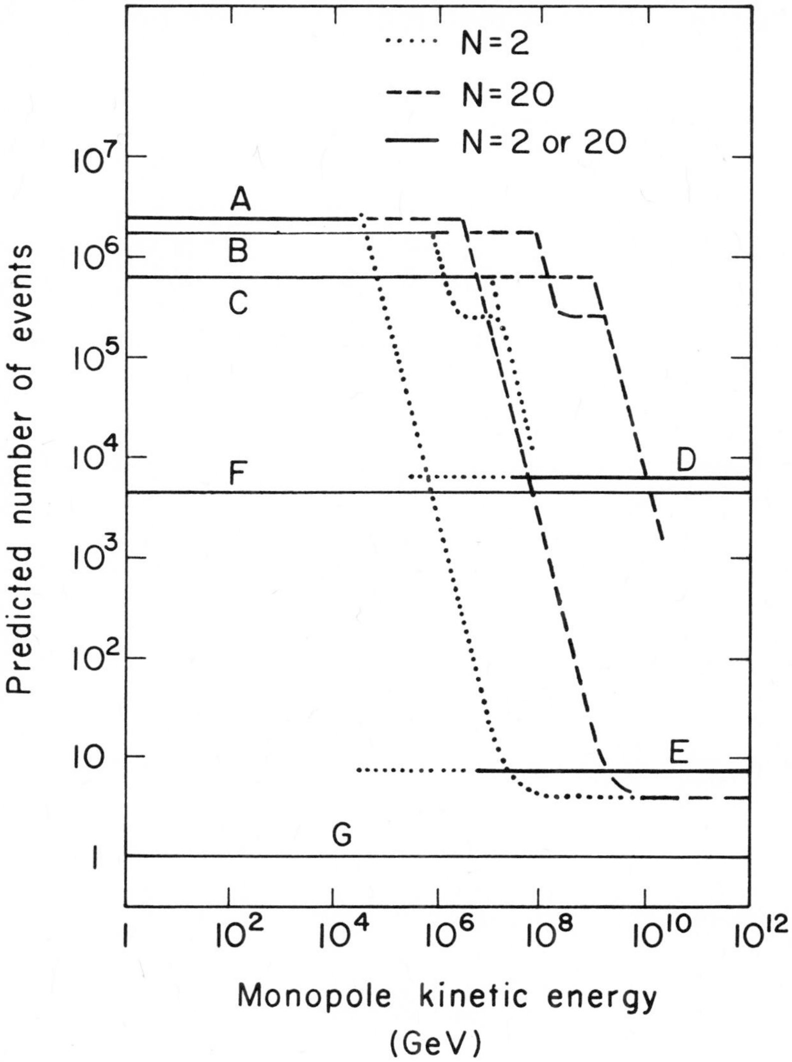

Figure 2 Predictions of the number of monopoles that would have been detected in previous monopole searches A, B, C, D, and E if the true flux of monopoles was $3.4 \times 10^{-13} cm^{-2} sec^{-1} sr^{-1}$. A, ref. 3.; B, ref.2.; C, ref.1.; D, ref. 4.; E, ref. 5.; F, ref. 10., see text for the significance of this "prediction;" G. ref. 6.

TABLE I

Positive Signature Experiments

Material Searched	Depth of Sample (g/cm^2)	Method of Detection	Limitations for Detecting $\nu=2$ Monopoles	Ref.
Managanese Nodules	$\sim 2.7 \times 10^5$	Extraction + Solid State Track Detectors	$M < \sim 2x10^4$ GeV	1
Deep Sea Sediments	$\sim 4.4 \times 10^5$	Extraction + Scintillator or Emulsion	$M < \sim 2x10^4$ GeV	2
Magnetic Extract from Georgia Clay Deposit	$\sim 2.5 \times 10^4$	Extraction + Scintillator or Emulsion	$M < \sim 2x10^4$ GeV	2
Lunar Soil and Rocks	10^3	Measure Magnetic Charge by Induction in a Coil	None	3

IV. CRITIQUE OF PREVIOUS SEARCHES

We classify the previous experiments in categories.

1. Positive signature experiments, in which the presence of a single monopole in the sample would be detected. Experiments of this type are listed in Table I, they are experiments A, B and C of Fig. 2.

2. Negative signature experiments, in which it can be determined with certainty that no monopole has traversed a sample by the absence of a signature due to a highly ionizing particle. If these experiments had seen a signal, there is in general no way to distinguish between a signal due to an electrically charged particle and a magnetically charged partical without further assumptions. Experiments of this type are listed in Table II, they are experiments D and E of Fig. 2.

3. Indirect evidence based on the predicted effect a flux of monopoles would have on other experimental observations in the universe. Evidence of this type is listed in Table III[9,10] and one result is shown as F on Fig. 2.

In the tables we have indicated the limitations of each of the experiments for the detection of $\nu = 2$ monpoles.

The positive signature experiments depend on the accumulation of monopoles in a sample over long periods of time. Monopole accumulation in the lunar sample depends on a) the slowing down and thermalization of monopoles guaranteed by their intense electromagnetic interaction with matter, and b) their being trapped in materials containing ferromagnetic components. They will either be trapped in the ordinary matter or, if not, then their trapping by ferromagnetic components is ensured by

TABLE II

Negative Signature Experiments

Material Searched	Depth of Sample (g/cm^2)	Limitations for Detecting $\nu=2$ Monopoles	Ref.
Mica	$\sim 2.0 \times 10^6$	Threshold for Detection only certain for $\nu \gtrsim 3$	4
Obsidian	$\sim 10^4$	Some loss of Solid Angle for $\nu=2$ (Threshold for $\nu=2$ has been questioned by one of the authors.[13])	4
Lexan	$\sim 10^3$	Same material as used in ref. 1	5

energy conservation.[11] No solid arguments have been advanced to throw doubt on these two assumptions. For the experiments using terrestrial samples, there is the additional assumption that the monopoles thermalized in the atmosphere or ocean will migrate along the earth's magnetic field lines to the ocean bottom before being trapped. This also seems a reasonable hypothesis, though maybe less certain than the previous two.

The monopoles would have been detected in the lunar samples by the current change they induced in a closed superconducting solenoid when they were passed through the solenoid.[12] This method does not alter the sample in any respect and the measurement can be repeated at will. Furthermore, the sensitivity of the detector to magnetic charge can be established without magnetic charge because of the identical roles played by current of magnetic charge and magnetic induction changes in the Maxwell equation

$$\vec{\nabla} \times \vec{E} = -\frac{\partial \vec{B}}{\partial t} - 4\pi \vec{J}_m \quad .$$

For the terrestrial samples of references 4 and 5, the detection of monopoles was to be accomplished by extracting them from the sample and accelerating them to high energy using a large magnetic field. The monopoles would then be detected by their large ionization. The necessity of extracting them and accelerating them to the detectors puts a limit on the monopole mass for which this technique would work. That limit is $M \lesssim 2 \times 10^4$ GeV.

From the positive signature experiments A, B and C on Fig. 2 alone, we conclude that for monopole kinetic energies less than 10^6 GeV and hence masses $< 10^6$ GeV for non relativistic monopoles like the event of ref. 6, the odds are greater than 1000 : 1 against the event of ref. 6 being a monopole.

For larger monopole energies and masses, the negative signature experiments give the most restrictive limits. The obsidian sample of ref. 4 should have detected about 8,000 tracks of the type reported in ref. 6. The authors indicate that calibration of obsidian threshold for identification of high charge should allow the detection of $\nu = 2$ monopoles. However, we understand that the threshold value has been questioned on the basis of new unpublished calculations of monopole energy loss that would lead to damage in crystals.[13] If this challenge can be substantiated, the only remaining experimental limit from negative signature experiments would come from ref. 5. Here the detection is with Lexan, the same detecting material used in ref. 6. The odds in this case are only 8 to 1 against the event of ref. 6 being a monopole.

In addition to the positive and negative signature experiment results, there is indirect evidence against such a large flux of monopoles from an argument in ref. 10 regarding the collapse of the galactic magnetic field. A flux of monopoles would cause the fields to decay with a time constant inversely proportional to the number density. By equating the decay time constant to the estimated time constant of generation of the fields ($\sim 10^8$ years), and assuming a velocity $\beta \approx 1$, the "sensitivity" limit shown as F on Fig. 2 was deduced. The meaning of this limit is as follows. If the true monopole flux in interstellar space was 7,000 times less than that implied by the one event of ref. 6, then the galactic magnetic fields would decay as fast as they were generated. Hence, this argument indicates odds of 7000 : 1 against the event being a monopole.

V. CONCLUSION

The results of previous monopole searches for a flux of monopoles in the cosmic rays have been reviewed in the light of the reported monopole candidate.[6] The results depend on the assumed mass and range of the monopole candidate. Assuming a range energy relationship based on N = 2 (dotted and solid curves of Fig. 2) and assuming the mass is of the same order of magnitude as or smaller than its kinetic energy (as was the case for the monopole candidate) we draw the following conclusions:

1. $M < 2 \times 10^4$. There are three independent experiments[1,3] that give odds $> 10^6$: 1 against the monopole interpretation.

2. $2\times10^4 < M < 10^6$ GeV. The Lunar experiment[3] gives odds greater than 1000 : 1 against the monopole interpretation.

3. $M > 10^6$. The absence of tracks in obsidian[4] gives odds of 8000 : 1 against the monopole hypothesis.

4. Any M. The existence of galactic magnetic fields gives odds of ~ 7000 : 1 against the monopole hypothesis.[10] Taking these published experiments before the reported event at face value, there are a priori odds greater than 8000 : 1 against the event reported ref. 6 being due to a magnetic monopole.

If the threshold for detection of $\nu = 2$ monopoles in the obsidian of ref. 4 can be effectively challenged and some way is found to either generate galactic

TABLE III

Indirect Evidence

Phenomenon Examined	Limitations for Detecting $\nu=2$ Monopoles	Ref.
Low Rate of Muon Poor Extensive Air Showers and Inverse Compton Scattering of 3°K Black Body Radiation on Monopoles	$M \lesssim 2 \times 10^{4}$ GeV. Beyond this Mass the Flux Limits are Less Restrictive than those of Other Experiments	9
Decay of the Galactic Magnetic Field	No Limitation is Discussed in ref. 10.	10

magnetic fields faster or to keep monopoles from destroying these fields, the a priori odds can be reduced to as low as 8 : 1 against the candidate for monopole energies and masses > 10^7 GeV.

Finally, the event was detected using a technique suited for a negative signature experiment. There is nothing in the experiment that demands that the track be due to a magnetically charged particle. The event could as well have been due to an electrically charged particle of high enough mass and may yet be explained in terms of a known high Z nucleus.[14] We conclude that the event is most probably not due to a magnetic monopole.

REFERENCES

1. R. L. Fleischer, I. S. Jacobs, W. M. Schwarz, P. B. Price, and H. G. Goodwell, Phys. Rev. 177, 2029 (1969) and R. L. Fleischer, H. R. Hart, Jr., I. S. Jacobs, P. B. Price, W. M. Schwarz, and F. Aumento, Phys. Rev. 184, 1393 (1969).
2. H. H. Kolm, F. Villa, and A. Odian, Phys. Rev. D 4, 1285 (1971).
3. P. H. Eberhard, R. R. Ross, L. W. Alvarez, and R. D. Watt, Phys, Rev. D 4, 3260 (1971) and R. R. Ross, P. H. Eberhard, L. W. Alvarez, and R. D. Watt, Phys. Rev. D 8, 698 (1973).
4. R. L. Fleischer, P. B. Price, and R. T. Woods, Phys. Rev. 1398 (1969) and updated in R. L. Fleischer, H. R. Hart, Jr., I. S. Jacobs, P. B. Price, W. M. Schwa and R. T. Woods, Jour. of Appl. Phys. 41, 958 (1970)
5. R. L. Fleischer, H. R. Hart, Jr., G. E. Nichols, and P. B. Price, Phys. Rev. D. 4, 24 (1971).
6. P. B. Price, E. K. Shirk, W. Z. Osborne, L. S. Pinsk Phys. Rev. Letters 35, 487 (1975).
7. P. A. M. Dirac, Proc. Roy. Soc. (London) A133, 60 (1931). We write magnetic charge $g = \nu g_o$ with $g_o = 137e/2$. g_o is the minimum charge predicted in Dirac's theory.
8. L. W. Wilson, Phys. Rev. Letters 35, 1126 (1975); E. V. Hungerfor, Phys. Rev. Letters 35, 1303 (1975); G. D. Badhwar, R. L. Golden, J. L. Lacy, S. A. Stephens, and T. Cleghorn, Phys. Rev. Letters 36, 12 (1976).
9. W. Z. Osborne, Phys. Rev. Letters 24, 1441 (1970).
10. E. N. Parker, The Astrophysical Jour. 160, 383 (1970

11. E. Goto, J. Phys, Soc. Japan 13, 1413 (1958); E. Goto, H. H. Kolm K. W. Ford, Phys. Rev. 132, 387 (1963); For an argument that is explicitly independent of the interaction of monopoles when they are inside the ferromagnetic material see P. H. Eberhard and R. R. Ross, LBL 4614, October 1975. Unpublished (submitted to Nuclear Phys. B).

12. L. W. Alvarez, M. Antuna, Jr., R. A. Byrns, P. H. Eberhard, R. E. Gilmer, E. H. Hoyer, R. R. Ross, H.H. Stellrecht, J. D. Taylor, and R. D. Watt, Rev. Sci. Instr. 43, 326 (1971); P. H. Eberhard, R. R. Ross, and J. D. Taylor, Rev. Sci. Instr. 46, 362 (1975).

13. P. B. Price, private communication.

14. P. B. Price, following paper at this conference.

STATUS OF THE EVIDENCE FOR A MAGNETIC MONOPOLE

P. B. Price

University of California

Berkeley, California 94720

In August, 1975, E. K. Shirk, W. Z. Osborne, L. S. Pinsky and I reported evidence[1] that we had detected a moving magnetic monopole, using a balloon-borne array of track detectors shown in Fig. 1. The Conference organizers have asked me to discuss the status of our evidence. I have agreed to do so, somewhat reluctantly since much remains to be done before the measurements of the accompanying ultraheavy cosmic rays are completed with all three types of detectors.

Our reasoning was straightforward. The very high, roughly constant ionization rate inferred from track etch rate measurements in the stack of Lexan detectors implies passage of a minimum-ionizing particle more highly charged than any known nucleus, yet the Cerenkov film detectors indicated a velocity less than ~0.68 c and the size of the track in the nuclear emulsion indicated a velocity ~0.5 c. At this velocity the ionization rate of a highly electrically charged particle would have changed dramatically with pathlength unless its mass to charge ratio were far greater than that of a nucleus.

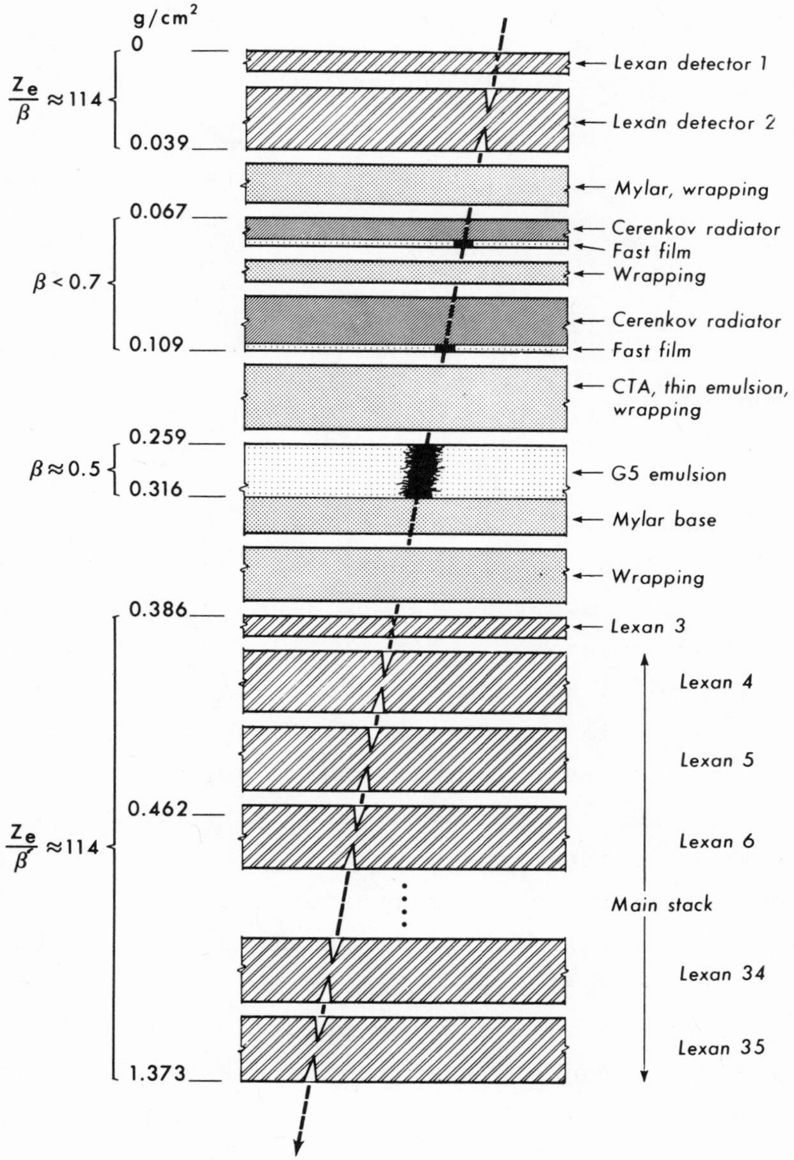

Figure 1. Detector array (schematic) with depths in g/cm² Lexan equivalent.

It has been known for many years that the ionization rate of a magnetic monopole is roughly independent of velocity. Bauer[2] and Cole[3] showed that the rate is given by replacing the quantity $Z_e e$ in the Bethe-Bloch equation with $g\beta$, the product of magnetic charge and velocity. (Z_e is the effective charge.) Assuming the sensitivity of our Lexan detectors to be the same as that of Lexan used in previous balloon experiments[4] and in a Skylab cosmic ray experiment,[5] we found that $Z_e/\beta \approx 137$ or that $g \approx 137$ e. The fit to the expected behavior of a monopole with twice the Dirac charge (and equal to the Schwinger charge) was so close that we were absolutely convinced of the validity of our evidence and decided to publish before carrying out the calibrations and analysis of the other events in the detector, which we knew would take nearly a year.

The Lexan data single out the monopole candidate as not just the end member of a smooth distribution of heavily ionizing cosmic ray nuclei but as a unique particle with qualitatively different behavior. This is obvious in Fig. 2, which shows the variation of track etch rate with depth in the Lexan stack for the monopole candidate and for the other particles found in the flight. Because etch rate is an increasing function of ionization rate, the curves in Fig. 2 are somewhat like Bragg curves. The data for the monopole candidate fit a horizontal line at an etch rate of ~2.9 μm/h, far above the other horizontal lines between about 0.3 and 0.8 μm/h that correspond to minimum-ionizing ($\beta \gtrsim 0.95$) nuclei with Z up to ~83 that were detected on the flight. Only particles with steeply rising etch rate curves, corresponding to slowing nuclei of lower velocity, reach etch rates as high as that of the monopole candidate. In none of our previous ultraheavy

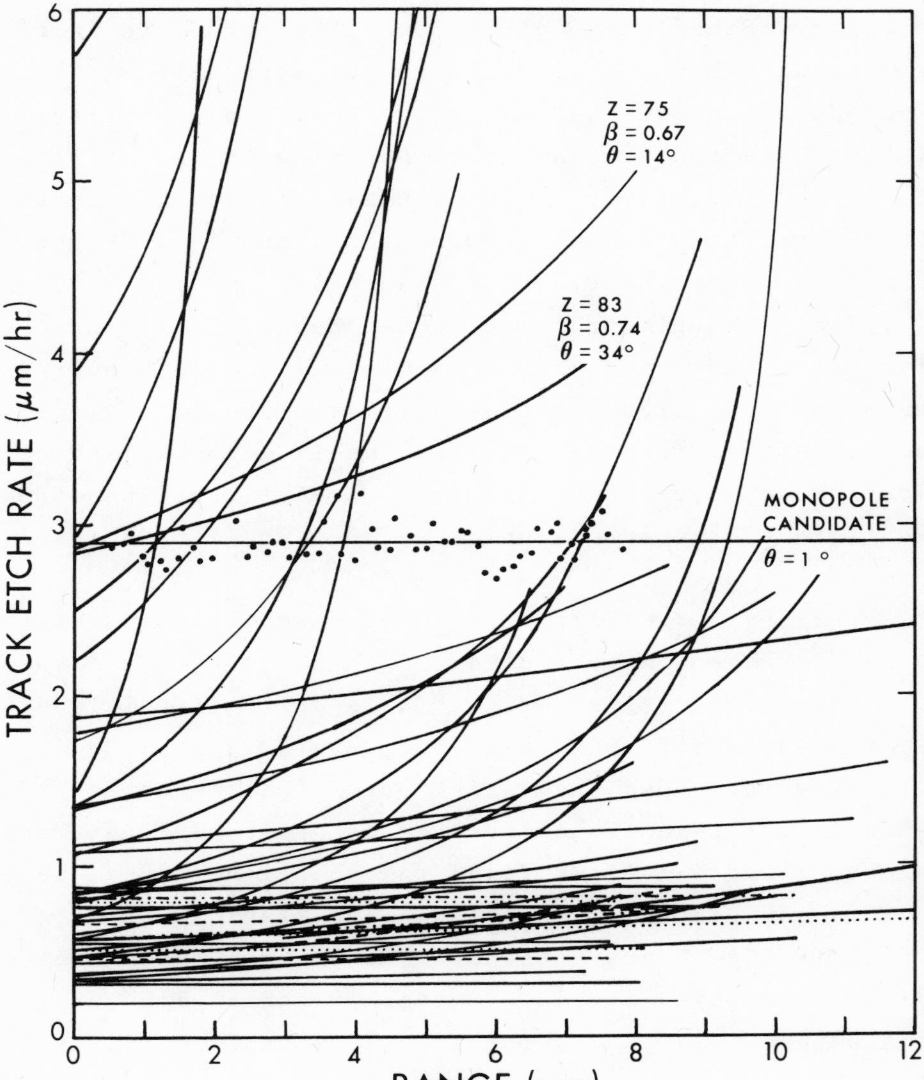

Figure 2. Response curves of the majority of the ultra-heavy particles from the Sioux City balloon flights. A few slow particles with very steep curves are not plotted

cosmic ray experiments had we seen events with constant etch rates higher than 1 μm/h.

After publishing the Letter reporting our evidence, we found that the Lexan used in that flight was slightly different in composition from that used in our previous experiments. It did not contain the trace of a UV-absorbing dye that is normally added to Lexan to retard its deterioration in sunlight. Instead of increasing with Z_e/β as $(Z_e/\beta)^{\alpha}$, with α in the range 3.5 to 4 as had been found previously,[4,5] the etch rate behaved as

$$v_T = 0.900(Z_e/90.18\ \beta)^{5.07}\ \mu m/h \quad . \qquad (1)$$

This required a downward revision of Z/β from ~137 to ~114. The higher value of the exponent meant that this Lexan was capable of detecting smaller changes in ionization rate than could the previous Lexan. Our first reaction was one of dismay that the revised ionization rate seemed to be significantly lower than expected for a monopole of strength 137. Steve Ahlen, a student of mine, then found that in a condensed medium the ionization rate of a monopole is not a constant but decreases continuously as it slows down. The old prescription[2,3] for finding dE/dx by replacing Ze by $g\beta$ in the Bethe-Bloch equation neglected the density effect. Using a restricted energy loss model of track formation, Ahlen[6] derived the curves in Fig. 3. The track etch rate in Lexan for a monopole of strength 137 e and velocity $\beta = 0.5^{+0.1}_{-0.05}$ is equivalent to that of a relativistic nucleus ($\beta \simeq 1$) with $Z_e = 121 \pm 2$. In view of the approximations used in Ahlen's treatment and of the crudity of the restricted energy loss model, this number is consistent with our revised estimate of $Z_e/\beta \simeq 114$ for the monopole candidate. Reasoning from the

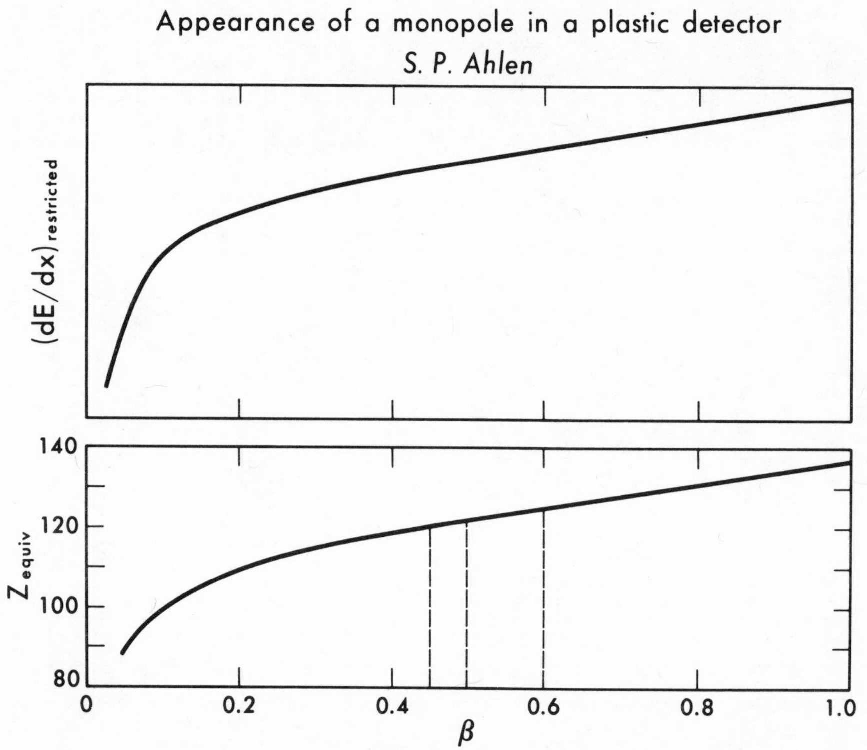

Figure 3. Effect of a monopole of strength g = 137 e in Lexan detectors, calculated by S. P. Ahlen (ref. 6). Upper curve shows velocity-dependence of energy loss to electrons with less than 350 eV, which produces etchable tracks in Lexan. Lower curve shows the equivalent charge of a highly relativistic (β=1) nucleus that would produce the same etch rate in Lexan as a monopole of velocity given by the abscissa.

observed numbers $Z_e/\beta \approx 114$ and $\beta = 0.5^{+0.1}_{-0.05}$, we now would infer a magnetic charge $g = 130^{+2}_{-4}$, with an additional uncertainty of at least ±5 charge units quoted in Ahlen's paper.

CRITICISMS

We expected and got a lively response to our paper[6-19] Some authors have critized our evidence and offered alternative explanations;[7-10] some have derived constraints on the properties or mode of production of the proposed monopole;[11-15] some have dealt with monopoles in general;[6,16,17] one reports a method of distinguishing a monopole from a nucleus by adding a linearly polarizing paint to a Cerenkov film detector;[18] and one reports a new negative search.[19] At the present stage of our calibrations, some of the criticisms of the evidence have become invalid, but some cannot be fully assessed until we are further along.

We and all our critics recognize that the constant, high ionization rate, together with the low velocity, would make a mundane explanation of the event impossible if the measurements were beyond reproach. Here are the criticisms:

1. There is a "glitch" in the Lexan data (see Fig. 4) that suggests that the ionization rate suddenly decreases and then increases gradually as would be expected if a fast nucleus underwent a nuclear collision in the Lexan, fragmenting into a slightly lighter nucleus.

2. The two data points in the upper sheet of Lexan can be rejected on the grounds that the sheet was separate and may have experienced a different mechanical, thermal and chemical history from the remainder of the stack.

3. The black points and triangles in Fig. 4 were

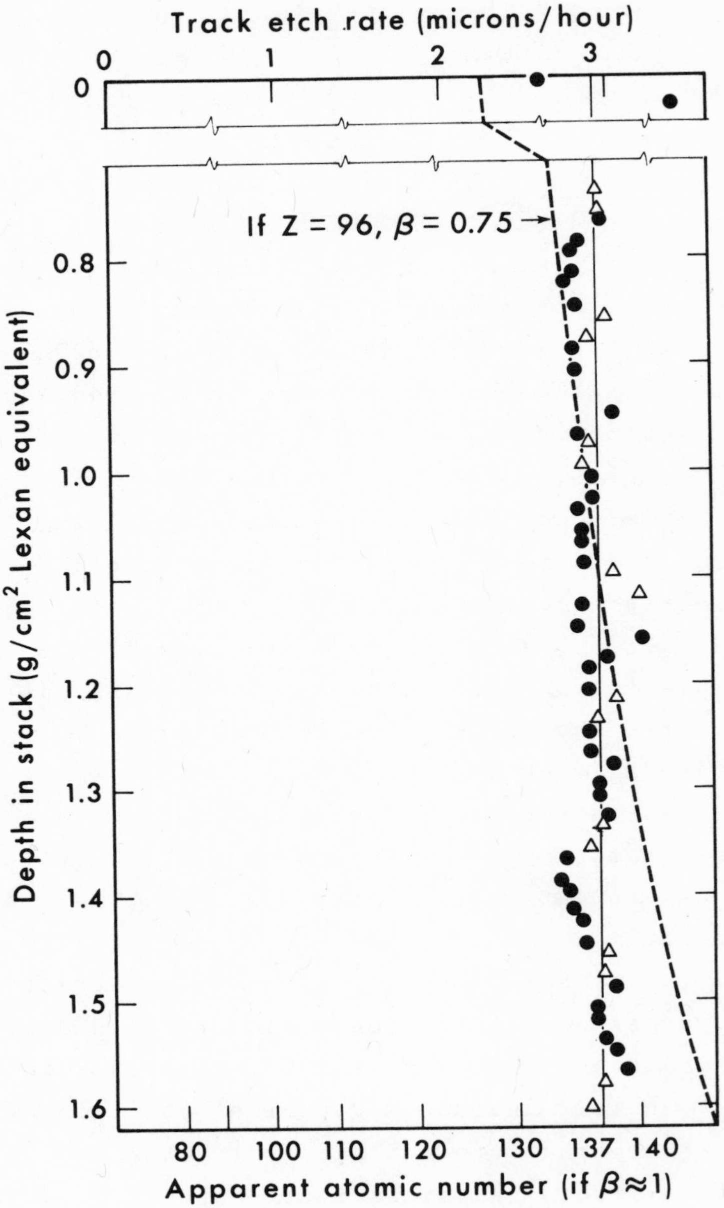

Figure 4. Original Lexan data for monopole candidate (Fig. 2 of ref. 1). Upper two points are from sheet 2; top two triangular points are from sheet 6; bottom two triangular points are from sheet 34.

obtained in sheets processed in two different etch tanks. A calibration was done only for the sheets corresponding to the black points; therefore, the triangular points can be rejected.

4. The method of velocity determination based on the track profile in nuclear emulsion has not been demonstrated to work. Further, in P. H. Fowler's model of track structure, it would not be possible unambiguously to distinguish the radial dependence of track structure of particles with $\beta \gtrsim 0.45$. Therefore, the information from the nuclear emulsion should be disregarded.

5. The thickness of material between the upper Lexan sheet and the main Lexan stack was labeled incorrectly in the paper. The actual thickness was less, reducing the difficulty of accounting for the data by a fragmenting nucleus.

Taking these points into account, the critics "explained" the event by a nucleus with $Z \approx 78$ or 79 that passed through the Cerenkov detectors with a velocity ~ 0.68 to 0.70 c, just below the velocity at which Cerenkov light would have produced a detectable number of photons. In order to maintain the right average ionization rate, the nucleus had to fragment twice in the main Lexan stack, losing about two charges each time. The second fragmentation is supposed to have occurred at the glitch in the data; the first fragmentation is not visible in the data.

6. To these published criticisms I shall add one of my own. Though the Cerenkov film technique has been discussed in detail in Pinsky's thesis[20] and measurements have been made of Cerenkov light images produced in the film by a few ultraheavy cosmic rays in a previous balloon flight,[4,20] the technique requires very exacting performance of Kodak's fastest experimental film and needs to be

tested thoroughly on the ensemble of particles that inclu the monopole candidate.

THE THICKNESS OF THE STACK

Not only did we overestimate the thickness of materi between the upper Lexan and the main Lexan stack, but we made a highly schematic drawing of the detector assembly that omitted two thin Lexan sheets, one of the Cerenkov detectors, a thin emulsion, a cellulose triacetate sheet, two Mylar sheets and the details of the layers of opaque wrapping paper around the emulsion and Cerenkov detectors We simplified the drawing in order to emphasize the main features of the experiment within the spatial confines c a Letter. Figure 1 of the present paper gives a more detailed breakdown of the stack showing all Lexan sheets, both Cerenkov detectors, the main emulsion, and the corre thicknesses in g/cm^2 Lexan equivalent, but still in some-what simplified form. In Fig. 2 of ref. 1, reproduced he as Fig. 4, we took the thickness of Cerenkov detectors, emulsion, and associated wrapping material to be 0.625 g/cm^2, whereas the correct thickness should be 0.347 g/cm Lexan equivalent. Referring to the correct Fig. 1 of the present paper, this material extends from the depth 0.039 g/cm^2 to 0.386 g/cm^2. The upper triangular data point in Fig. 4 corresponds to sheet 6. It was plotted at ~0.74 g/cm^2 but should be at 0.462 g/cm^2. All lower points in that figure will appear at the proper depth if 0.278 g/cm is subtracted. Our overestimate of the stack thickness is equivalent to a change in velocity of ~0.02 c for a nucleus with Z~78 and an initial velocity of ~0.68. For an initial velocity of 0.73 c, it is equivalent to a change in velocity of only 0.015 c. As we shall see in the next section, when all the Lexan data unjustifiably omitted by Alvarez are included (having now been calibrated), they rule out fragmenting nuclei with velocities as high as 0.74 c. Whether one starts with a nucleus at $\beta=0.68$ or 0.70 is thus irrelevant, and the error in stack thickness is unimportant provided either the emulsion or Cerenkov detectors can rule out velocities appreciably higher than 0.74 c.

The next four sections include a discussion of the remaining criticisms, which must be shown to be invalid before worrying unduly about other difficulties such as the negative results of other monopole experiments of much greater collecting power.

DATA AND CALIBRATION OF THE LEXAN DETECTORS

The principles and applications of nuclear tracks in dielectric solids are treated in a recently published book.[21] of all track-recording solids, Lexan plastic is the kind most used for identifying charged particles. Because of its low cost, high resolution, and insensitivity to lightly ionizing particles, it is ideal as a detector of large collecting power to study the rare, ultraheavy cosmic rays and to search for hypothetical, heavily ionizing particles. In a solution of a suitable chemical reagent, material along the trajectory of a heavy particle is etched out at a rate that depends on the ionization rate, leaving cone-shaped etch pits whose lengths can be measured in a microscope. The track etch rate, v_T (defined at etch pit length divided by etch time), increases as some power of Z_e/β that must be determined for each batch of Lexan and exposure history. A single expression fits values of v_T extending over at least three orders of magnitude for $Z \gtrsim 20$ and $\beta \gtrsim 0.2$.

Figure 5 illustrates schematically how we determined

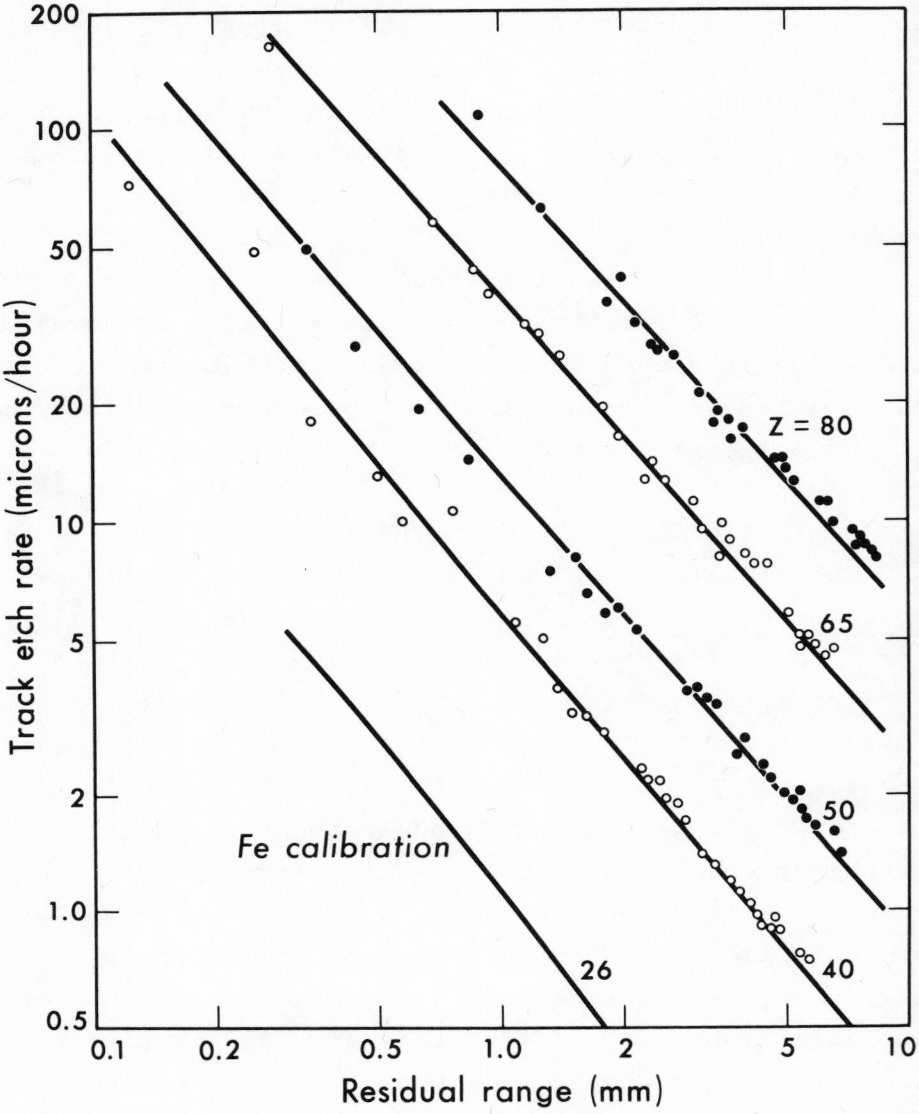

Figure 5. Response curves of several stopping ultraheavy nuclei as a function of residual range, along with the curve resulting from measurements of numerous stopping Fe nuclei. The curves of Fig. 2 become nearly straight when plotted with residual range as abscissa, using log-log pa

the two constants in the power law relation for v_T. A scanning criterion was adopted that favored the selection of events with $20 \lesssim Z \lesssim 30$. Because of the pronounced cosmic ray abundance peak at Z=26 (iron), the measurement of 50 to 100 events, each comprising several pairs of etch pits in consecutive Lexan sheets, sufficed to define a surve of etch rate vs. residual range for Fe. In this short account we show only the result, a curve labeled "Fe calibration." To first order, this curve, together with a table of range-energy relations, enabled us to determine both constants in eq. 2. The density of stopping Fe nuclei was sufficiently high that we were able to carry out the calibration in the very sheets containing the monopole candidate. The criticism in point 3 is invalid because we calibrated the sheets etched in both tanks with Fe tracks and found the same values for the constants in eq. 2 for both etchings.

Out of some 600 candidates found in a stereomicroscopic scan of the entire nuclear emulsion, we have thus far verified that 64 of them have $Z \gtrsim 40$ and we have measured their etch pits in the Lexan sheets. Fourteen of them came to rest in the Lexan stack, producing tracks with extremely high etch rates near the ends of their ranges. Data for four stopping particles are shown in Fig. 5. The requirement that the data for these 14 particles of known range have the correct slope on the graph of v_T vs. range is a stringent check on the exponent in eq. 1.

We searched through the data for the 64 events with $Z \gtrsim 40$ for evidence of Lexan sheets with higher or lower sensitivity than given by eq. 1. We found that sheet 2 (in the notation of Fig. 1) was systematically only about 0.94 times as sensitive as the sheets in the main stack.

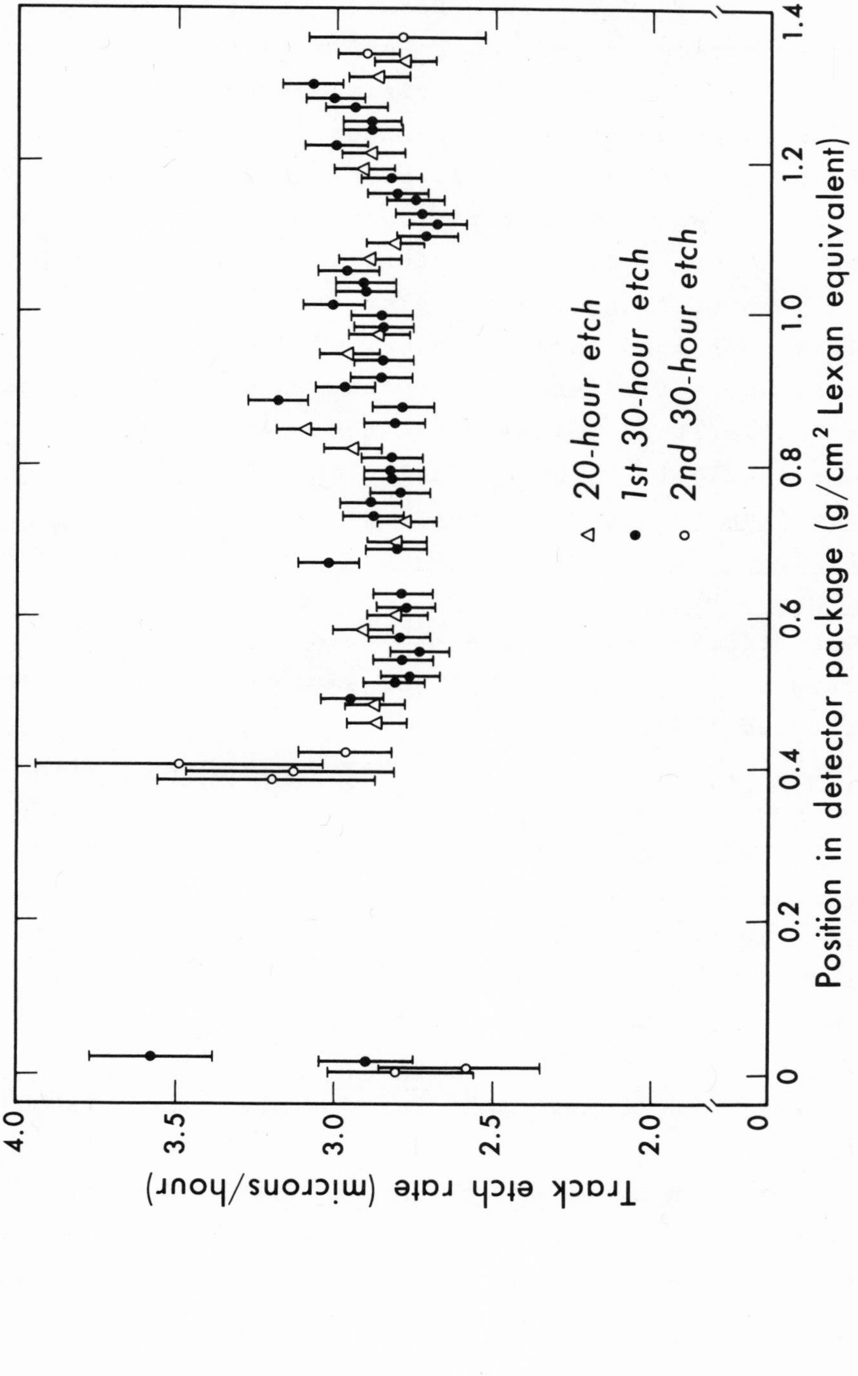

Figure 6. Calibrated Lexan data for monopole candidate.

However, the data in sheet 2 showed no larger dispersion than did data for sheets in the main stack, so that the criticism in point 2 is invalid.

In order to increase our lifting power, we flew part of the stack ($10m^2$) on September 18, 1973, and 20 m^2 of the stack on a second balloon launched on September 25, both from Sioux City. Both portions stayed at float altitude (3 g/cm^2 and ~4.5 g/cm^2 respectively) for 60 hours. Our calibrations show that both portions have the same sensitivity.

Figure 6 shows the calibrated Lexan data for the monopole candidate. The data in sheet 2 are raised by the factor $(0.94)^{-1}$ and given error bars that represent the standard deviation about the factor 0.94 for this sheet based on the measurements for all 64 cosmic rays. No data exist for sheets 5 and 12, which had been etched for a long time (160 h) to form holes that allowed the event to be initially found by ammonia scanning. We initially set aside sheets 1,3,4, and 35, but after our published evidence had been criticized we etched these sheets in a third tank and calibrated their sensitivity individually using the 64 cosmic rays. The results for the monopole candidate, with error bars, are shown in Fig. 6. The main Lexan stack, comprising sheets 4 through 35, was bolted together as a unit. We found that the outer surfaces of the stack (top of sheet 4, bottom of sheet 35) were somewhat more sensitive than the inner surfaces, and a correction has been applied to those two data points in Fig. 6. Not surprisingly, the thin Lexan sheets (1 and 3), having been manufactured in a different batch from the other sheets, required slightly different constants in the etch rate equation. A detailed account of the calibrations will appear in a future paper.

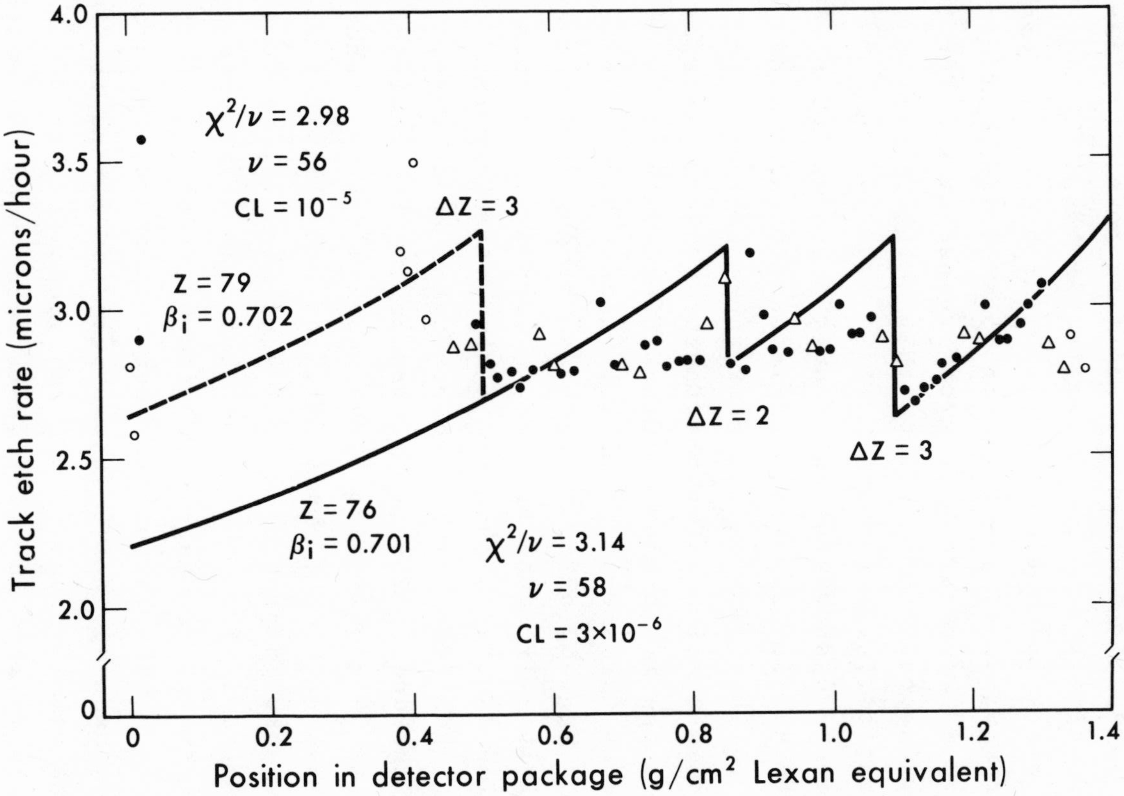

Figure 7. Best fits for doubly and triply fragmenting nuclei with $\beta = 0.7$ at the Cerenkov detector.

Figures 7 through 11 show various attempts to fit the Lexan data with fragmenting nuclei having initial velocities $\beta_i c$ (in sheet 1) ranging from 0.7 c up to 0.98 c. I used eq. 1 and a range-energy table to generate the curves, trying in each figure to minimize the square error by judicious choices of Z, β_i, and ΔZ. In Figs. 7 and 8 I worked backward from the glitch.

For each curve I have listed the statistic χ^2, the number of degrees of freedom, and the confidence level for the fit. To compute χ^2 one needs to know σ. I want to test the hypothesis that one of the curves in Figs. 7 through 10 gives as good a fit as the line of zero slope at the average etch rate 2.88 μm/h in Fig. 11. For the main Lexan stack (excluding sheets 4 and 35), assuming a normal distribution of measured etch rates about the average rate, I calculate a fractional σ of 0.0337. Including the separately determined σ's for sheets 1 to 4 and 35, I get a root mean square σ_{rms}=0.0356 for the monopole fit. This is quite a reasonable choice; about half of the fractional σ's of the data from the optimum curves from eq. 2 for the 62 cosmic rays with $Z \gtrsim 40$ fall between 0.03 and 0.04. This procedure of course insures that $\chi^2/\nu \approx 1$ for the line in Fig. 11 and thus avoids the error common in particle physics experiments of underestimating σ. (See Rosenfeld's discussion[22] of the Particle Data Group's use of a Scale Factor to inflate the quoted σ's in experiments so that $\chi^2/\nu \approx 1$.)

We now wish to find the confidence levels associated with the larger values of χ^2 that I calculate for the curves in Figs. 7 to 10. The F-test is suited for comparing the variances of two curves through a set of data. The statistic F is defined as the ratio of reduced chi-squares for the two curves. Based on the F-test, in the

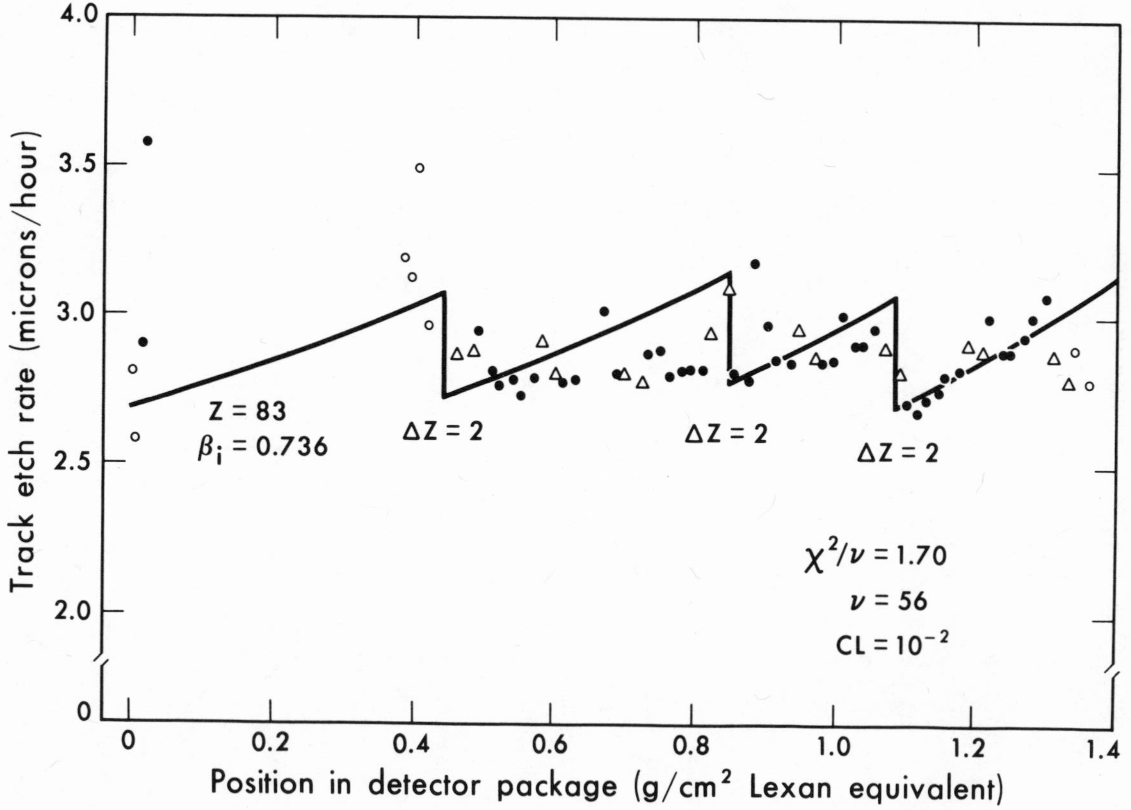

Figure 8. Best fit for a triply fragmenting bismuth nucleus with β = 0.736 at the Cerenkov detector.

figures and in column 6 of Table 1 I have listed the confidence levels that the curves in Fig. 7 to 10 are as good a fit to the data as is the straight line in Fig. 11. The values are more conservative (higher) by about a factor 10 than would be the values computed with a χ^2 test.

The doubly fragmenting nucleus with $Z \approx 78$ hypothesized by Alvarez[9] and by Fowler[10] has been widely publicized. I believe the Lexan data rule out that hypothesis and also the one shown in Fig. 8. When the number of degrees of freedom is very large, a reduced χ^2 as low as 2 or 3 leads to extremely low confidence levels. Figure 12, which compares the error distributions for the curve with two interactions in Fig. 7 and for the straight line fit in Fig. 11, makes the point quite clearly. In the case of the fragmenting nucleus, not just one or two but many points lie outside the Gaussian error envelope derived from the σ_{rms} of 0.0356 for the straight line fit. The Lexan data alone cannot rule out a fast nucleus of uranium, curium, or a superheavy element (Figs. 9-11). Only if the emulsion or Cerenkov measurements show that the velocity could not have been as high as 0.82 c or 0.86 c, respectively, can these scenarios be ruled out.

FRAGMENTATION AND THE "GLITCH" IN THE LEXAN DATA

In computing the overall confidence level for the fragmenting nuclei in Figs. 7 to 9, we must consider not only the fit to the Lexan data but also the product of two quantities: the probability of a given number of fragmentations with just the right decrease of charge to follow the Lexan data, and the total number of nuclei in all balloon flights that entered the stack with initial ionization rates and velocities that could have simulated

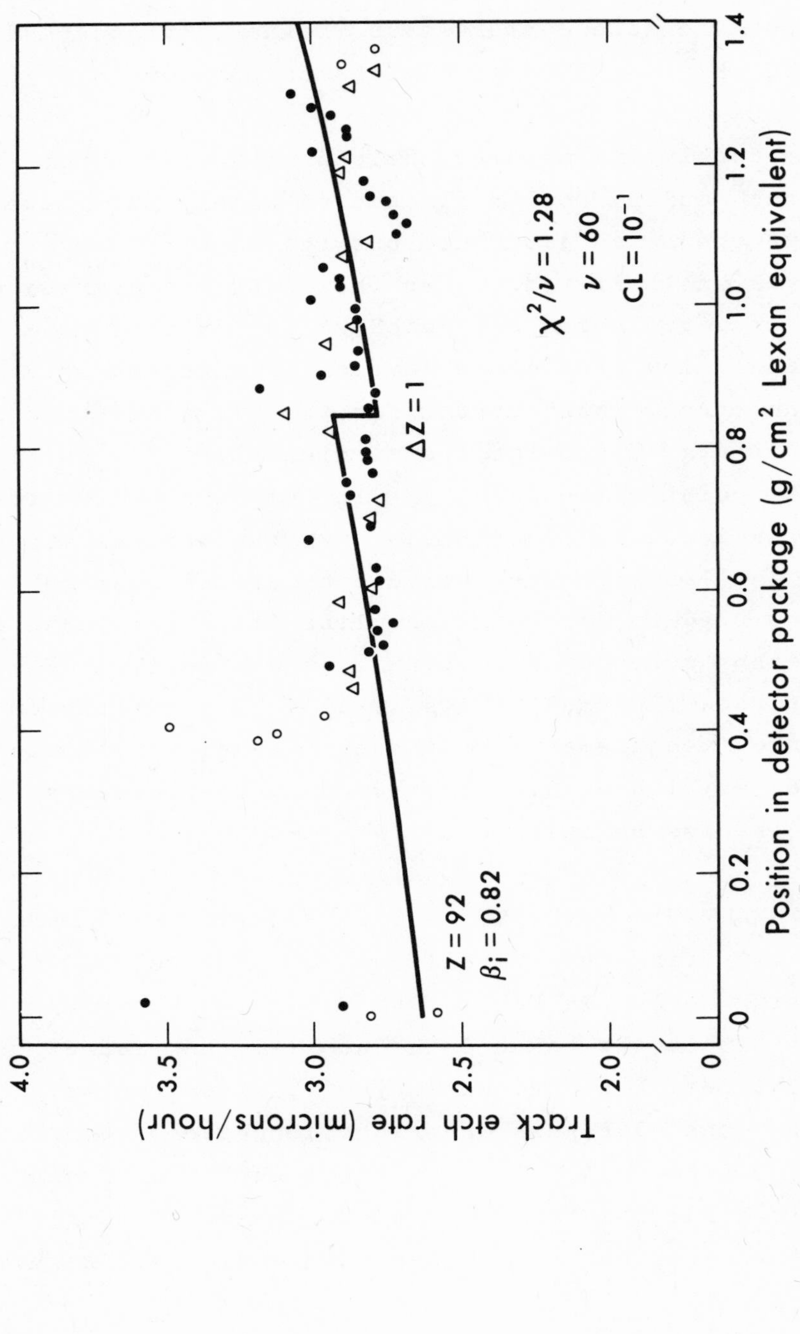

Figure 9. Best fit for a once-fragmenting uranium nucleus with β = 0.82 at the Cerenkov detector.

a monopole if the fragmentations occurred.

Alvarez[9] assumed "several hundred" nuclei and a total probability of order unity for a doubly fragmenting platinum nucleus to have been seen in some flight. Fleischer and Walker[8] did a more realistic calculation. They considered nuclei with three possible velocities at the emulsion--0.7 c, 0.65 c and 0.6 c--and concluded that at the highest velocity a fragmenting nucleus would be a reasonable interpretation, whereas at the lowest velocity only a monopole could account for the data. To fit the data in the main stack (ignoring the data in the upper sheet) they assumed 2,3, and 8 fragmentations, each with $\Delta Z \approx 2$ to 4, for $\beta=0.7$, 0.65, and 0.6, occurring with probabilities they calculated to be $\sim 10^{-3}$, 2.4×10^{-15} per incoming nucleus. For the three cases they assumed 14, 13, and 8 nuclei in the right range of Z and β and arrived at total probabilities of 0.017, 3×10^{-4}, and 6×10^{-13} for $\beta=0.7$, 0.65, and 0.6.

I have followed the procedure of Fleischer and Walker to calculate the numbers in column 5 of Table 1, making two changes to make the calculations more realistic.

(1) Shirk and I examined all previous ultraheavy cosmic ray experiments to see how many nuclei were detected in a suitable range of Z and β. Flights launched from the southern U. S. could collect none because the geomagnetic cutoff rigidity excludes nuclei with $\beta \lesssim 0.8$ to 0.85. Flights from the northern U. S. fall into two categories. Those by the Bristol-Dublin collaboration employ very thick stacks (~ 5 g/cm^2) with enough material to detect velocities less than ~ 0.85 c with no difficulty. In our Minneapolis experiment[4] we detected no particle in a suitable range of Z and β. In our Skylab experiment[5] we detected one lead nucleus (Z=82) with $\beta=0.68$ and with

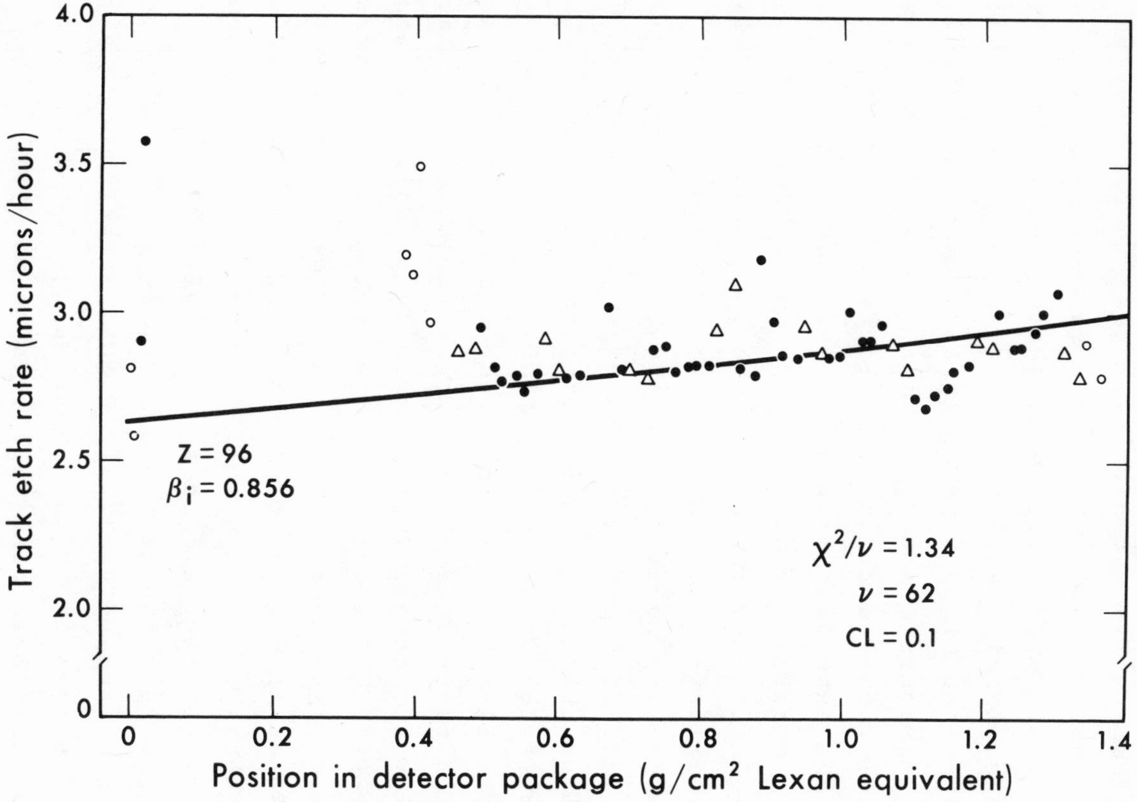

Figure 10. Best fit for a curium nucleus with β = 0.856 at the Cerenkov detector.

Z/β increasing from 121 to 153 through the stack. In our Sioux City flights we detected two nuclei with initial Z/β near that of the monopole candidate. Their etch rate curves are labeled in Fig. 2. One of them actually fragments, but with a loss of 34 charges. Figure 13 shows the data for that event, plotted with the same scale as in Fig. 6 for the monopole candidate. Thus, instead of the 13 candidates assumed by Fleischer and Walker, we use the observed number of four particles (including the monopole candidate) that should multiply the probability of a sequence of fragmentations by a single particle.

(2) I assumed the same fragmentation mean free path as did Fleischer and Walker, but with a window in ΔZ that was two instead of three units wide.

Is the glitch in the Lexan data an "obvious fragmentation," as claimed by Alvarez? If it were, then the above estimates are irrelevant, this paper is irrelevant, and I would immediately go back to the research I was doing before last July. ("Monopoles don't fragment.") Without having seen other Lexan data, it is quite natural to interpret the glitch as a sudden loss of charge. However, correlated variations in etch rate occurring over several consecutive sheets are not uncommon. Some show upward glitches; most of them must be attributed to the chemistry and physics of the plastic and of the etching process, not to nuclear or atomic processes. Figure 13 is an example of large fluctuations in the data that appear to the eye to be correlated.

In the course of our studies of ultraheavy nuclei we have seen four definite fragmentations with $\Delta Z \approx 3$ to 6 and another ten with larger ΔZ, including the one in Fig. 13. The data following the fragmentation have a <u>shallower</u> slope than those preceding the fragmentation, for the

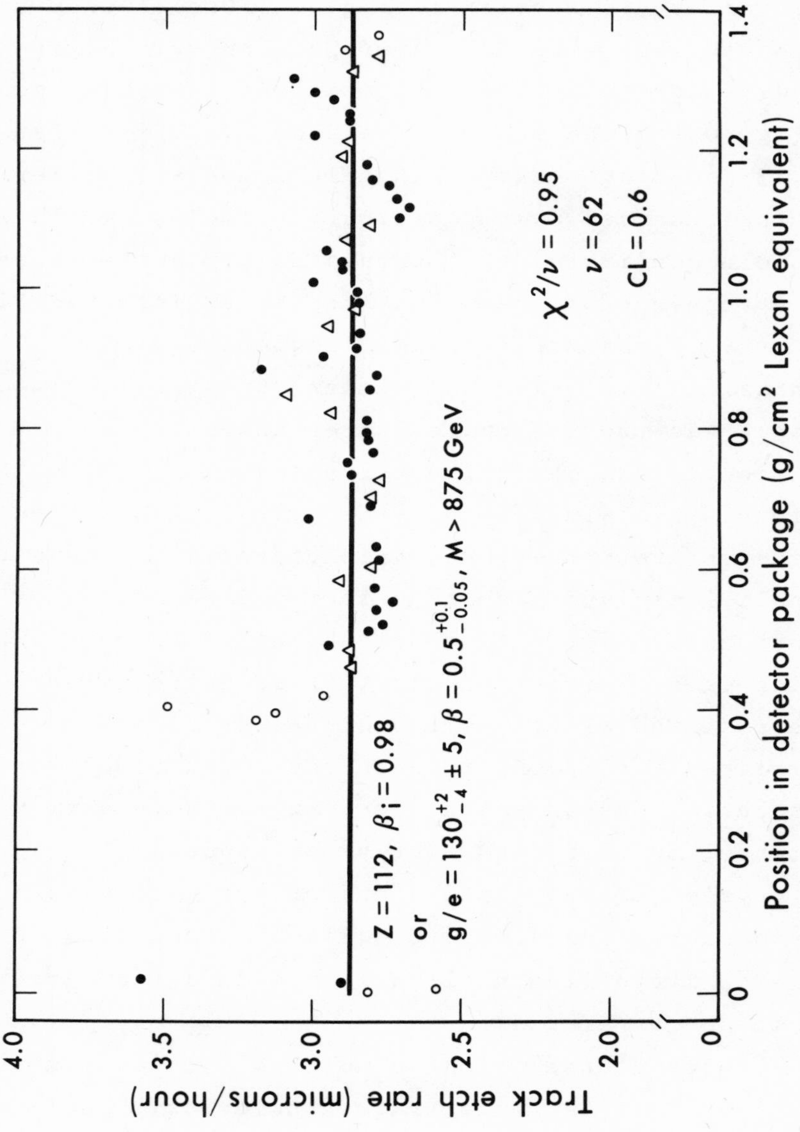

Figure 11. Best fit for a straight line of zero slope.

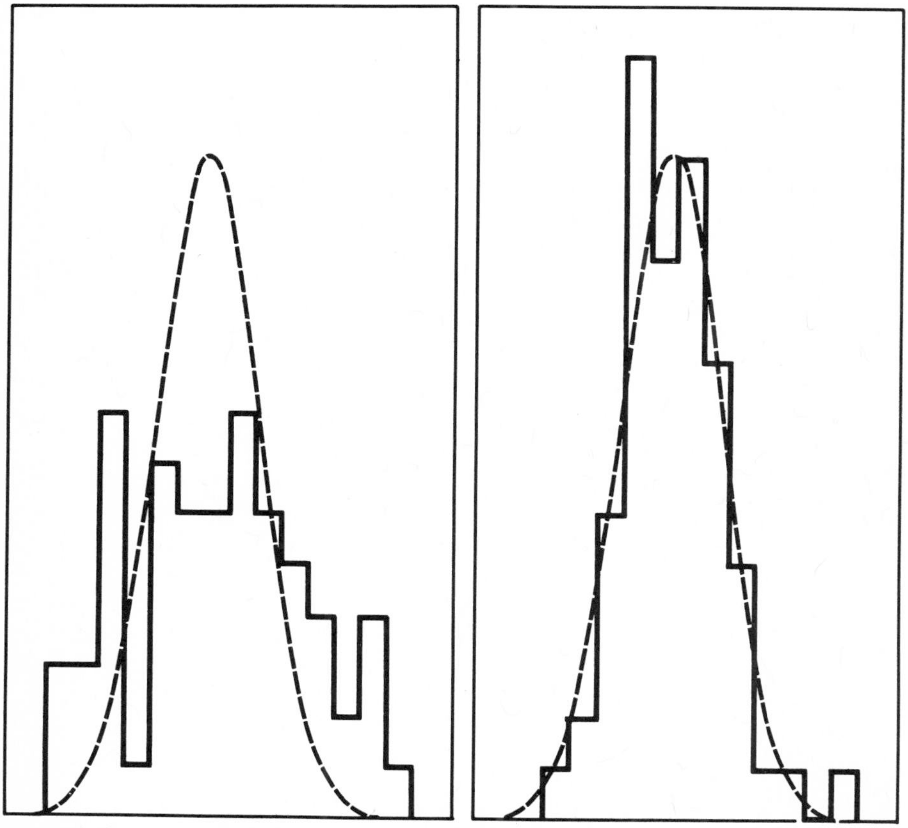

Twice-fragmenting nucleus with $Z = 76$, $\beta = 0.70$.

$\chi^2/\nu = 3.14$, with $\nu = 58$

Straight line of zero slope.

$\chi^2/\nu \approx 1$, with $\nu = 62$

Figure 12. Error distributions for (a) a twice-fragmenting nucleus with $\beta = 0.7$ at the Cerenkov detector (Fig. 7) and (b) a straight line of zero slope (Fig. 11). The curves are identical Gaussians with $\sigma = 0.035\ \bar{v}_T$ (see text). The confidence levels are 3×10^{-6} for the fragmenting nucleus and ~1 for the straight line.

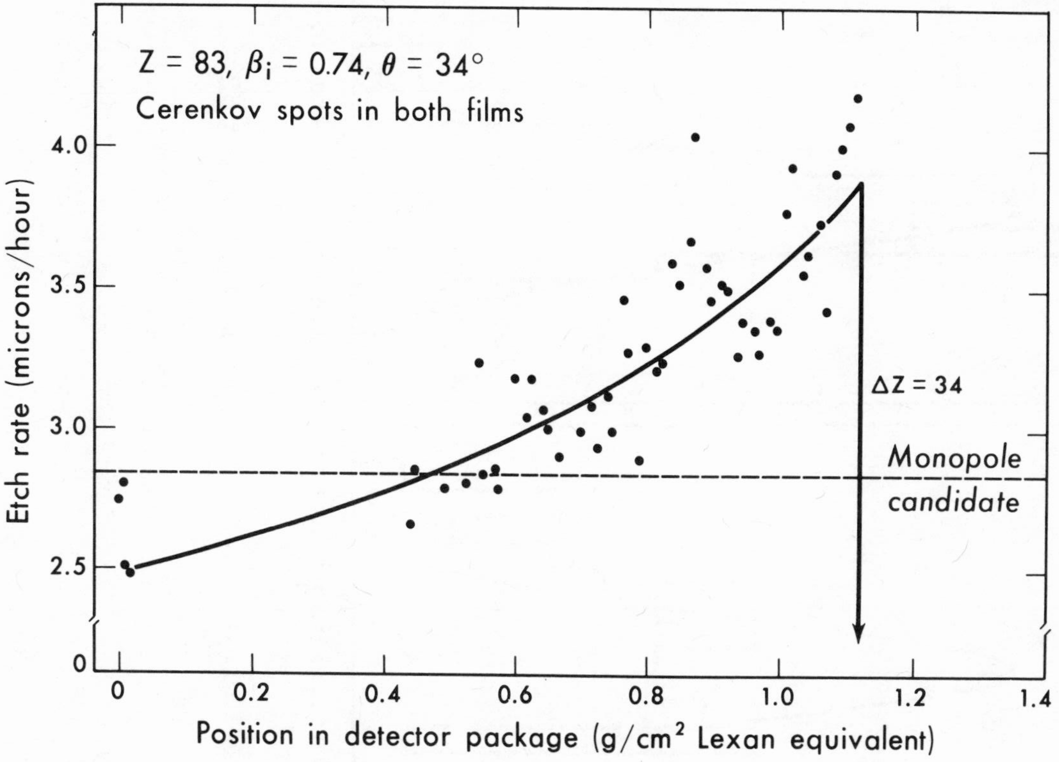

Figure 13. Data for the nucleus that comes closest to simulating the monopole candidate. Both the emulsion and the Cerenkov film indicated that it had $\beta>0.7$. It fragmented with loss of 34 charges at 1.1 g/cm^2.

simple reason that a fragmenting nucleus loses charge and mass but continues on at about the same velocity and thus has a greater range and a smaller gradient to its Bragg curve than it would have had. The glitch in the data for the monopole candidate is different and unphysical in that the data following the step have a much higher slope than the data preceding the step.

MEASUREMENTS AND TESTS OF THE NUCLEAR EMULSIONS

As early as 1969 W. Z. Osborne had the idea that a single layer of nuclear emulsion could be used to estimate both Z and β of a heavy particle, for velocities between ~0.3 c and ~0.7 c. As a first test of his method we exposed a stack of Lexan below a layer of emulsion in a spectacularly long balloon flight (14 days) launched from Minneapolis in 1970.[4] The results, though encouraging, have not been thoroughly analyzed until recently and have not been published even though the Lexan data were published several years ago.[4] It is thus true that the method must be regarded as untested. Here I give a brief account of it and show results for 32 cosmic rays with $Z>50$ from the Minneapolis flight and for 77 cosmic rays with $26 \leqslant Z \leqslant 83$ from the Sioux City flights. The procedure in all the flights was to scan all the emulsions in a stereomicroscope at Houston, locating tracks with large core and halo radii (defined below) that might correspond to cosmic rays with $Z>26$. Coordinates, azimuth and zenith angles, and core and halo radii were recorded and sent to Berkeley. We etched the Lexan sheets, followed these tracks until they either ended or penetrated the entire stack, and determined Z and β for the heaviest events and for a number of the Fe tracks.

In G-5 emulsion the track of a heavy nucleus consists

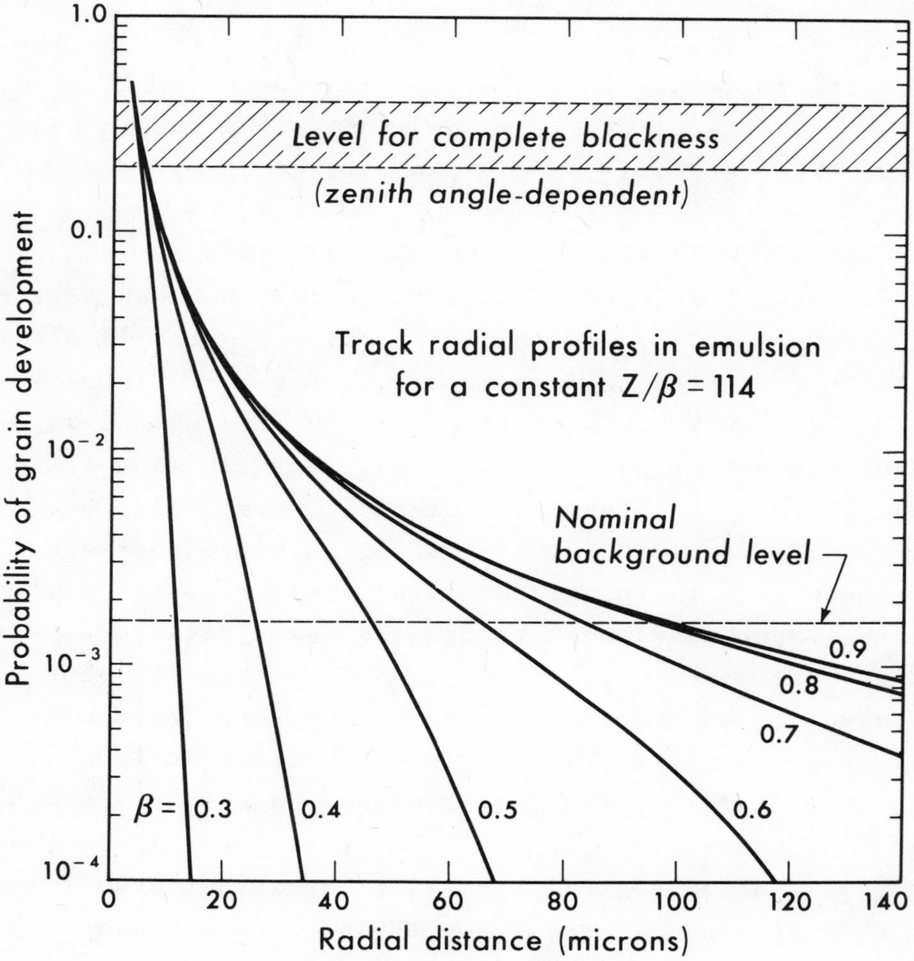

Figure 14. Probabilities of grain development around the track of a particle with $Z/\beta=114$, calculated by W. Z. Osborne.

of a solid core of fully developed silver grains, extending to a radial distance that depends on Z/β virtually independently of β, surrounded by a halo of silver grains whose density decreases radially until it is indistinguishable from the background grain density. The radial distribution of silver grains is determined by the energy and angular distribution of δ-rays, which depend on Z and β of the incoming particle, and by the radial transport and energy deposition of these δ-rays. Osborne has used the model of Katz and co-workers to compute the probability of grain development as a function of Z, β and radial distance, using the Mott cross section instead of the less accurate Rutherford cross section. Figure 14 shows a set of Osborne's radial profiles for various velocities at a constant value of $Z/\beta=114$ pertinent to the monopole candidate. The probabilities corresponding to an opaque core and to the background gray level are marked. For higher or lower Z/β the curves move up or down.

It is probably fair to say that the dependence of core radius on Z and β is uncontroversial, because at distances of less than a few microns from the particle's trajectory most of the blackening is caused by electrons of low energy, for which the assumption of diffusive transport due to the intense multiple Coulomb scattering in nuclear emulsion is valid. The dependence of core radius on Z and β has been studied in a series of papers[24] by a Swedish emulsion group, who find that the model of Katz and co-workers fits their measurements of cosmic ray track widths over a wide range of β and charges up to 26.

Figure 15 shows our measurements of core radius, made by eye with a reticle and an oil immersion objective, as a function of Z/β. Here β refers to the velocity at the emulsion as determined from the value of Z and β measured

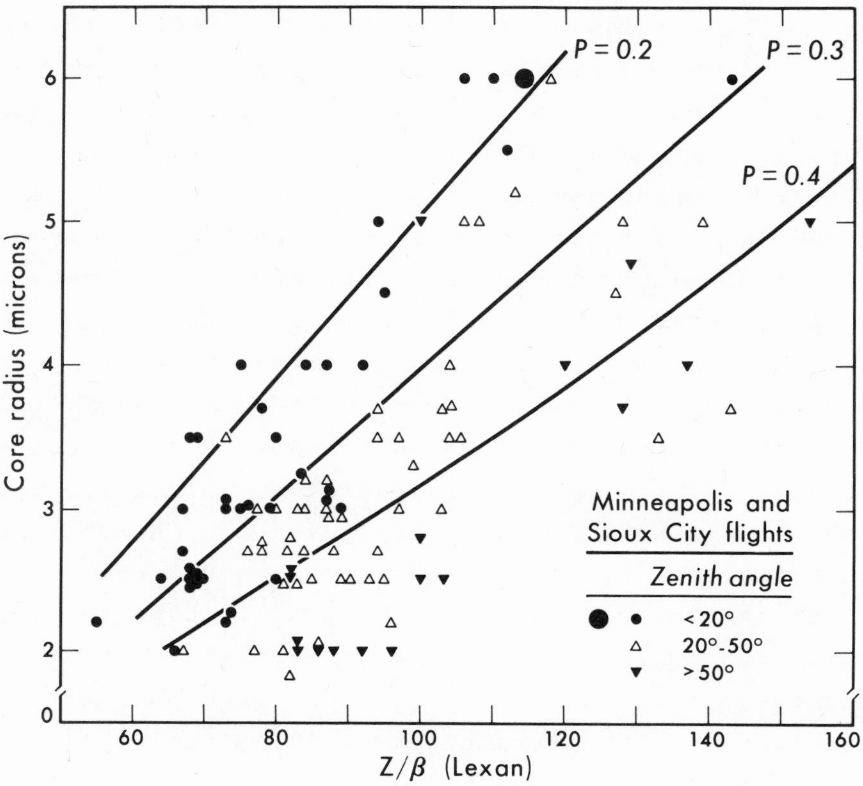

Figure 15. Measurements of core radius in emulsion for particles with various zenith angles and values of Z/β inferred from Lexan data. The curves for different probabilities of grain development were calculated by Osborne. The large black circle is for the monopole candidate.

for the same event in the Lexan stack. In agreement with the Swedish group, we find a pronounced zenith angle effect: steep tracks have an apparently larger core width than do shallow tracks with the same Z/β. For the extremely heavily ionizing events we have studied, two effects may contribute. (1) When looking down a nearly vertical track, it appears black out to a greater distance, corresponding to a smaller probability of grain development (note the curves in Fig. 15 corresponding to probabilities of 0.2, 0.3, and 0.4), than does a shallow track. (2) During fixing, the undeveloped silver halide grains are removed, the emulsion shrinks in thickness, and the solid mass of silver grains in the core, being incompressible, may be displaced outward for a very steep track more than for a shallow track.[7]

The monopole candidate, which came in at a zenith angle of 11°, is plotted in Fig. 15 with the same (but enlarged) symbol as are other events with zenith angles from 0 to 20°. From the fact that it follows the trend with Z/β of the other steep events (near the curve P=0.2), one cay say that its core radius of 6 μm is consistent with its having a value Z/β between ~100 and ~140. Thus, I conclude that the portion of nuclear emulsion traversed by the monopole candidate was neither anomalously sensitive nor insensitive compared to the other emulsions in the flights.

At low probabilities of grain development, corresponding to transport and energy deposition of δ-rays at distances of many tens of microns out from the trajectory, the shapes of the curves in Fig. 14 are disputed. Using a simple diffusion model and additional simplifying assumptions, Fowler[10] was able to integrate his expression for the energy deposition by δ-rays as a function of Z, β,

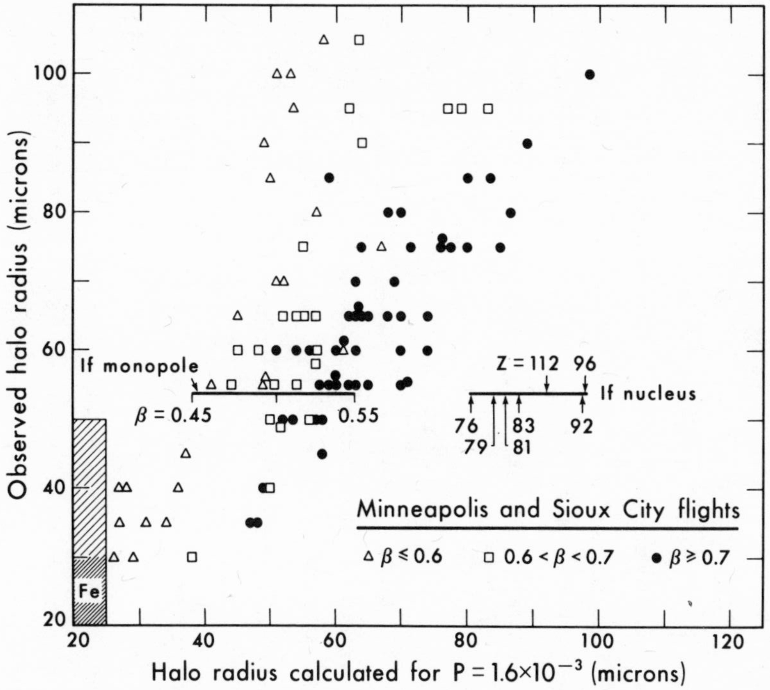

Figure 16. Measurements of halo radius in emulsion for particles with various zenith angles. The abscissa gives halo radius calculated using Osborne's model with $P = 1.6 \times 10^{-3}$ for the edge of the halo and using Z and β measured with the Lexan stack. The line segments at an observed halo radius of ~55 μm show where the point should be plotted if our event was a monopole at various velocities or one of the nuclear candidates.

and radial distance. For values of $\beta \gtrsim 0.45$ he has claimed that his curves are so close together that one can tell nothing about the velocity of the particle (point 4 of the criticism). They are so different from Osborne's curves that at least one of the two models must be wrong. Fowler's statement that Osborne's method cannot work at $\beta \gtrsim 0.45$ has been widely publicized and has been cited by Alvarez[9] as his justification for rejecting our emulsion evidence that $\beta \approx 0.5$ for the monopole candidate.

Osborne has pointed out that Fowler's own published data[25] on radial profiles of ultraheavy cosmic rays are inconsistent with his diffusion model. Alvarez has privately expressed doubts to Fowler that his random walk model is valid for the more energetic electrons. Ray Hagstrom (LBL) has shown that all of Fowler's simplifying assumptions act in the same direction to underestimate the velocity-dependence of the radial distribution of the energy deposited by fast electrons. He concludes that Fowler's model is invalid, and he is developing his own model of track profiles. Of course, ultimately the test of correctness of a model is the extent to which it agrees with experiment. Osborne is now testing a computer-driven image-recognition system that records the positions of all silver grains outside the core region and calculates a radial profile. Until we have such profiles for the events in the Sioux City flights, we must use measurements made by eye. The eye cannot recognize quantitatively the probability of grain development, P, but it can estimate the radial distance at which the halo of grains around a track fades into the background of randomly developed grains. The background typically corresponds to $P \approx 10^{-3}$ For the Minneapolis and Sioux City flights we use the value $P = 1.6 \times 10^{-3}$ as the level at which the eye sees the "edge" of the halo.

Independent observers at Houston and at LBL have measured the halo radius of the monopole candidate, obtaining values ranging from 50 to 55 μm. These values imply a velocity ~0.5 c if the curves in Fig. 14 are correct, if the dispersion about the expectation value is small and if the eye correctly locates the radius at which $P=1.6 \times 10^{-3}$.

To assess these questions, in Fig. 16 I have plotted Osborne's observed halo radius as a function of the value calculated from the model, using as inputs the values of Z and β (at the emulsion) determined in the Lexan stack and $P=1.6 \times 10^{-3}$. <u>I believe this figure contains the mos important new results since our original publication</u>.

Let us examine this comparison of experiment with theory for any trends. First of all, I find that the "errors" are uncorrelated with zenith angle. One of Fowler's[10] criticisms of Osborne's model was that, due to the escape of high-energy δ-rays from the surface of an emulsion of finite thickness (the transition effect), the measured halo radius should depend on zenith angle. Experimentally we are unable to detect such an effect. (Re call what we did for the <u>core</u> radius.)

Second, I find that the distributions of errors for the Minneapolis and Sioux City flights are indistinguishable. This is a reassuring result, showing that data taken four years ago, long before the monopole candidate was found, follow the same trend as the recent data, usin the same value $P=1.6 \times 10^{-3}$. Let me point out that the measurement most susceptible to subjective judgment, rely ing wholly on the human eye, is made <u>first</u>, without any information from the Lexan, followed by a set of ~60 etch rate measurements in the Lexan.

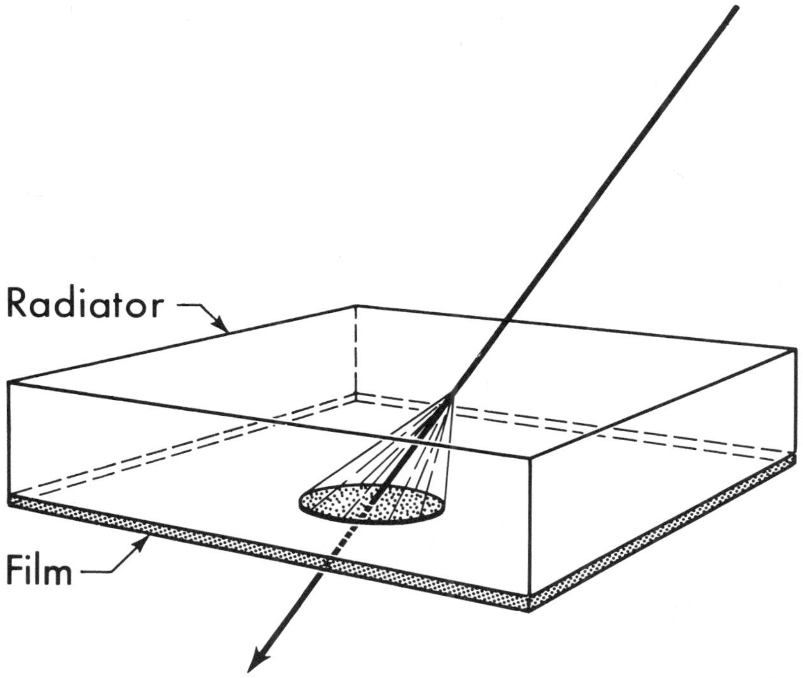

Figure 17. Cerenkov method of Pinsky (ref. 20). Two radiator-film units were used in our Sioux City flights. The two plastic radiators were 100 μm and 200 μm thick; the Kodak 2485 film was 12 μm thick.

Third, notice the correlation of errors with velocit Events with $\beta \gtrsim 0.7$ lie within a tight band, about ±10 μm wide, with a sharp edge at low observed halo radii, below which there are no stragglers. Events with lower velocit tend to lie higher and show a large dispersion toward positive errors. Consider, for example, the shaded area labeled "Fe." A conscious effort was made to reject the 10^5 to 10^6 Fe tracks in order to concentrate on the track of rare, heavier nuclei, yet many of the events with halo radii between 30 and 50 μm, thought to have $Z \gtrsim 35$, turned out to be Fe when measured in the Lexan. They tended to be at small zenith angles, which meant that their core radii were fatter than for shallow tracks (Fig. 15). Thi together with their larger than average halo radii, caused them to be recorded as candidates for $Z>35$.

The large positive errors for the particles with lowest velocities cannot be strictly a physiological defect of the human eye. The event with a halo radius of 105 μm and a calculated radius of only 58 μm was measured with Peter Fowler's photodensitometer in Bristol and verified to have a light-absorbing halo extending out to more than 100 μm.

One or both of two possibilities seem likely: (1) The theory underestimates the <u>average</u> radial distance to which electrons ejected by particles with $\beta<0.7$ diffuse. (2) The theory does not take into account <u>fluctuations</u> in the distance diffused. It seems intuitively reasonabl that very steep radial profiles for low β (Fig. 14) are more vulnerable to positive fluctuations in radial distan by δ-rays than are the shallow profiles for high β. The essence of diffusion is to reduce concentration gradients It would be very unphysical to have a large dispersion toward <u>lower</u> observed halo radii. Inward fluctuations of t

Table 1. Confidence Levels for Nuclear Explanations of the Monopole Candidate

Z	Mass (amu)	β	No. of frags.	Total prob. to occur in some flight	Conf. level for fit to Lexan data[1]	Conf. level for fit to emulsion data[2]	Figure of merit
				Nuclear Explanations			
76	192	0.70	2	2×10^{-3}	3×10^{-6}	$<<10^{-2}$	$<<6\times10^{-11}$
79	197	0.70	3	3×10^{-5}	10^{-5}	$<<10^{-2}$	$<<3\times10^{-12}$
81	205	0.74	2	2×10^{-3}	3×10^{-3}	$<<10^{-2}$	$<<6\times10^{-8}$
83	209	0.74	3	3×10^{-5}	10^{-2}	$<<10^{-2}$	$<<3\times10^{-9}$
92	238	0.82	1	10^{-1}	10^{-1}	$<<10^{-2}$	$<<10^{-4}$
96	247	0.86	0	1	10^{-1}	$<<10^{-2}$	$<<10^{-3}$
112	296	$\geqslant0.98$	0	?	1	$<<10^{-2}$	$<<10^{-2}$
				Hypotheticl Particles			
$g/e=137$	>875	~0.5	0	?	1	1	?
$Q/e\approx60$	$\gtrsim2000$	~0.5	0	?	1	1	?

[1] Based on F-test. The χ^2 test gives ~10 times lower confidence level.

[2] Based on measurements of halo radii for 110 nuclei.

few electrons at the edge of a halo would be swamped by outward fluctuations of the more numerous electrons from regions closer to the core. Note that a complete radial profile would not be so sensitive to fluctuations in diffusion distance of those few electrons that travel to the edge of the halo. This is so because fast electr cause the greatest blackening near the end of their range and the distribution of δ-rays energies decreases as $(\text{energy})^{-2}$. A quantitative model of these effects is being developed by Hagstrom.

Where should the point for the monopole candidate appear in the figure? The horizontal lines at an observe halo radius of ~55 μm indicate the values calculated for the various nuclear scenarios shown in Figs. 7 to 11 and for a monopole of velocities 0.45 c to 0.55 c. Recall that the Lexan data are incompatible with fragmenting nuclei with Z=76 to 83. A nuclear explanation of the event would require an extremely large negative fluctuation in electron diffusion distances not exhibited by any of the data in Fig. 16. The emulsion evidence provides strong support for the claim that the event is unique. I would appear to be compatible, within the framework of Osborne's model, with a monopole of velocity ~0.45 to ~0.6 c.

We now need to assess the confidence level that the measured halo radius is compatible with a nuclear interpretation. A complete physical model would allow a reali tic error distribution to be computed, one that is clearl asymmetric about a 45° correlation line. Even at this stage we could construct a Gaussian distribution of error that would clearly err on the conservative side because of its symmetric shape. I shall be even more conservativ and say that the hypothesis that the event was a nucleus has been tested at the level N^{-1}, where N=110, the number of events studied. In column 7 of Table 1 I assigned a confidence level "less than 10^{-2}" to the consistency of the emulsion measurement with the various nuclear hypotheses. A confidence level based on the magnitude of the negative error would appear to be far lower.

MEASUREMENTS AND TESTS OF THE CERENKOV DETECTORS

Figure 17 illustrates the principle of the Cerenkov method developed by L. S. Pinsky.[20] A particle with $\beta>\beta_c=n^{-1}$ (where the refractive index $n\approx 1.51$) generates a cone of Cerenkov photons along its path in a plastic radiator coated on the bottom by a layer of Eastman Kodak film 2485, the fastest film currently available. For the simplest case of vertical incidence this light falls on a circular area of radius $T\cdot\tan\theta_c$, where T=radiator thickness and $\theta_c\equiv\arccos(n\beta)^{-1}$ is the angle at which the photons from each element of pathlength are emitted.

At a radial distance r, the number of photons per unit area that reach the film is given by

$$I(r)=\frac{\alpha Z^2\Delta k(1-n^{-2}\beta^{-2})}{2\pi r\tan\theta_c}=\frac{\alpha Z^2\Delta k\sin\theta_c\cos\theta_c}{2\pi r} \qquad)2)$$

where α=fine-structure constant and $\Delta k=2\pi[\lambda_2^{-1}-\lambda_1^{-1}]$, the band pass for a particular film and radiator.

For an extremely heavy nucleus the region in which to look for the Cerenkov image is pinpointed by the solid black ionization spot that fills the depth of the 12 μm film and has a radial extent from a few to 30 μm, depending on Z/β. If this ionization spot cannot be found, it is difficult to locate the Cerenkov halo, because the coordinates of the track are precise only to a few mm. The Cerenkov halo has a much lower grain density than the

Table 2. Performance of Cerenkov Film Detector

Minneapolis flight (refs. 4,20); single radiator and film			Sioux City flight (in progress); two separate radiator film combinations			
Z (Lexan)	β (Lexan)	β (Cerenkov)	Z (Lexan)	β (Lexan)	Image in 200 μm Cer. detector?	Imag in 100 μm Cer. detect
90±3	.70±.01	.720±.013	83	0.95	yes	yes
80±8	.76±.05	.701±.011	83	0.76	yes	yes
76±2	.79±.01	.829±.012	82	0.93	yes	yes
74±6	.72±.04	$.741^{+.028}_{-.022}$	82	0.66	yes	yes
>60	>.65	.684±.005	77	0.86	yes	yes
>60	>.75	.793±.005	77	0.77	yes	yes
>60	>.7	$.717^{+.039}_{-.026}$	76	0.93	yes	yes
>60	>.7	>.95	76	0.82	weak	weak
>60	>.75	>.95	75	0.68	yes	yes
>50	>.7	$.810^{+.124}_{-.074}$	74	0.73	yes	yes
80±3	.56±.01	no spot	68	0.71	yes	yes
69±1	.70±.03	" "	59	0.77	yes	weak
68±7	.70±.03	" "	58	0.70	weak	weak
66±6	.60±.02	" "	58	0.64	yes	yes
65±3	.60±.05	" "	49	0.62	yes	yes
>65	>.6	" "	81	C.60	no	no
64±6	.74±.04	" "	65	0.80	no	no
63±3	.70±.05	" "	62	0.67	no	no
63±2	.63±.04	" "	61	0.60	no	no
60±4	.69±.02	" "	61	0.75	no	no
58±3	.58±.02	" "	60	0.71	no	no
56±2	.53±.02	" "	57	0.67	no	no

Table 2. Performance of Cerenkov Film Detector (Cont.)

Minneapolis flight (refs. 4,20); single radiator and film			Sioux City flight (in progress); two separate radiator film combinations			
Z (Lexan)	β (Lexan)	β (Cerenkov)	Z (Lexan)	β (Lexan)	Image in 200 μm Cer. detector?	Image in 100 μm Cer. detector?
55±5	.63±.03	" "	53	0.67	no	no
54±6	.67±.03	" "	52	0.70	no	no
52±3	.52±.02	" "	monopole candidate		no	no
51±2	.58±.01	" "				
50±4	.55±.02					

ionization spot. Within a series of rings around the ionization spot Pinsky counts developed grains, corrects for the background grain density, and computes I(r). In favorable cases (large Z, intermediate β) he sees a sudden drop in intensity that directly gives him the Cerenkov angle and therefore the velocity. If β is very high, the angle will be so large that I(r) will decrease to the background level at a radial distance less than T tan θ_c. He can still estimate β from the radial variation of I(r), solving eq. 2 for sin θ_c cos θ_c, which is single-valued for θ_c=0 to 45°, corresponding to velocitie from 0.66 c to 0.94 c, and is roughly 0.5 for higher velocities.

Table 2 summarizes the measurements Pinsky has made on detectors from the Minneapolis and Sioux City flights. Because the Minneapolis payload crashed and was dragged miles across country, some of the Cerenkov films were destroyed. Data from the usable films are shown. The results from this flight were encouraging. Where he saw no Cerenkov image, the velocity determined in the Lexan was consistent with the inequality β<0.68 except for one particle with Z=64±6, β=0.74±0.04. In the ten cases wher he saw a Cerenkov image, his estimate of β was consistent with that from the Lexan.

At this writing only a few observations have been ma of the detectors from the Sioux City flights. No quantit tive determinations of β have been made. Though the qual tative observations of Cerenkov images <u>in both films</u> for 14 events in Table 2 are encouraging, I believe it is too early to use the absence of Cerenkov images at the monopole candidate to further lower the confidence level for nuclear interpretation. The detection of a Cerenkov imag requires establishing the existence of a small signal abo a large background of developed grains. To make quantitative profiles of grain density around the ionization spots of the events, Pinsky plans to use the same computer-operated image-recognition system Osborne will use on the emulsion.

The Cerenkov data will be most convincing in assessing confidence levels for the nuclei with largest Z and β. At a given radial distance in the Cerenkov film, the photon intensity for the three nuclear candidates with Z=92, 96, and 112 would exceed that for a nucleus with Z=65 and β=0.7 by factors of 3.1, 3.5, and 4.8 respectively. From the qualitative results in Table 2 it appears that signals from nuclei with $Z \geq 65$ at $\beta \geq 0.7$ are detectable. The response curve of Kodak film 2485 is such that one would expect signals greater than three times the minimum detectable signal to be impossible to miss unless one argued that the film was locally damaged or locally abnormally insensitive. The existence of two independent Cerenkov radiator-film combinations, each in its own protective wrapping, would require a critic to argue that both films were locally damaged or locally abnormally insensitive. The Cerenkov film thus provides a constraint that complements the constraint imposed by the Lexan data.

DISCUSSION

Table 1 gives, I believe, a conservative view of the status of our work. The last column is simply the product of the numbers in the previous three columns. I use the undefined term "figure of merit" to warn the reader that one should not literally interpret the number as a probability or an overall confidence level. It is a convenient way of summarizing the relative merits of the various nuclear scenarios.

The rows labeled "hypothetical particles" indicate that there are two classes of particles that are equally consistent with all the data. A monopole has the attractive features that the charge of $g \approx 130$ e inferred from ou data (if $\beta \approx 0.5$) is consistent with the predicted value $g = 137$ e, and the lower limit of 875 amu for its mass, inferred by Ahlen[6] from the absence of a negative slope to the Lexan data, is consistent with 't Hooft's theoretical model[26] in which monopoles exist with mass$\approx 137\ M_W \approx 10^4$ amu, where M_W is the mass of the intermediate vector bosc A monopole has the unattractive features that it has not been detected in experiments with up to a million times greater collecting power, and it is hard to account for i low velocity without rather contrived assumptions. Most previous experiments would have missed seeing monopoles i they have masses greater than $\sim 10^4$ amu, which is consistent with our lower limit. For example, the collecting power of the lunar experiment of Alvarez and co-workers[27] decreases rapidly for monopoles of large mass, which bury themselves at great depths instead of in the shallow subsurface soil. Let it suffice to say that the history of physics shows that theorists have a way of explaining apparent conflicts with nature if sufficiently compelling experimental evidence requires it.

We cannot rule out the second hypothetical particle, one with electric charge given by $Q/e = \beta \cdot (Z/\beta) \approx 0.5 \times 114 \approx 60$. In order that its Bragg curve, at $\beta \approx 0.5$, not rise any faster than the Lexan data permit, its mass must exceed ~ 2000 amu. Such a particle has the attractive feature that its flux does not conflict with flux limits set by other experiments that have sought highly electrically charged particles. Yock[28] has proposed that hadrons consist of "subnucleons" with large mass and strong electric charge, bound by Coulomb forces. His heaviest subnucleon is consistent with the charge and mass that we require.

It seems conceivable that a "collapsed" or "abnormally dense" nuclear particle, as discussed by Bodmer[29] and by Lee and Wick,[30] might have a huge mass and a charge of ~60. Bodmer has pointed out that, if the potential well is deep enough, the state of lowest energy may be one in which some of the nucleons convert into neutral hyperons with their own Fermi levels, so that Z/A is far less than that for normal nuclei.

To bring this discussion back to reality, let me close by affirming what all scientists believe, that science advances by criticism, painful as it may seem to those on the receiving end. In the absence of strong criticism we might have pressed ahead with plans for a further series of balloon experiments, neglecting the critical measurements of the other events on the Sioux City flights. Through the efforts of Steve Ahlen, Ray Hagstrom, and others, we are learning more about the expected behavior of monopoles and about the capabilities of nuclear emulsions and Cerenkov film detectors. It is possible that, when we have generated radial profiles of all the tracks in the emulsion and made quantitative measurements of the Cerenkov images, the question of the uniqueness of this event may be settled. It cannot be <u>proved</u> to have been produced by a monopole, but if it can be shown at a high confidence level not to have been produced by any nucleus, future experiments of expanded scope will be justified.

ACKNOWLEDGMENTS

I am indebted to my colleagues Edward Shirk, Zack Osborne, and Larry Pinsky for continuing the investigation of this event. The notoriety and pressure which have made it difficult for us to carry out our research with untroubled minds are a direct consequence of my own eagerness to announce what I thought was strong evidence for a monopole. Following my return from the Munich Cosm Ray Conference, there was strong pressure to publish a retraction and not to carry out calibrations and further measurements. A number of friends, among them Sumner Davis, Ed McMillan, John Reynolds, Emilio Segré, and Charles Townes, supported me in my determination to complete the investigation with an open mind. I am grateful to Steve Ahlen, Brian Cartwright, Fred Goldhaber, Maurice Goldhaber, Ray Hagstrom, and Ed McMillan for numerous valuable ideas and suggestions. It was a great pleasure to have had several conversations with Paul Dirac while at the Conference.

REFERENCES

1. P. B. Price, E. K. Shirk, W. Z. Osborne, and L. S. Pinsky, Phys. Rev. Letts. 35, 487 (1975).
2. E. Bauer, Proc. Camb. Phil. Soc. 47, 777 (1951).
3. H. J. D. Cole, Proc. Camb, Phil. Soc. 47, 196 (1951).
4. E. K. Shirk, P. B. Price, E. J. Kobetich, W. Z. Osborne, L. S. Pinsky R. D. Eandi, and R. B. Rushing, Phys. Rev. D7, 3220 (1973).
5. P. B. Price and E. K. Shirk, Proc. 6th Inter. Cosmic Ray Conf., Vol. 1, p. 268, Munich, 1975.
6. S. P. Ahlen, "Monopole Track Characteristics in Plastic Detectors," submitted to Physical Review.
7. M. W. Friedlander, Phys. Rev. Lett. 35, 1167 (1975).
8. R. L. Fleischer and R. M. Walker, Phys. Rev. Lett. 35, 1412 (1975).
9. L. Alvarez, LBL Report No. 4260 (1975).
10. P. H. Fowler, Proc. 14th Inter. Cosmic Ray Conf., Vol. 12, p. 4049, Munich, 1975.
11. L. W. Wilson, Phys. Rev. Lett. 35, 1126 (1975).
12. E. V. Hungerford, Phys. Rev. Lett. 35, 1303 (1975).
13. G. D. Badhwar, R. L. Golden, J. L. Lacy, S. A. Stephens, and T. Cleghorn, Phys. Rev. Lett. 36, 120 (1976).
14. S. A. Bludman and M. Ruderman, "Theoretical Limits on Interstellar Magnetic Poles Set by Nearby Magnetic Fields," submitted to Phys. Rev. Lett.
15. P. H. Eberhard and R. R. Ross, "Are Monopoles Trapped by Ferromagnetic Material," preprint.
16. G. 't Hooft, "Magnetic Charge Quantization and Fractionally Charged Quarks," preprint, Univ. of Utrecht.
17. G. Kalman and D. ter Haar, Nature 259, 467 (1976).
18. Ray Hagstrom, Phys. Rev. Lett. 35, 1677 (1975).

19. R. A. Carrigan, F. Z. Nezrick, and B. P. Strauss, "Search for Misplaced Magnetic Monopoles," Fermilab preprint.

20. L. S. Pinsky, Ph.D. Thesis, available from Johnson Space Center as NASA report TMX-58102 (1972).

21. R. L. Fleischer, P. B. Price, and R. M. Walker, Nuclear Tracks in Solids: Principles and Applicatior Univ. of California Press, Berkeley (1975).

22. A. H. Rosenfeld, Ann. Rev. Nuc. Sci. 25, 555 (1975).

23. R. Katz and J. J. Butts, Phys. Rev. 137, B198 (1965); R. Katz and E. J. Kobetich, Phys. Rev. 186, 344 (1969 R. Katz, S. C. Sharma and M. Homayoonfar, Nucl. Instr Meth. 100, 13 (1972).

24. R. Andersson, L. Larsson, and O. Mathiesen, Rept. LUIP-CR-73-02, Univ. of Lund, Sweden (1973); L. Larsson, R. Andersson, and O. Mathiesen, Nucl. Instr. Meth. 114, 35 (1974); M. Jensen and O. Mathiesen, Rept. LUIP-CR-75-03, Univ. of Lund. Sweden (1975);S. Behrentz, Rept. LUIP-CR-75-05, Univ. of Lund, Sweden (1975).

25. P. H. Fowler, V. M. Clapham, V. G. Cowen, J. M. Kidd, and R. T. Moses, Proc. Roy. Soc. London, Ser. A. 318, 1 (1970).

26. G. 't Hooft, Nucl. Phys. B79, 276 (1974).

27. P. H. Eberhard, R. R. Ross, L. W. Alvarez, and R, D. Watt, Phys. Rev. D4, 3260 (1971); R. R. Ross, P. H. Eberhard, L. W. Alvarez, and R. D. Watt, Phys. Rev. D8, 698 (1973).

28. P. C. M. Yock, Int. J. Theor. Phys. 2, 247 (1969); Annals of Physics 61, 315 (1970); Phys. Rev. D, in press.

29. A. R. Bodmer, Phys. Rev. D4, 1601 (1971).

30. T. D. Lee and G. C. Wick, Phys. Rev. D9, 2291 (1974).

THE GAUGING AND BREAKING OF SIMPLE SUPERSYMMETRIES*

Peter G. O. Freund
The Enrico Fermi Institute
University of Chicago, Chicago, Ill. 60637

It is widely believed that the strong, electromagnetic, weak, and maybe even gravitational interactions can all be described by one unified gauge theory with spontaneous symmetry breaking. Although no such unified theory has been fully constructed as yet, it is clear that among its ingredients there must be

i) spin 1/2 Fermi fields
ii) spin 1 and 2 gauge fields
iii) spin 0 Nambu-Goldstone and Higgs fields.

Depending on the interplay of these three types of fields one may consider three categories of theories.

A) Phenomenological Gauge Theories introduce all three categories of fields as "fundamental" (i.e., in the lagrangian). Such theories have no problem achieving essentially "by hand" the desired degree of symmetry

*Work supported in part by the NSF: contract No. PHYS74-08833 A01.

breaking. The price paid for this, as a rule, is a loss of asymptotic freedom on account of the large number of scalar and Fermi fields. This has led to the consideration of

B) Pure Gauge Theories. In these theories only Fermi and gauge fields are postulated, the symmetry breaking is produced (or more accurately hoped to be produced) dynamically and correspondingly the scalar fields are composite and harmless. The approach, thus assumes Fermi and gauge fields to be more "fundamental" than Nambu-Goldstone or Higgs fields.

So far we considered renormalizable quantum field theories, at least while disregarding gravitation. In the presence of gravitation no renormalizable theory has been put forward as yet[1], and were one, following Cartan[2], to gauge via gravitation the full Poincaré group, one would pick up an explicit A-type four-Fermi interaction as shown by Weyl.[3] On the face of it this appears highly non-renormalizable. Were one therefore to abandon renormalizability altogether, one could consider a third type of fundamental theory involving only Fermi fields with both scalar and gauge fields as composites. This is essentially

C) A relativistic version of the BCS theory and an example in this direction is the Nambu - Jona-Lasinio model.

Whichever of the approaches A) - C) one follows, a number of arbitrary elements invariably turn up: the specification of the gauge group, the representation to which the basic Fermi and/or scalar fields are assigned and the scalar couplings, (of course no ambiguity is involved in the representation assignments, and coupling of gauge fields). It is therefore important to develop a method for removing the arbitrary features connected with the scalar and Fermi fields. If they only could be gauge fields, their group assignments and (via the principle of minimal coupling) their couplings would also be determined. We ask is it possible that all fields (scalar and Fermi fields included) are gauge fields? Alternatively, is it possible that all fields (gauge and Fermi fields included) are Nambu-Goldstone (NG) fields? Finally, are there theories in which every field is simultaneously a gauge and a Nambu-Goldstone field? I shall provide affirmative answers[4] to these questions and as you will see, supersymmetry will be of the essence. One may find this somewhat surprising, as we are all familiar with examples of NG fields that are not gauge fields (the $\vec{\pi}$ fields in the σ-model) or of gauge fields that are not NG fields (the O(3) Yang-Mills fields in the absence of further sources). Yet, for quite some time, two examples are known of fields that at the same time are gauge fields of a group G and NG fields of a larger group $\bar{G}$. Indeed for the electromagnetic[6] field G = U(1), $\bar{G}$ = Lorentz group, and for the gravitational field[7] G=Lorentz group and essentially* $\bar{G}$ = GL(4,R).

Of these two cases the gravitational one turns out to be readily generalizable so as to include nonabelian gauge fields - and eventually fermions - as NG fields. Let us see how this comes about[8]. As a first step consider a space with one time-like and 3+N space-like

*More precisely, $\bar{G}$ is the semigroup of general coordinate transformations in four dimensions but according to Ogievetsky's theorem[7] these can all be obtained by suitably iterating infinitesimal GL(4,R) transformations and conformal boosts. Spontaneously breaking the conservation of conformal boosts then only adds a four vector NG field which automatically plays the role of a Weyl-type gauge field of dilatations.

dimensions where N is the dimension of a internal symmetry group G. Consider the $(4+N)^2$-parameter group GL(4+N,R) and the (4+N) (3+N)/2-parameter Lorentz group O(3+N,1). Suppose we were to spontaneously break GL(4+N,R) down to O(3+N,1). In the process $(4+N)^2 - (4+N)(3+N)/2 = 10 + 4N + N(N+1)/2$ NG fields are picked up. Of these, 10 are just components of the vierbein of gravitation in a symmetric gauge, N(N+1)/2 are scalars of the four dimensional Lorentz group, that transform as components of a tensor under the internal symmetry G and 4N are Lorentz four-vectors in the adjoint N-dimensional representation of G. Using the standard technique of nonlinear group realizations one can find the transformation law of the NG fields under general coordinate transformations in the 4+N dimensional space. In particular under the transformations that leave the four ordinary space-time coordinates unaffected, while performing essentially a space-time dependent G-transformation on the N internal coordinates, the 4N components of the N four-vector NG fields transform precisely as gauge potentials. One thus proves the gauge invariance of the theory and identifies 4N NG-fields as gauge fields. To this point the trouble is that the theory is played in 4+N-dimensional space with N-unobservable dimensions. One can avoid this problem by imposing, as in Kaluza-Klein theory[9], a cylindricity condition in the N internal coordinates. While this may seem somewhat artificial, it has the virtue of leading to a unique theory[4]. Indeed choosing the 4+N general invariant action as the 4+N-space integral of the (4+N)-dimensional scalar curvature density, it is found that with the cylindricity condition the internal space integration can be carried out explicitly and the action reduces to the sum of the ordinary 4-dimensional Hilbert-

Einstein action for gravitation, the Yang-Mills action for the gauge fields and a specific but rather complicated action for the scalar field. There are some differences in that internal indices are contracted with an internal metric tensor expressed in terms of the scalar fields and not with the Kronecker symbol as usual, and that the whole lagrangian is multiplied with the square root of the determinant of the internal metric; like in Jordan-Brans-Dicke theory one has a variable gravitational coupling. In fact, this is just a nonabelian version of Jordan-Brans-Dicke theory. Yet, were all scalar fields to be constant over space and time, the theory would reduce <u>precisely</u> to the coupled Einstein-Yang-Mills theory. While I must refer you to the original paper (Ref. 4) for the full details, I hope that the discussion so far clearly conveys the idea that we are dealing here with a theory in which <u>every</u> field is a NG field, even the spin 1 and 2 gauge fields. The converse (every field is a gauge field) is not yet true at this stage. In addition, one sees that there is an intimate relationship between the Hilbert-Einstein and Yang-Mills actions as pieces of the same 4+N-dimensional scalar curvature density.

It is at this point that supersymmetry enters the problem. Indeed assume the additional "internal" dimensions to be fermionic rather than bosonic (i.e. anticommuting rather than commuting). This would be advantageous for a number of reasons. First, the problem of the observability of the internal dimensions would go away by itself. On account of the anticommutation of the internal coordinates, fields can depend on them only in a trivial polynomial way. Dynamics determines then only the coefficients of this polynomial which are ordinary fields depending only on the four space-time variables. This is essentially the so-called superfield concept of Salam and Strathdee.[10] Second, were we able to implement a geometrical structure analogous to the 4+N-Bose-dimensional case, we would now have <u>all</u> fields (fermions included!) as both gauge and NG fields. There would be no arbitrariness left in the choice of the lagrangian once the number of internal Fermi dimensions is given and the condition of a Hilbert-Einstein type action of the spin 2 field imposed. To properly understand such a theory some mathematical footwork is required. What are the supersymmetric generalizations of the GL (4+N, R), O(3+N, 1), and conformal algebras? To answer these questions one is advised to go for the more general question: what are all simple graded Lie algebras (GLA's)? The answers to these questions are contained in recent work by V. G. Kac[11], Freund and Kaplansky[12], Djoković and Hochschild[13], and also of Corwin, Ne'eman and Sternberg[14], Nahm and Scheunert[15], Srivastava[16], and probably others.

I will first introduce various families of GLA's and then give the full classification. Consider a mod2 graded vector space V with m Fermi and n Bose dimensions (in order for every sector in V to have a definite grade one has to impose the usual superselection rule of forbidding superpositions of Fermi and Bose elements). Consider all complex linear transformations of V. They are obviously graded. In the matrix representation, the bracket [ ] of two such transformations A and B is defined as their commutator except when both A and B are odd (Fermi) in which case it is their anticommutator. The bracket operation is bilinear, grade additive (mod 2), graded antisymmetric and obeys a graded Jacobi identity.

Therefore these transformations span a graded Lie algebra: $GL(m|n)$. It has invariant graded Lie subalgebras (ideals), so it is not simple.

From the condition

$$tr([A,B]) = 0 \quad ,$$

the trace of a transformation A is defined as the difference of its traces over the Fermi and Bose sectors of V. For $m \neq n$ all traceless (in the sense just defined) transformations in $GL(m|n)$ form a simple $(m + n)^2-1$ dimensional GLA $SL(m|n)$ with Bose sector $SL(m) \times GL(n)$. For $m = n$ the unit matrix is traceless and the center has to be divided out before $SL(m|m)$ becomes the simple $(m+n)^2-2$ dimensional GLA $PSL(m|m)$.

Endow the vector space V with 2r Fermi and s Bose dimensions with a metric G antisymmetric (symmetric) on its Fermi (Bose)part:

$$(x,y) = x^T Gy \quad . \qquad \text{for } x,\ y \in V$$

G can always be brought to the form

$$G = \begin{pmatrix} C & 0 \\ 0 & 1 \end{pmatrix} \begin{matrix} 2r \\ s \end{matrix} \quad , \qquad C = \begin{pmatrix} i\sigma_2 & & \\ & \ddots & \\ & & i\sigma_2 \end{pmatrix} \quad .$$

(columns of G: 2r, s)

The elements of $GL(2r|s)$ that leave the metric (x,y) invariant obey

$$(x,Ty) + (Tx,y) = 0 \quad ,$$

except when x and T are both Fermi when

$$(x,\ Ty) - (Tx,\ y) = 0 \quad ,$$

as the order of x and T in the two terms is reversed. They span a simple $r(2r+1) + s(s-1)/2 + 2rs$-dimensional GLA OSp $(2r|s)$ whose Bose Sector is the ordinary Lie algebra $Sp(2r) \times O(s)$.

Consider all pairs (A, ε) where A is an anti-hermitian traceless n x n matrix and ε can take the two values $\varepsilon = 1$ (Fermi). The bracket is defined as

$$[(A,\varepsilon), (A',\varepsilon')] = \begin{cases} (AA' - A'A, \ \varepsilon + \varepsilon') & \text{if } \varepsilon\varepsilon' = 0 \\ (i(AA' + A'A - \frac{2}{n}\mathrm{Tr}(AA')1), \ 0) & \text{if } \varepsilon\varepsilon' = 1 \end{cases}$$

(for n=3 we recognize these to be the f- and d- products respectively). Again we have a ($2n^2 - 2$ - dimensional) simple GLA FD(n) (first discussed by Michel and Radicati, see ref. 14.)

Now consider[11,13] the GLA P(n), graded by the integers -1, 0, +1 such that its (Bose) null sector P_0 is SL(n+1), its (Fermi) + 1 sector P_+ contains all symmetric $(n+1) \times (n+1)$ matrices and its (Fermi) - 1 sector P_- contains all antisymmetric $(n+1) \times (n+1)$ matrices. The graded Lie algebra is defined by

$$[a,b] = \begin{cases} ab-ba & a,b\varepsilon P_0 \\ 0 & a,b\varepsilon P_+ \text{ or } a,b\varepsilon P_- \\ ab & a\varepsilon P_0,\ b\varepsilon P_- \\ ab+(ab)^T & a\varepsilon P_0,\ b\varepsilon P_+ \\ ab-(ab)^T & a\varepsilon P_-,\ b\varepsilon P_0 \end{cases}$$

The grading by the integers induces, of course, a mode 2 grading on P(n).

Next, consider an n-dimensional space. The general coordinate transformations on this space have the generators $(x_1)^{m_1} \ldots (x_n)^{m_n} \partial_\mu$ $\mu=1,\ldots,n$. On an ordinary space with all dimensions bosonic, the number of these generators is infinite ($m_n = 0,1,2,\ldots$), and they generate one of Cartan's four types of infinite pseudogroups. If, on the other hand, the space is purely fermionic (i.e., all dimensions are fermionic) then each m_n can only take the values 0 and 1 so that there are only a finite number, $n2^n$, of such generators and they span[11,12,14] a simple GLA W(n), for $n \geq 2$. There are three further types of Cartan algebras which all are infinite in the ordinary (pure Bose) case but become finite-dimensional[11,12] GLA's: S(n+2), $\tilde{S}$(n+2), H(n+3) in the pure Fermi case. We shall not describe them in detail here (see ref. 11).

Finally, Kaplansky[12] has found exceptional GLA's: a doubly infinite family Γ(A, B, - A - B) (with A, B arbitrary real numbers) of 17-dimensional GLA's all with the same Bose part $S_p(2) \times O(4)$, a 31-dimensional GLA Γ_2 with $G_2 \times SL(2)$ as its Bose part and finally, a 40-dimensional GLA with $O(7) \times SL(2)$ as its Bose part.

<u>There are no simple GLA's other than those just described: the special linear, orthosymplectic, exceptional, FD(n), P(n) and the four Cartan families.</u> A more detailed picture of the distinctive properties of these various classes can be gained from the following table.

SIMPLE GRADED LIE ALGEBRAS

<table>
<tr><td>SL(m|n)</td><td>OSp(2r|s)</td><td colspan="2">Exceptional</td><td>PSL(m|m)</td><td>OSp(2r|2r+2)</td><td>FD(n)</td><td>P(n)</td><td colspan="2">Cartan type</td></tr>
<tr><td>$m \neq n$</td><td>$s \neq 2r+2$</td><td>Γ_2; Γ_3</td><td>$\Gamma(A,B,C)$</td><td>$m \geq 2$</td><td></td><td>$n \geq 3$</td><td></td><td>W(n+1)</td><td>S(n+2);
$\tilde{S}$(n+2);
H(n+3)</td></tr>
<tr><td colspan="3">Killing: Killing form nondegenerate</td><td colspan="7">Anti-Killing: Killing form vanishes</td></tr>
<tr><td colspan="6">Classical: admit nondegenerate metric form</td><td colspan="4">Hyperexceptional: not classical</td></tr>
<tr><td colspan="9">Bose reducible: the representation of the Bose sector on the Fermi sector is fully reducible</td><td>Cartan type
not Bose reducible</td></tr>
</table>

Of these GLA's only those of the OSp(2r|1) family are fully reducible.[13]

We further notice that the Bose parts of most simple GLA's are not simple (they are simple only for OSp(2r|1)). This by itself may have importance for physics. The fact that the ordinary symmetry of hadrons is not simple: $SU(N)_{flavor} \times SU(3)_{color} \times U(1)_{baryon\ or\ number}$ is unsatisfactory and has led to the search of a larger "unified" simple symmetry Lie-algebra. But SU(N) x SU(3) x U(1) is precisely the Bose part of the compact real form SU(N|3) of SL(N|3). So maybe simplicity is to be sought not at the symmetry but at the supersymmetry level. SU(N|3) might be related to a world with "subquarks" N of which are Fermi and carry flavor and 3 of which are Bose and carry color (or with Fermi and Bose interchanged)[17]. Of course this is not realistic as spin is not yet included. One simple way of including spin is to place spin-flavor independence characterized by SU(2N) (like the Gürsey-Radicati-Sakita SU(6) for N=3) instead of the flavor independence $SU(N)_{flavor}$ factor and obtain SU(2N|3). This is however, not yet relativistic and should be treated just like SU(6). As a rule its representations will contain colored along with color singlet states, but the former may be confined by "infrared slavery" as long as exact color gauge invariance prevails.

We now return to our sheep. The obvious generalization of the ideas above (i.e., of a Riemannian geometry) to the mixed Bose-Fermi case is to gauge an orthosymplectic "supergroup"[18]. This can be done and one has to consider OSp(8N|3,1). The reasoning for this is straightforward, the 3, 1 in OSp(8N|3, 1) corresponds to the real form of OSp(8N|4) in which the symmetric metric has signature +---, the 8N corresponds to the fact that the Fermi sector must be built of Majorana spinors, there being four (real) components to such a spinor and two such spinors for a Dirac spinor. So N is the number of Dirac spinors. The Bose sector of OSp(8N|3,1) is Sp(8N) x O(3,1) the latter factor being the Lorentz algebra. The Sp(8N) shuffles all spin components of the 2N Majorana spinors, somewhat in the manner of the SU(6) -supermultiplet theory. The internal symmetry part (Lorentz scalar generators) of this Sp(8N) is O(2N) the orthogonal transformations among the 2N Majorana spinors. Instead of the expected SU(N) we are thus dealing with the somewhat larger internal symmetry O(2N) (its rank is larger by one and it has roughly twice as many generators). This is a Pauli-Gürsey type symmetry and should be investigated independently of this geometric framework. Ogievetsky's theorem again generalizes and all fields (spins 0, 1/2, 1, 3/2, 2) in this theory are both gauge and NG fields (essentially of GL(8N|4,R) Fermi and scalar fields. One may be somewhat surprised to find scalar gauge fields, but this follows quite naturally from gauging Fermi generators over the Fermi part of superspace. The dynamical details of such a theory are quite complicated and work along these lines is pursued at Northeastern and Chicago. Zumino[19] has recently proposed a graded non-riemannian geometry with torsion connected with the Wess-Zumino rather than the orthosymplectic algebra.

I believe the most remarkable features of these theories to be the merger of the NG and gauge field concepts and the ensuing specification of all interactions, without introducing extra "matter" fields.

Since it is time-honored to think of NG fields as composite, one may consider all fields in the gauge lagrangian as composite, this just being a "phenomenological" lagrangian for a grandiose spontaneous symmetry breakdown in a more fundamental theory whose agents (fields?) prefer to stay in the background. It is not hard to construct models, say on a lattice[20], with a fundamental interaction that vanishes fast with the lattice spacing (as a power law) in such a manner that the interaction of the composite NG fields it leads to ("our" quarks, leptons, gauge fields and, NG and Higgs fields) vanish much more slowly (logarithmically) so that for lattice spacings of the order of the Planck length the original lagrangian is practically null while the interaction of the NG fields is "left over" and describable phenomenologically by a gauge theory. In any case, distinguishing various levels of "fundamentality" among Fermions, gauge, and NG fields appears artificial on the basis of the work I just reported.

REFERENCES

1. Formal arguments suggest the possibility that certain acceptable Weyl-invariant theories of gravitation may be renormalizable, see P.G.O. Freund, Ann. of Physics 84, 440 (1974); see also, S. Deser in Proceedings of the XVII International Conference on HEP, London 1974, G. Manning et al., editors, Rutherford Lab 1974, p. 1264.
2. See T. W. B. Kibble, J. Math. Phys. 2, 212 (1961).
3. H. Weyl, Phys. Rev. 77, 699 (1950).
4. P. G. O. Freund and Y. M. Cho, Phys. Rev. D12, 1711 (1975).
5. P. Nath and R. Arnowitt, Phys. Lett. B56, 177 (1975), and to be published; P. Nath, talk at the 1975 Conference on Gauge Theories and Modern Field Theory at Northeastern University.
6. J. D. Bjorken, Ann. of Phys. (N.Y) 24, 174 (1963); G Guralnik, Phys. Rev. B1404, B1417 (1964); Y. Nambu, Progr. Theor. Phys. Suppl. Extra No. 190 (1968).
7. V. I. Ogievetsky, Lett. Nuovo Cimento 8, 988 (1973), V. I. Ogievetsky and A. B. Borisov, Dubna report E2-7684.
8. The following argument is taken from ref. 4.
9. T. Kaluza, Berl. Ber. 966 (1921); O. Klein, Helv. Phys. Acta Suppl. IV, 58 (1956). See also Y. M. Cho, J. Math. Phys. (in press); L. N. Chang, F. Mansouri, G. Macrae, to be published. J. Scherk and J. Schwarz, to be published.

10. A. Salam and J. Strathdee, Nucl. Phys. B76, 477 (1974).

11. V. G. Kac, Functional Anal. and Suppl. 9, 91 (1975).

12: P. G. O. Freund and I. Kaplansky, J. Math. Phys. Feb. 76; I. Kaplansky, to be published; P. G. O. Freund, J. Math. Phys., Feb. 76.

13. D. A. Djoković and G. Hochschild, to be published.

14. L. Corwin, Y. Ne'eman and S. Sternberg, Rev. Mod. Phys. 45, 573 (1975).

15. M. Scheunert and W. Nahm to be published.

16. P. P. Srivastava, to be published.

17. O. W. Greenberg, Phys. Rev. Lett. 35, 1120 (1975); J. Pati, A. Salam and J. Strathdee, Phys. Lett. 59B, 265 (1975).

18. See ref. 5 and the last paper in ref. 12.

19. B. Zumino, talk at the 1975 Conference on Gauge Theories and Modern Field Theory at Northeastern University.

20. This was realized by R. Carlitz and the author.

CHARGE SPACE, EXCEPTIONAL OBSERVABLES AND GROUPS*

Feza Gürsey

Physics Department

Yale University, New Haven, Ct. 06520

ABSTRACT

It is shown that a class of exceptional quantum mechanical spaces represented by octonionic matrices and first introduced by Jordan, von Neumann and Wigner are suitable for representing the states of basic fermionic constituents (leptons and quarks) of elementary particles. In these exceptional spaces, the transformation groups that leave scalar products invariant are the exceptional groups. A gauge field theory based on E_7 is given as an example for the unification of weak, electromagnetic and strong interactions.

I. INTRODUCTION

In the early thirties, when physicists were confronted with the newly discovered phenomena of nuclear physics, Jordan, von Neumann and Wigner[1] (JNW)

---

*Research (Yale Report COO-3075-136) supported in part by the U. S. Energy Research and Development Administration under contract no. AT(11-1)3075.

investigated a generalization of Quantum Mechanics as a possible framework for the understanding of subatomic physics. They found that the projection operator form of Quantum Mechanics in which a state associated with the ket $|\alpha>$ is represented by the hermitian idempotent projection operator $P_\alpha = |\alpha><\alpha|$ can be generalized if P_α is a 3×3 hermitian octonionic matrix. It was later shown by Chevalley and Shafer[2] that these octonionic projection operators can be transformed by means of the exceptional group F_4 in the same way as a ket in n-dimensional space is transformed by the unitary group U(n). Probabilities are then invariant under such transformations. Other exceptional groups E_6, E_7, E_8 act on projection operators made of ordinary matrices and 3×3 Jordan octonionic matrices[3]. Since nuclear phenomena can be described in a conventional Hilbert space we may be tempted to see if there is a correspondence between the space of internal degrees of freedom of leptons and quarks (generalized charges such as leptonic isospin, strong isospin, strangeness, color, charm, etc.) and the space of exceptional quantum mechanical states. If such a correspondence exists the basic fermion or vector meson states must correspond to irreducible representations of exceptional groups.

In previous papers[4] the color group $SU^c(3)$ was identified with the subgroup of the automorphism group G_2 of octonions which leaves one octonionic imaginary unit, say e_7, invariant. That direction e_7 then appears in the unitary representations of the Poincaré group that act in an ordinary complex Hilbert space. The group F_4 has $SU(3) \times SU^c(3)$ as maximal subgroup with a new SU(3) interchanging the 3 octonions appearing in the 3×3 Jordan matrix. Thus the new SU(3) can be interpreted as acting on 3 flavors [5,6]. Higher exceptional groups correspond to more flavors [7,8]. The maximal subgroups relevant for the color-flavor classification of degrees of freedom in exceptional Hilbert spaces are

$$F_4: \quad SU(3)\times SU^c(3),$$

$$E_6: \quad SU(3)\times SU(3)\ SU^c(3),$$

$$E_7: \quad SU(6)\times SU^c(3),$$

$$E_8: \quad E_6\times SU^c(3).$$

Hence E_6 and E_7 correspond to 6 flavors and 3 colors. F_4, E_6 and E_7 all possess a fundamental representation smaller than their adjoint representation which can be used for representing basic fermions. The smallest representation of E_8 is its 248 dimensional adjoint representation which is very large compared to observed quark and lepton degrees of freedom. E_7 is also the largest exceptional group with a pseudoreal fundamental representation of dimension 56. Accordingly all the 56 components are complex and the overall phase of this representation which is not part of the group can be used as a global fermion number transformation that is not associated with a massless vector boson. But E_7 being a real group we have the advantage of not generating Adler-Bell-Jackiw anomalies in a gauge theory based on E_7. It follows that E_7 is a good candidate for a gauge group acting in charge space if exceptional observables are associated with the internal degrees of freedom of elementary particles. That is why we shall give examples of the phenomenology implied by an E_7 gauge group for the unification of elementary particle interactions.

II. EXCEPTIONAL QUANTUM MECHANICAL SPACES

A conventional quantum mechanical state is represented either by the ket $|\alpha>$ or the projection operator $P_\alpha = |\alpha><\alpha|$.

The probability for a transition $\alpha \to \beta$ is

$$\Gamma_{\beta\alpha} = |<\beta|\alpha>|^2 = <\beta|\alpha><\alpha|\beta> \ . \tag{2.1}$$

Thus we can write

$$\Gamma_{\alpha\beta} = \mathrm{Tr}(|\alpha><\alpha|\beta><\beta|) = \mathrm{Tr}\ P_\alpha\ P_\beta = \mathrm{Tr}\ P_\alpha \cdot P_\beta , \tag{2.2}$$

where

$$P_\alpha \cdot P_\beta = \tfrac{1}{2}(P_\alpha\ P_\beta + P_\beta\ P_\alpha), \tag{2.3}$$

which defines a Jordan product for the hermitian matrice P_α and P_β.

The transition probabilities are invariant under th unitary transformations

$$P_\alpha \to U\ P_\alpha\ U^{-1}, \qquad P_\beta \to U\ P_\beta\ U^{-1}. \tag{2.4}$$

Because of the associativity of the complex number multiplication and matrix multiplication, this implies

$$|\alpha> \to U|\alpha>, \qquad |\beta> \to U|\beta>, \ (U\ U^\dagger = 1) \tag{2.5}$$

which is the usual transformation law in conventional Hilbert space. For normalized states we have

$$\mathrm{Tr}\ P_\alpha = 1, \qquad \mathrm{Tr}\ P_\beta = 1, \qquad P_\alpha^2 = P_\alpha,\ P_\beta^2 = P_\beta. \tag{2.6}$$

Then, using (2.2) and (2.6) we can write

$$d^2_{\alpha\beta} = \frac{1}{2}\,\mathrm{Tr}(P_\alpha - P_\beta)^2 = 1 - \Pi_{\alpha\beta}. \qquad (2.7)$$

Now in a projective geometry in which we represent the normalized idempotent P_α as a point with n homogeneous and n-1 inhomogeneous coordinates $d_{\alpha\beta}$ would be the invariant distance between points α and β. We see that it is simply related to the transition probability. When $\alpha=\beta$ their relative distance vanishes while the transition probability equals unity. A subgroup H of $G=U(n)$, (in this case $H=U(n-1)\times U(1)$) will leave the idempotent P_α invariant. This is the stability group in the corresponding projective geometry leaving the point unchanged. The transformations of the coset G/H will change the point, hence the state. The set of all transformed states is a homogeneous space of dimension dim G - dim H and forms our quantum mechanical space in which positive definite transition probabilities can be defined from the invariant distance between two separate points.

Note that in this algebraic formulation of quantum mechanics only Jordan multiplication enters. Indeed, an infinitesimal transformation

$$U = 1 + iA, \qquad (2.8)$$

where iA is antihermitian acts on states in a way which can be expressed solely by Jordan products. We show that by writing

$$U\,P_\alpha\,U^{-1} = P_\alpha + i[A, P_\alpha] = P_\alpha - \frac{1}{4}\,[[H_1, H_2], P_\alpha], \qquad (2.9)$$

where we have expressed $-4iA$ as a commutator of two hermitian matrices H_1 and H_2: Now we have the identity

$$- \frac{1}{4} [[H_1, H_2], P_\alpha] = (H_1 \cdot P_\alpha) \cdot H_2 - H_1 \cdot (P_\alpha \cdot H_2)$$

$$= (H_1 \; P_\alpha \; H_2), \qquad (2.10)$$

where the right hand side is just the associator of the Jordan product. For a finite unitary transformation we obtain

$$T \, P_\alpha = P_\alpha + (H_1 \; P_\alpha \; H_2) + \frac{1}{2!}(H_1, (H_1 \; P_\alpha \; H_2), H_2) + \cdots, \qquad (2.11)$$

an expression involving multiple Jordan associators. Thus states, transition probabilities, unitary transformations in the algebraic formulation of Quantium Mechanics involve only hermitian matrices and their Jordan product.

From the preceding consideration it follows that if one can generalize hermitian matrices, projection operators and Jordan products to the case in which complex numbers are replaced by octonions a generalized Quantum Mechanics becomes possible. This is exactly what (JNW) did by introducing 3×3 matrices J hermitian with respect to octonionic conjugation.

If we set

$$J = \begin{pmatrix} \alpha & c & \bar{b} \\ \bar{c} & \beta & a \\ b & \bar{a} & \gamma \end{pmatrix}, \qquad (2.12)$$

where α, β, γ are scalars and a, b, c, octonions with the bar denoting octonionic conjugation, J is hermitian and depends on 27 parameters. We have

$$\mathrm{Tr}\, J = \alpha + \beta + \gamma, \qquad (2.13)$$

$$\mathrm{Det}\, J = \alpha\beta\gamma - \alpha(a\bar{a}) - \beta(b\bar{b}) - \gamma(c\bar{c}) + (ab)c + \bar{c}(\bar{b}\bar{a}). \qquad (2.14)$$

The "hermitian" matrices J are closed under the Jordan product. Thus the transformations (2.11) remain meaningful. When H_1 and H_2 are traceless Jordan matrices (2.11) represents the 52 parameter group F_4 when applied on J. We can take as a projection operator a Jordan matrix with trace unity and determinant zero. TrJ, TrJ^2 and Det J are invariant under F_4. Transition probabilities can be defined by means of (2.7) and are left invariant by F_4 which replaces the unitary transformation for the usual finite Hilbert spaces.

If P_α is a Jordan matrix of the form (2.12) which satisfies the relations (2.6), it is a projection operator corresponding to a pure state. It is left invariant by the subgroup SO(9) of F_4. Thus SO(9) plays the same role as a generalized phase group in a unitary geometry. The quantum mechanical space of all the states corresponds to the elements of the coset space F_4/SO(9) which is known to be a Moufang projective sphere. SO(9) is the stability group of a point in that projective space and correspondingly is a generalized phase transformation that leaves a quantum mechanical state unchanged. If such a state corresponds to a particle state with a given momentum a translation should not change the state so that the translation operator should result in a particular SO(9) transformation on that state. In a previous work this transformation was shown to single out a special direction e_7 in octonionic space. Thus, when we consider F_4 together with the Poincaré transformations the stability group that leaves e_7 invariant will be relevant. This is just the subgroup $SU^c(3)\times U(1)\times SU(s)$ of SO(9) which has the same structure as the direct product of the color group with the Weinberg-Salam group.

When the Jordan matrix J is complex it is a 27 dimensional representation of E_6. Then J^* corresponds to the $\overline{27}$ representation. $Tr(J \cdot J^*)$ and Det J are invariant under E_6. In this case the projection operator belongs to the adjoint representation (78) contained in $27\times\overline{27}$. The stability group is $SO(10)\times U(1)$ and the projection operator P_α for a pure state is in correspondence with the coset $E_6/(SO(10)\times U(1))$ which has 32 real or 16 complex components.

Finally, the 56-dimensional complex set

$$\Psi:(\rho,\ J,\ K,\ \sigma), \tag{2.15}$$

where ρ and σ are complex numbers invariant under E_6, J transforms like 27 and K like $\overline{27}$, form the fundamental representation of E_7 which is a group with 133 parameters. The dual set

$$\tilde{\Psi}:\ (\sigma^*,\ -K^*,\ J^*,\ -\rho^*) \tag{2.16}$$

also transforms like Ψ under E_7. The quadratic form which has symplectic structure in terms of Ψ and $\tilde{\Psi}$ is also invariant and reads

$$\Psi^\dagger\Psi = \rho^*\rho + \sigma^*\sigma + \tfrac{1}{2}Tr(J^*\cdot J + K^*\cdot K). \tag{2.17}$$

The stability group in the geometry based on E_7 is $SO(12)\times SO(3)$. This is the group that will leave a projection operator invariant. Again fixing the e_7 octonion direction leads to the subgroup $SU^c(3)\times SU(4)\times SU(2)\times U(1)$ that will act as a generalized phase group on a pure state.

It is interesting to note that the generalized projective geometries associated with quantum mechanical spaces in which the transformation groups are F_4, E_6, E_7 and E_8 are exceptional in the sense that Desargues' theorem is not valid for them. This theorem is the result of the embeddability of a quantum mechanical space in a higher one of the same kind and plays an important role in measurement theory in connection with the building of new states in a larger space by taking direct products of states in lower dimensional spaces. The failure of this property in exceptional quantum mechanical spaces must lead to new properties for the observability of quark states which are not Desarguesian. Such a measurement theory in exceptional quantum mechanical space has not yet been worked out. On the other hand color singlets are ordinary complex numbers that lie in a Desarguesian subset and should enjoy all the observability properties of conventional quantum mechanics. Leptons and hadrons will then be associated with color singlet parts of given representations of exceptional groups.

III. E_7 AS A UNIVERSAL GAUGE GROUP

Elsewhere[5] we have shown how F_4 can be regarded as the transformation group on a set of 4 Dirac leptons and 3 sets of colored quarks. Two representations of E_6 have been used[8] to describe a set of 6 colored quarks and 9 Dirac leptons. Finally we have also considered[9] a model based on E_7 with 6 colored quarks and 10 Dirac leptons. Since the 56 representation of E_7 is pseudoreal it can be taken as purely left handed and accommodate two kinds of left handed spinors

$$\psi_L = \frac{1}{2}(1+\gamma_5)\psi \text{ and } \widehat{\psi}_R = \frac{1}{2}(1+\gamma_5)\psi^c. \qquad (3.1)$$

Using split octonion units[10]

$$u_0 = \frac{1}{2}(1 + ie_7), \quad u_j = \frac{1}{2}(e_j + ie_{j+3}), \quad (j=1,2,3) \qquad (3.2)$$

and their complex conjugates obtained by changing the sign of i, we write explicitly the various components of the 56 representation in terms of basic fermions. Under $SU(6)\times SU^c(3)$ we have

$$56 = (20,1) + (6,3) + (\bar{6},\bar{3}). \qquad (3.3)$$

The color singlet part (20.1) is a SU(6) tensor $L^{\alpha\beta\gamma}$ completely antisymmetrical in its quark indices α, β, γ and represents 20 Majorana or 10 Dirac leptons. The left handed quarks are represented by Q_α^j where α is a quark index with 6 flavors and j is the color index. Then $(Q_\alpha^j)^*$ will transform like antiquarks and will be denoted by $Q^\alpha{}_j$. The charge operator is taken to be a member of the octet of the SU(3) subgroup and a singlet under $SU^c(3)$. Then the charges of the quarks are 2/3,-1/3,-1/3,2/3,-1/3,-1/3 for $\alpha = 1,\ldots,6$. The upper indices correspond to opposite charges. Thus the leptons $L^{\alpha\beta\gamma}$ are split into 6 neutral leptons and 4 charged ones of charge unity. If $(\alpha\beta\gamma)$ = (145), (164), (124), (314), are particles of charge -1, then (356), (256), (236), (235), are antiparticles with charges +1. Taking the 56 dimensional representation for the basic fermions we we make the following identifications:

$$\rho = L^{456}, \qquad \sigma = L^{123}, \qquad (3.4)$$

$$J = u_0\lambda + u_0^*\lambda^T + u_j^* Q_L^j + u_j \hat{Q}_R^j, \qquad (3.5)$$

$$K = u_0\theta + u_0^*\theta^T + u_j^* C_L^j + u_j \hat{C}_R^j, \qquad (3.6)$$

where T denotes transposition and

$$\lambda = \begin{pmatrix} L^{156} & L_+^{256} & L_+^{356} \\ L_-^{164} & L^{264} & L^{364} \\ L_-^{145} & L^{245} & L^{345} \end{pmatrix}, \quad \theta = \begin{pmatrix} L^{234} & L_+^{235} & L_+^{236} \\ L_-^{314} & L^{315} & L^{316} \\ L_-^{124} & L^{125} & L^{126} \end{pmatrix}, \qquad (3.7)$$

$$Q_L^j = \begin{pmatrix} 0 & Q_3^j & -Q_2^j \\ -Q_3^j & 0 & Q_1^j \\ Q_2^j & -Q_1^j & 0 \end{pmatrix}, \quad C_L^j = \begin{pmatrix} 0 & Q_6^j & -Q_5^j \\ -Q_6^j & 0 & Q_4^j \\ Q_5^j & -Q_4^j & 0 \end{pmatrix}, \qquad (3.8)$$

and similarly for $\hat{Q}_R^j$ and $\hat{C}_R^j$. Under the $SU(3)\times SU(3)\times SU^c(3)$ subgroup of $SU(6)\times SU^c(3)$ we have the representations

$\rho=(1,1,1)$, $:(\bar{3},3,1)$, $Q_L^j:(3,1,3)$, $\hat{Q}_R^j:(1,\bar{3},\bar{3})$, (3.9)

$\sigma=(1,1,1)$, $:(3,\bar{3},1)$, $C_L^j:(1,3,3)$, $\hat{C}_R^j:(\bar{3},1,\bar{3})$, (3.10)

so that the light quarks Q_L^j and the charmed quarks C_L^j together form the $(6,3)$ representation of $SU(6)\times SU^c(3)$ and $\hat{C}_R^j$ together $\hat{Q}_R^j$ form the $(\bar{6},\bar{3})$ representation. Thus, if the 6 quarks are denoted u d s c b b' both $(u_L\ d_L\ s_L\ c_L\ b_L\ b'_L)$ and $(c_R\ b_R\ b'_R\ u_R\ d_R\ s_R)$ are in the $(6,3)$ representation.

For the charged weak current we choose the color singlet

$$J_0^+ = (J_0)_2^1 + (J_0)_3^4. \qquad (3.11)$$

To this current contribute both the left handed and right handed quarks so that the hadronic part reads

$$J_0^{\dagger had} = u_L^\dagger d_L + c_L^\dagger s_L + c_R^\dagger b_R + u_R^\dagger b'_R \ . \quad (3.12)$$

The mass matrix will cause mixings for the physical quarks. A simple mixing involving a Cabibbo angle θ and an angle ϕ will replace d_L and s_L by $d_L(\theta)$, $s_L(\theta)$ and b_R and b'_R by $b_R(\phi)$, $b'_R(\phi)$.

Note that this current is parity violating and the corresponding neutral current is both parity violating (parity violation is in d, s, b and b') and has no $\Delta S=1$ part.

The overall structure of the current (3.12) is such that it will give a $\Delta I=\frac{1}{2}$ rule[11] and give rise to anomalies in $\bar{\nu}$ scattering when the b' channel starts opening[9,12].

The phenomenological implications of this current are in good agreement with recent experiments.

Under the SU(2) group generated by $J_\pm$ and J_3, we have the following structure

$$\text{Doublets:} \quad \begin{pmatrix} u_L \\ d_L \end{pmatrix}, \begin{pmatrix} c_L \\ s_L \end{pmatrix}, \begin{pmatrix} u_R \\ b'_R \end{pmatrix}, \quad \begin{pmatrix} c_R \\ b_R \end{pmatrix} , \quad (3.13)$$

$$\begin{pmatrix} L^{123} \\ L_-^{124} \end{pmatrix}, \begin{pmatrix} L^{234} \\ L_-^{314} \end{pmatrix}, \begin{pmatrix} L_+^{356} \\ L^{456} \end{pmatrix}, \quad \begin{pmatrix} L^{256} \\ L^{156} \end{pmatrix}. \quad (3.14)$$

Singlets:

$$d_R,\ s_R,\ b_L,\ b'_L, \tag{3.15}$$

$$L^{125}, L^{126}, L^{364}, L^{365},\ \frac{1}{\sqrt{2}}(L^{245}+L^{315}),\ \frac{1}{\sqrt{2}}(L^{264}-L^{316})\ . \tag{3.16}$$

Triplets:

$$\begin{pmatrix} L_+^{236} \\ \frac{1}{\sqrt{2}}(L^{264}+L^{316}) \\ L_-^{164} \end{pmatrix} \qquad \begin{pmatrix} L_+^{235} \\ \frac{1}{\sqrt{2}}(L^{245}-L^{315}) \\ L_-^{145} \end{pmatrix}. \tag{3.17}$$

In principle e_L^- and μ_L^- could belong either to doublets or to triplets. An analysis of the mass matrix seems to favor the triplet assignment with $L_-^{145}=e_L$, $L_-^{164}=\mu_L$. Also $L_+^{256}=\hat{\mu}_R$, $L_+^{356}=\hat{e}_R$.

In a gauge theory based on E_7 with the charge operator chosen as a color singlet generator of SU(3) the unrenormalized Weinberg angle θ_W corresponding to the current (3.11) has the value 3/4. The theory also has one unrenormalized gauge coupling constant g" which we can take as the coupling constant for massless gluons associated with the color group. Using the methods of Georgi, Quinn and Weinberg[13] we find the following renormalized values of g" and θ_W:

$$\ln\frac{M}{\mu} = \frac{3(4\pi)^2}{22}\ \frac{1}{c^2+3c'^2}\ \left(\frac{1}{e^2} - \frac{c^2+c'^2}{g''^2}\right), \tag{3.18}$$

$$\sin^2\theta_W = \frac{c^2}{c^2+3c'^2}\left(1 + 2c'^2 \frac{e^2}{g''^2}\right), \qquad (3.19)$$

where, for E_7 we have $c^2=2$, $c'^2=2/3$, μ is the renormalization point and M is a mass scale set gy the Higgs mechanism in spontaneous symmetry breaking. Using the value $g''^2/4\pi = 0.2$ at $\mu = 3$ GeV from a fit to the ψ-decay rate we find $\sin^2\theta_W = .52$ and $M = 3.10^{23}$ GeV. In general in such a unified field theory the baryons will decay into leptons plus mesons in second order. The decay is only suppressed by such a high mass scale M. In this case the proton life-time is about 10^{63} years For a larger g" the Weinberg angle is smaller, M higher and the proton even more stable. The minimum acceptable value of M is of the order of the Planck mass $G^{-1/2}$. The predicted Weinberg angle is also in reasonable aggrement with neutral current data[9].

Finally in ratio $R = (e^+e^- \rightarrow \text{hadrons})/(e^+e^- \rightarrow \mu^+\mu^-)$ turns out to be 4 from the quark contributions alone, 5 if one heavy charged lepton contributes and 6 if both charged heavy leptons contribute. Since the threshold for at least one heavy lepton is passed, if we assume that 5 of the quarks are already produced we obtain a minimum of 4.67 for R and a maximum of 6. These values are also in good agreement with the Spear data.

IV. CONCLUDING REMARKS

We have seen that the association of the internal degrees of freedom of elementary particles with the exceptional Quantum Mechanical spaces that are coset spaces related to exceptional Lie groups provides us with the lepton and quark degrees of freedom, including color, that are needed for the phenomenology of weak, electromagnetic and strong interactions. The group E_7 emerges as a reasonable condidate for a spontaneously broken gauge theory provided the Higgs mechanism introduces a very high mass scale of the order of the Planck mass. Then most of the old weak interaction phenomenology together with ν, ν high energy scattering data and e^+e^- annihilation experiments are accounted for. The color containment problem is not yet solved but our identification of color with an automorphism group of octonions suggests an algebraic origin related to non-Desarguesian properties of colored states.

Another problem is the complete working out of the the mass matrix which upon diagonalization will yield the various values of mixing angles such as the Cabibbo angle. Two mass scales are needed for the spontanteous symmetry breaking, one M of the order of the Planck mass and the other of the order of lepton and quark masses. This suggests that we may need two Higgs fields χ and ϕ. If χ belongs to the 56 representation it cannot be regarded as a bound state of the basic fermions and it cannot have a renormalizable coupling to the quarks and leptons in 56. Thus it can only give mass to some vector mesons. This must be the Higgs field associated with mass scale M. The other Higgs fields ϕ which gives masses to the basic fermions are contained in the symmetric part of 56×56 and can only belong to representations 133 and 1539. These fields could be of dynamical origin. If we have a scalar and a pseudoscalar (133) associated with a mass scale of a few GeV range, through their Yukawa couplings to basic fermions they can give the heavy leptons masses of the same order as charmed quark masses. The light quark masses become comparable with light lepton masses. Here the problem is to see if the neutrinos can remain massless.

A last remark concerns CP violation which can also be generated spontaneously in the six quark model implicit in the E_7 theory without interfering with CP conservation in the light quark sector. There the problem is to obtain an effective superweak theory rather than a milliweak theory.

I am very grateful to my collaborators M. Günaydin, P. Ramond and P. Sikivie with whom I carried out most of this work and to G. Domokos, S. Kövesi-Domokos, M. Gell-Mann, L. Michel from whom I learned a great deal at various stages of the development of the ideas discussed above. It is also with great pleasure that I thank my colleagues at Yale for discussions and valuable suggestions.

REFERENCES

1. P. Jordan, J. von Neumann and E. P. Wigner, Ann. Math. 35, 29 (1934), J. Segal, Ann. Math. 48, 930 (1947), S. Sherman, Ann. Math. 64, 592 (1956).
2. C. Chevalley and R. D. Schafer, Proc. Nat. Acad. Sci. U. S. 36, 137 (1950).
3. See for example J. Tits, Proc. Colloq. Utrecht, p. 175 (1962) and H. Freudenthal, Advances in Mathematics, vol. I, p. 145 (1965).
4. M. Günaydin and F. Gürsey, Lett. Nuovo Cimento 6, 401 (1973) and Phys. Rev. D9, 3387 (1974). F. Gürsey, in Johns Hopkins Workshop on Current Problems in High Energy Particle Theory, p. 15 (Johns Hopkins University, 1974).
5. F. Gürsey, in Kyoto International Symposium on Mathematical Problems in Theoretical Physics, ed. H. Araki, p. 189(Springer 1975).
6. F. Gürsey, in Proc. 4th Int. Colloquium on Group Theoretical Methods in Physcis (U. Nijmegen, 1975).
7. F. Gürsey, P. Ramond and P. Sikivie, Phys. Rev. D12, 2166 (1975).
8. F. Gürsey, P. Ramond and P. Sikivie, Phys. Lett. 60B, 177 (1976).
9. F. Gürsey and P. Sikivie, E_7 as a Universal Gauge Group, Yale preprint, January 1976, to be published.
10. M. Günaydin and F. Gürsey, J. Math. Phys. 14, 1651 (1973).
11. R. V. Mohapatra, Phys. Rev. D6, 2023 (1972), A. de

Rujula, H. Georgi and S. L. Glashow, Phys. Rev. Lett. 35, 69 (1975).

12. See M. Barnett, A. Review of Models with More than Four Quarks, Fermilab prepring 75/71-THY (1975).
13. H. Georgi, H. Quinn and S. Weinberg, Phys. Rev. Letters 33, 451 (1974).

SEMICLASSICAL QUANTIZATION METHODS IN FIELD THEORY

A. Neveu

California Institute of Technology

and

The Institute for Advanced Study

Princeton, New Jersey

This is an overview of the semiclassical WKB method which has been developed by R. DASHEN, B. HASSLACHER and myself in Ref. 1 and which can be applied to finding solutions to field theories which are inaccessible to perturbation techniques.

In particular, it is possible to find solutions to the full nonlinear interacting classical equations of motion of various models, which behave like bound, stable field configurations in space-time, with particle properties. The question arises as to whether these solutions survive the process of second quantization. In Ref. 1 we give a method for answering that question, the accuracy of which depends both on how much one knows about the classical problem, and the strength of the coupling constant, in direct proportion.

Our method is based on the works of KELLER, GUTZ-WILLER and MASLOV, who developed a general semiclassical formalism for use in atomic physics. These techniques are directed toward the computation of energy levels - or particle masses in field theory. We approach the problem through the quantum action principle in the Feynman path integral representations, since this provides the most natural connection between the classical problem and its second quantized analogue. Also, since we start from a Lagrangian formalism, any divergences that emerge can be handled by standard renormalization techniques.

For weak coupling, it was found that time-independent classical solutions are interesting[1,2]. In the weak coupling limit, our WKB quantization of static solutions to classical field equations is equivalent to a number of other schemes. The differences enter when on contemplates classical motions which cannot be reduced to a time independent field. That such solutions are interesting should be obvious from the fact that the Bohr orbits of hydrogen are not time-independent solutions to classical equations of motion but rather are motions which are periodic in time. The real power of th WKB method is the quantization of motions analogous to Bohr orbits. To find an example of how the semiclassical method works in field theory, we have studied the sine-Gordon equation in one space and one time dimension.[3] It is defined by the Lagrangian

$$L = -\frac{1}{2}(\partial\mu\phi)^2 + \frac{m^4}{\lambda}\left[\cos\left(\phi\frac{\sqrt{\lambda}}{m}\right) - 1\right] \qquad (1)$$

and is completely solvable at the classical level: there exists an algorithm[3] from which all solutions to the Lagrange equations for ϕ can be constructed. In particular, to apply our quantization method, we look for classical solutions which become particles when quantized. There are two types of these:

First, there is the soliton (and the antisoliton) which is a solution that is time independent in its rest frame. The other, which we call the doublet, is a soliton-antisoliton bound state. In its rest frame, the doublet field oscillates periodically in time. Doublet solutions exist for a continuous range of classical energies. The WKB method will quantize the doublet energies, yielding a discrete spectrum of particle masses.

The particle spectrum of the sine-Gordon Hamiltonian turns out to be the following: the soliton and anti-soliton have a mass $M = 8m/\gamma'$, where $\gamma' = [\lambda/m^2] / [1 - \lambda/8\pi m^2]$. The doublet produces the remaining series of states at masses

$$M_n = \frac{16m}{\gamma'} \sin \frac{n\gamma'}{16} \quad ,$$

$$n = 1,2,3 \ldots\ldots < 8\pi/\gamma' \quad . \tag{2}$$

The original "elementary particle" of the theory is the $n = 1$ state in eq. (2). As $\lambda \to 0$ γ' vanishes, and one easily sees that M approaches the weak coupling mass, $m + O(\lambda^2)$, of the elementary particle. Notice that according to eqs. (1) and (2) there is a finite number of doublet states. As the coupling γ' increases the states disappear one by one. What happens is that they decay into soliton-antisoliton pairs. This may be seen by observing that when the nth state disappears, M_n is just $16\ m/\gamma'$, or twice the soliton mass. At $\gamma' = 8\pi$, the $n = 1$, or "elementary particle" state itself breaks up and disappears from the spectrum; only solitons and antisolitons remain.

The weak coupling behavior of M_n is quite interesting. Expanding, one finds:

$$M_n = nM_1 - \frac{M_1}{6} \left(\frac{\lambda}{16m^2}\right)^2 (n^3-n) + 0\,(\lambda^3)\ ,$$

$$M_1 = \frac{m}{16\gamma'} \sin \frac{\gamma'}{16} = m\left[1 - \frac{1}{6}\left(\frac{\lambda}{16m^2}\right)^2\right] + 0(\lambda^3)\ , \qquad (3)$$

which corresponds to a nonrelativistic n-body bound state made up of n particles with physical mass M.

This is the same as one finds upon solving the n-body Schrödinger equation with δ-function potential obtained from the ϕ^4 term in the interaction Lagrangian. Thus for weak coupling, the doublet states can be thought of as bound states of n "elementary particles". Of course, n cannot be too big. When $\gamma'n$ is greater than 8π, the state breaks up into a soliton-antisoliton pair. In fact, for $\gamma'n$ large (but less than 8π), the states are probably best thought of as soliton-antisoliton bound states.

The semiclassical calculation suggests that all states with $\gamma'n$ less than 8π are stable. The mass ratio as given by eq. (2) and the symmetry of the Lagrangian under $\phi \to -\phi$ account for the stability of the n = 1,2,3 states. It takes further symmetry to keep the n = 4 state from decaying into two n = 1 states. At a classical level, the sine-Gordon equation has an infinite number of nontrivial conserved quantities[3]. If, as conjectured, these survive in the quantum theory, they would provide enough quantum numbers to stabilize all the bound states: the S-matrix, as conjectured in Ref. 4, would consist of pure phases.

We have also extended our work on the ϕ^4 theory in two dimensions. This system is not exactly solvable. For small coupling, however, one can find the analogue of the sine-Gordon doublet states. We obtain a formula like (3) with a different coefficient of $n^3 - n$. The interpretation is the same except that we no longer know what happens for strong coupling. It is a reasonable speculation, however, that for large λn the states break up into a kink - antikink pair. Although our results for the ϕ^4 theory are neither as complete nor as elegant as those for the sine-Gordon case, we regard this calculation as important. It shows that the method is not restricted to special, classically solvable equations like the sine-Gordon system.

Coleman[5] has obtained the remarkable result that the sine-Gordon system can be mapped into the massive Thirring model. The relationship between the sine-Gordon coupling λ and the four-fermion coupling g of the Thirring model is $\frac{\lambda}{4\pi m^2} = \frac{1}{1+g/\pi}$, or $\gamma' = 8\pi/(1 + 2g/\pi)$. What are the fermions? They are almost certainly the solitons. To see this, we observe that at $\gamma' = 8\pi$, the Thirring model coupling g vanishes. This is just the point where the n = 1 state unbinds. For γ' slightly less than 8π, the four-fermi coupling is weak and attractive. There will then be one non-relativistic fermion-antifermion bound state. Summing diagrams in the Thirring model, one finds that through order g^3, the mass M_B of the bound state is given in terms of the fermion mass M_f by

$$\frac{2 M_f - M_B}{M_f} = g^2 - \frac{4g^3}{\pi} + O(g^4) \quad . \tag{4}$$

Identifying M_B with M_1, and M_f with the soliton mass

$8\,m / \gamma'$, we compare this to

$$\frac{2M\ (\text{soliton}) - M_1}{M\ (\text{soliton})} = 2(1 - \sin\frac{\gamma'}{16}) = g^2 - \frac{4g^3}{\pi} + O(g^4)\ ,$$

where we have used Coleman's identification of the coupling constants. It is remarkable that both the g^2 and g^3 terms agree. We have not computed beyond order g^3 in the Thirring model. For $\gamma' > 8\pi$, the four-fermion coupling is repulsive and there is no bound state.

Coleman also finds that the theory is singular at $\lambda/m^2 = 8\pi$. At this point, γ' goes to infinity and it is evident that our semiclassical solution is also singular.

The agreement between our approximation and Coleman's precise results suggests that WKB may be exact for the mass spectrum of the sine-Gordon equation. This is not beyond the realm of possibility. Recall that the Bohr-Sommerfeld quantization conditions give the energy levels of hydrogen exactly. To investigate this question, we have gone to the weak coupling regime, and carried out an exact calculation of M_2/M_1 through order $(\lambda/m^2)^4$. This is done by summing Feynman diagrams in a way which is equivalent to solving the Bethe-Salpeter equation. The exact result is

$$\frac{2M_1 - M_2}{M_1} = \left(\frac{\lambda}{16m^2}\right)^2 + \frac{4}{\pi}\left(\frac{\lambda}{16m^2}\right)^3 + \left(\frac{12}{\pi^2} - \frac{1}{12}\right)\left(\frac{\lambda}{16m^2}\right)^4 + O(\lambda^5). \tag{6}$$

One can easily calculate the same quantity using eq. (2) for M_1 and M_2. Expanding, one finds that the coefficients of λ^2, λ^3 and λ^4 are identical. This is a highly nontrivial result: to get the exact order λ^4 term, one has to keep two-loop diagrams in the kernel of the

Bethe-Salpeter equation. We can show that the agreement in order λ^4 is special to the sine-Gordon equation, and will not occur in the generic case.

As argued above, we conjecture that eq. (2) gives the mass ratios of Lagrangian (1) exactly to all orders of perturbation theory. It does not, however, give the absolute masses exactly as a function of the bare mass, as can already be seen in lowest nontrivial order (order λ^2).

We have also investigated a model which contains fermions and developed a general method for handling them in semiclassical calculations.

Specifically, we have used a WKB method to compute the particle spectrum of the Gross-Neveu model. It is in two dimensional space-time and is defined by the Lagrangian,

$$L = \sum_{k=1}^{N} i\, \bar{\psi}^{(k)} \not{\partial} \psi^{(k)} + \frac{g^2}{2} \left(\sum_{k=1}^{N} \bar{\psi}^{(k)} \psi^{(k)} \right)^2 . \qquad (7)$$

The model thus contains N fermions coupled symmetrically through a scalar-scalar interaction. We will generally surpress the particle type indices k and use the notation

$$i\, \bar{\psi}\, \not{\partial}\, \psi = \sum_k i\, \bar{\psi}^{(k)}\, \not{\partial}\, \psi^{(k)} ,$$

$$\bar{\psi}\, \psi = \sum_k \bar{\psi}^{(k)}\, \psi^{(k)} . \qquad (8)$$

The model is renormalizable (g is dimensionless), γ_5 invariant and formally scale invariant. For large N one can sum the leading sets of diagrams and establish that in this limit the model is asymptotically free. Gross and Neveu[6] also found that $\bar{\psi}\psi$ develops a vacuum expectation value so that γ_5 invariance is spontaneously broken. In the process the dimensionless coupling constant g is traded for an arbitrary dimensional parameter $g \langle\bar{\psi}\psi\rangle$ and disappears from the theory. The end result is that the theory contains no dimensionless parameter other than the number of fermions N. Consequently, any physical dimensionless quantity such as the ratio of two particle masses can depend only on N. This rather striking phenomenon, whose ultimate origin is the renormalization group, will be present in our WKB calculations. We can take this as an indication that semiclassical methods are compatible with renormalization group ideas.

Following Gross and Neveu, we find it useful to replace (7) by

$$L = \bar{\psi}\, i\, \partial\!\!\!/ \psi + g\sigma\bar{\psi}\psi - \frac{\sigma^2}{2} \quad , \tag{9}$$

where we have used the notation of (8) and have introduced a neutral scalar field σ. Using the equation of motion

$$\sigma = g\, \bar{\psi}\psi \quad , \tag{10}$$

the Lagrangian in (9) becomes equivalent to that in (1.1). Our WKB method is based upon the evaluation of certain functional integrals by a stationary phase approximation. It is not obvious how to use a stationary phase method when there are integrations over anticommuting fermion fields. The advantage of the Lagrangian in(9)is that the fermi fields enter bilinearly and can be integrated out of the problem leaving an effective action containing only the boson field σ. We then do the σ-integration by stationary phase. To do this we must find space-time dependent fields σ around which the effective action is stationary. This effective action is nonlocal and highly nonlinear but it turns out to be possible to find stationary points. The first such example was found by Callan, Coleman, Cross and Zee.[7] It is analogous to the kink in the ϕ^4 theory or the soliton in the sine-Gordon equation, i.e., it is a particle-like solution which is time-independent in its rest frame and which has a peculiar topology. We have found a large number of further stationary points of the effective action. In particular, we find solutions which are particle-like but have a nontrivial time dependence in the rest frame. The WKB method then quantizes these classical solutions, producing a spectrum of particle masses.

The kink-like solutions produce an exotic sort of particle which probably has no counterpart in four dimensions. However, the vast majority of our solutions are not kinks. They correspond to less exotic objects such as the original fermion, fermion-antifermion bound states or multifermion bound states. Such states surely exist in four-dimensional theories and we would conjecture that in four, as well as in two dimensions, there is a correspondence between classical field configurations and particle states. Assuming this to be so, it remains to be seen if such a correspondence can be effectively exploited.

Below we will describe the particle spectrum of the model as given by our WKB calculation. To interpret this spectrum we will need to know something about the symmetries of the model. The Gross-Neveu model has an obvious U(N) internal symmetry. Actually it has an O(2N) symmetry of which U(N) is a subgroup. This may be seen as follows. Choose a Majorana representation for the γ matrices $\gamma^0 = \sigma^y$, $\gamma^1 = i\,\sigma^x$ and write

$$\psi^{(k)} = \psi_1^{(k)} + i\,\psi_2^{(k)} \quad , \tag{11}$$

where $\psi_1^{(k)}$ and $\psi_2^{(k)}$ are hermitian two component spinors. The Lagrangian then takes the form

$$L = \sum_k i(\psi_1^{(k)} \frac{\partial}{\partial t} \psi^{(k)} + \psi_2^{(k)} \frac{\partial}{\partial t} \psi_2^{(k)} + \psi_1^{(k)} \sigma_z \frac{\partial}{\partial x} \psi_1^{(k)} + \psi_2^{(k)} \sigma_z \frac{\partial}{\partial x} \psi_2^{(k)})$$

$$- g\,\sigma \sum_k (\psi_1^{(k)}\,\sigma_y\,\psi_1^{(k)} + \psi_2^{(k)}\,\sigma_y\,\psi_2^{(k)}) - \frac{\sigma^2}{2} \quad , \tag{12}$$

which is hermitian and nonvanishing because the ψ's anticommute. When written in the form(12),it is clear that the Lagrangian is invariant under orthogonal transformations on the 2N component vector $\psi_j^{(k)}$, $k = 1,2,\ldots$ $\ldots N$, $j = 1,2$. The fermion number operator $Q \equiv \int \psi^+\psi\, dx$ has nontrivial commutation relations with other generators of O(2N). Therefore a nontrivial representation of O(2N) will contain states with more than one value of Q. Hence we may expect, for example, that some fermion-antifermion states will be degenerate with fermion-fermion states. The σ field is an O(2N) scalar while ψ is an O(2N) vector. The only other O(2N) representations which we will encounter are the totally antisymmetric O(2N) tensors of rank $n_0 < N$. The number of states in a multiplet corresponding to such a tensor is $n_0!\,(2N - n_0)!/(2N)!$. Scalars and O(2N) vectors are special cases of completely antisymmetrical tensors of rank $n_0 = 0$ and $n_0 = 1$ respectively.

Because of our inability to evaluate certain Gaussian functional integrals we have not been able to carry through a complete WKB calculation in the Gross-Neveu model. What we have been able to do is a sort of zeroth order calculation which, in ordinary potential theory, is analogous to using the quantization rule $\oint pdq = 2n\pi$ rather than the more accurate $\oint pdq = (2n+1)\pi$. (In the sine-Gordon equation the analogous approximation is equivalent to setting $\gamma' = \frac{\lambda}{m^2} \times [1- \frac{\lambda}{8\pi m^2}]^{-1} \approx \frac{\lambda}{m^2}$). Even with this approximation our results should become exact in the limit of large N and are probably <u>qualitatively</u> correct for any N greater than 2 or 3.

We find the particle spectrum shown in Fig. (1). There is a large, unexpected degeneracy beyond that required by O(2N) symmetry. This degeneracy might be real or it may be an artifact of our zeroth order calculation. There are supermultiplets listed by a "principle quantum number" $n = 1,2,... < N$. The common mass of the states in the nth supermultiplet is

$$M_n = g\ \sigma_0 \frac{2N}{\pi} \sin\left(\frac{n}{N}\frac{\pi}{2}\right) \quad , \qquad n = 1,2,... < N \quad , \tag{13}$$

where σ_0 is the vacuum expectation value of σ. We see that ratios of masses are independent of g as they should be. If n is odd the supermultiplet is composed of fermions and contains O(2N) representations corresponding to all completely antisymmetrical tensors of rank $n_0 = 1, 3, 5... \leq n$. For example, the n = 1 state is a fermion belonging to a vector representation of O(2N). This is the "elementary particle" of the theory. For large N,

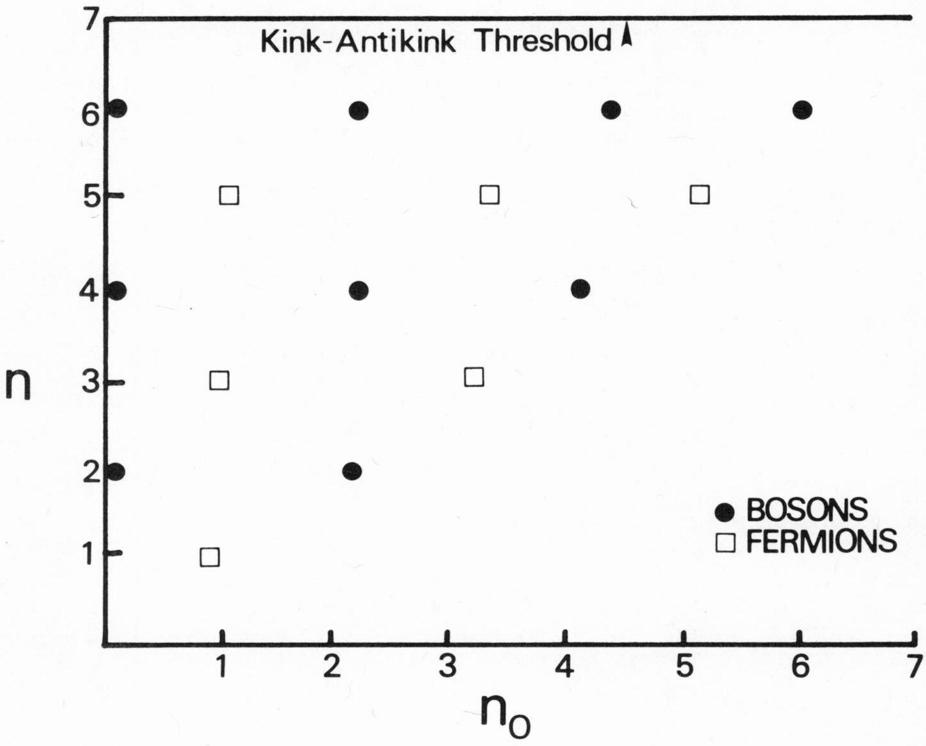

Figure 1

Particle Spectrum (see text)

$$M_1 \approx g\,\sigma_0 \quad , \tag{14}$$

which agrees with the result of Gross and Neveu. The $n = 3$ supermultiplet contains an O(2N) vector which is some kind of excited state of the elementary particle and a completely antisymmetrical O(2N) tensor of rank 3. The latter is a bound state of three fermions and/or antifermions. The supermultiplets with n even are composed of bosons and contain O(2N) antisymmetrical tensors of rank $n_0 = 0, 2, 4 \ldots \leq n$. For example, $n = 2$ contains an O(2N) scalar and an antisymmetric tensor of second rank. The tensor is a set of two body bound states with fermion-fermion, antifermion-antifermion and fermion-antifermion quantum numbers. The O(2N) scalar is a different sort of object. It may be thought of as a particle associated with the σ field. At the $n = 4$ level there is an excited σ, a state which can be thought of as an excitation of the second rank tensor at $n = 2$ and a new state corresponding to a completely antisymmetrical tensor of rank 4. This new object is a bound state of 4 fermions and/or antifermions analogous to the 2 and 3 particle states found at levels $n = 2$ and 3. The pattern continues in the same way for $n = 5,6\ldots$ on up to N.

The quantum numbers of the states in our spectrum are not unexpected. In the limit of large N the leading exchange is a sum of bubbles.

In the nonrelativistic limit, this exchange produces an attractive δ-function potential. Such a potential will produce bound states only in channels where the spatial wave function is completely symmetrical. For fermions this means that the O(2N) wave function must be completely antisymmetrical, i.e., an O(2N) antisymmetric tensor.

For large N the bubble exchange is weak[6] and a Schrödinger equation calculation is valid. In this way one finds a binding energy which agrees with that computed from (13)

$$|E_B| = nM_1 - M_n = M_1 \left(\frac{\pi}{2N}\right)^2 (n^3-n) + O\left(\frac{1}{N^3}\right) \quad (15)$$

to the indicated order in N^{-1}. These nonrelativistic bound states correspond to the states with $n_0 = n$. They are the lowest states with given O(2N) quantum numbers and are consequently stable. Eq. (15) is valid only if n/N is small. For n and N both large the binding energy per particle is, in units of M_1

$$\frac{n\, M_1 - M_n}{n\, M_1} = 1 - \frac{2N}{\pi n} \sin\left(\frac{n}{N}\frac{\pi}{2}\right) , \quad (16)$$

which for $n/N \sim 1$ shows binding by a finite fraction of the rest mass. Thus strong binding can occur even for large N.

The bubble exchanges are not the only important interaction for large N. For fermion-antifermion interactions in an O(2N) single state the annihilation bubbles are dominant. The sum of these bubbles leads to an interaction which is marginally attractive. In leading order in N, Gross and Neveu[6] found a σ bound state at the fermion-antifermion threshold. It is presumably the n=2, O(2N) singlet state discussed above. We find that it is bound in the next order in N^{-2}. This disagrees with a detailed diagrammatic calculation by Schoenfeld[8] who finds that the bound state remains at threshold to this order. We do not understand the origin of this discrepancy. In any case there is a weak attraction between fermion-antifermion pairs in an O(2N) singlet state. One might therefore imagine that the particles in the model will be made up of a number of fermions and antifermions paired into O(2N) singlet states plus further "valence" fermions and antifermions in an antisymmetrical tensor state. Our particle spectrum is consistent with such a picture.

The particle spectrum ends at n = N, where the mass is $M_N = 2Ng\sigma_0/\pi$. The mass of the Callan-Coleman-Gross-Zee kink is (in our zeroth approximation) $M_{kink}=Ng\sigma_0/\pi$. Thus the Nth state is just at the kink-antikink threshold. Higher mass states would be unstable against decay into kink-antikink pairs.

There is a striking similarity between the sine-Gordon equation and the Gross-Neveu model. In the zeroth order WKB approximation the particle spectrum of the sine-Gordon theory is given by $M_n \approx [m\ 2\xi/\pi]\ \sin(\pi n/2\xi)$, where $\xi = 8\pi m^2/\lambda$, plus a soliton at mass $M_{soliton} \approx m\xi\pi$. With the identification $N \to \xi$ the energy levels are identical to those of the Gross-Neveu model. The particle content of the levels is of course very different in the two theories. There is no doubt an underlying reason for this correspondence between the theories but we do not know what it is. However, we can use this correspondence to try and guess what would happen if we could do a complete WKB calculation. In the sine-Gordon equation the result of the complete calculation is simply to replace λ/m^2 in the zeroth order formula by $\frac{\lambda}{m^2} \times [1 - \frac{\lambda}{8\pi m^2}]^{-1}$ which is equivalent to making the replacement $\xi \to \xi - 1$. The analogous replacement in the present model would be to replace N by N-1 in Eq. (1.7) and in the formula for the kink mass. The theory would then be singular at N=1. One expects such a singularity since at N=1 the Gross-Neveu model can be Fierz transformed to the usual Thirring model which contains a single massless fermion. Our zeroth order calculation is certainly not valid for N as small as 1.

If it were to turn out that a full WKB calculation differs from the present one only by changing N to N-1, then the extra degeneracy in the mass spectrum would presumably be real and a consequence of some underlying dynamical symmetry. Another possibility is that in a complete WKB calculation the masses within a supermultiplet will be split by terms of order N^{-2}. If this happens, the n=2 singlet state might remain at threshold to order N^{-2} in agreement with Schoenfeld.

While the finer details of our approximate semiclassical spectrum are clearly not to be taken too seriously, the qualitative picture of a rich spectrum organized into some kind of supermultiplets is almost certainly correct. This unexpected wealth of particle states seems to be a consequence of the asymptotic freedom of the theory. The detailed form of the classical σ field which corresponds to a quantum bound state suggests that the binding mechanism is not a direct interaction between the bound fermions but rather is some kind of vacuum polarization effect. The fact that the theory is unstable in the infrared is most likely the reason for this.

REFERENCES

1. R. DASHEN, B. HASSLACHER, and A. NEVEU, Phys. Rev. D10, 4114, 4130, 4138 (1974), D11, in press, and IAS preprint COO 2220-46 (May 1975). For a review and more explicit details, see R. RAJARAMAN, Physics Reports, to be published.
2. J. GOLDSTONE, and R. JACKIW, Phys. Rev. D11, 1486 (1975), J.L. GERVAIS and B. SAKITA, Phys. Rev. D11, 2943 (1975), C. CALLAN and D. GROSS, Nucl. Phys. B93, 29 (1975).
3. There is a considerable literature on the sine-Gordon equation. A review of this and related equations is given in A. SCOTT, F. CHU and D. Mc LAUGHLIN, Proceedings of the IEEE, 61, 1443 (1973). See also L. FADDEEV and L. TAKHTAJAN Theor. and Math. Phys. 21, 160 (1974) and 22 (in press).
4. L. FADDEEV, IAS preprint: "Quantum theory of solitons" and references therein.
5. S. COLEMAN, Phys. Rev. D11, 2088 (1975).
6. D. GROSS and A. NEVEU, Phys. Rev. D10, 3235 (1974).
7. Private communication
8. J. SCHONFELD, Princeton University preprint (1975).

ON THE THEORY OF THE DIRECT FOUR-FERMION INTERACTION

Yu.M. Makeenko, K.A. Ter-Martirosyan,
A.B. Zamolodchikov
(Presented by K.A. Ter-Martirosyan)
Institute for Theoretical and Experimental Physics
Moscow, U.S.S.R.

ABSTRACT

The solution of a coupled system of integral equations for the direct four-fermion interaction theory is obtained in the $d=2+\varepsilon$ dimension space. It exists only for the SU(2) symmetric interaction and for the positive coupling constant G. It is scale invariant and corresponds to a stable fixed point of Gell-Mann - Low equations.

The real four dimensional case $d=4$ is considered. Here the same solution with a small change in the indices is shown to be approximately valid and all integrals in the equations are found to be convergent. The solution so obtained allows one to estimate the so-called "nonparquet" terms omitted in the set of equations. They are found to be numerically small.

A general iteration procedure with an obtained

solution as a zero order approximation is suggested.

1. INTRODUCTION

It is known that direct four-fermion V-A Hamiltonian

$$H = G_o(\bar{\nu}O_\alpha e)(\bar{e}O_\alpha \nu) \tag{1}$$

with $O_\alpha = \gamma_\alpha \frac{1+\gamma_5}{2}$ describes very well, in the first Born approximation, the weak interaction at low energies. However, since the cross-section grows with increasing energy, $\sigma \sim G^2E^2$, and only s-wave is scattered, the contradiction with unitarity always occurs at $E \sim 10^3$ GeV. Hence, the higher orders in the coupling constant G come into the play[1] and should be considered. For this purpose Feynman perturbation theory could not be used, as the interaction (1) is not renormalizable in the usual way. This is why the renormalizable Weinberg-Salam[2] model of the weak interaction gained popularity during recent years. Unfortunately, it requires the introduction of a number of new particles and is not simple. The same is true for the strong interaction scheme with intermediate non-Abelian vector gauge fields, which is also not simple. All other types of renormalizable interactions (e.g. Yukawa $\pi\bar{N}N$ or electromagnetic $\gamma\bar{e}e$ interactions, meson $\lambda\phi^4$ interaction, etc.) lead to the well-known "zero charge" difficulty[3] connected with the vanishing of physical charge in the limit of a point, i.e. local, interaction. It is revealed also in the increase of the "effective charge" $g(\tilde{p}^2)$ at given momentum $\tilde{p}^2$ in these theories with increasing $\tilde{p}^2=\vec{p}^2-p_o^2$, unlike the behavior in the asymptotically free gauge theories, where it decreases as $\tilde{p}^2 \to \infty$.

In principle there could be the so-called "fixed

point" theories with $g(\tilde{p}^2)\to g_1$ = const as $\tilde{p}^2\to\infty$. It is just the case which we shall consider below for the four-fermion interaction (1).

In a number of papers the weak interaction has been investigated by the dispersion relations approach. This enables one to express corrections of order G^2 and G^3 (ref. 4) to the Born amplitudes, at energies less than 10^3 GeV, through some constants connected with cross-sections of physical processes and to obtain limitations on the weak cross-section values[5] at high energies.

The weak interaction (1) is considered below with the help of the set of the so-called "parquet" integral equations for the four-fermion vertex part stated by one of the authors[6] in 1958 (see also ref. 7 for the boson case). They are just Bethe-Salpeter chains developed simultaneously along s, t and u channels. Iterations of these equations with the Born term as a zero approximation reproduce exactly the usual perturbation theory result with quadratic unrenormalizable divergences. However, beyond perturbation theory the equations can admit the finite solutions[8,9]. In the paper by Abrikosov, Galanin et al[10] the solution leading to the vanishing of the four-fermion charge in the local limit was found. To find this solution, the "two-cutoff technique" was used. However, it violates the Pauli principle, i.e. the antisymmetry of the fermion vertex part.

Another possibility is considered below. The integral equations can have in fact another solution, rapidly decreasing with rising particle momenta and leading to the convergent integrals. Hence, the cutoff dependence disappears entirely and instead of the charge vanishing situation, the "fixed point" solution appears.

In Section 2 (and Appendix 1) the "parquet" integral

equations for V-A interaction (1) are listed. Section 3 contains general arguments[11-13] showing that they could have finite solution in the asymptotic scaling region. The exact solution in the $d= 2 + \varepsilon$ dimensional space at $\varepsilon << 1$ is obtained in Section 4. For this purpose the spinor matrices and space integrals are continued[14,15] to the noninteger d.

To first order in ε the solution is found both by summing leading graphs by the renormalization group method[16], and by solving the set of "parquet" integral equation determined for any noninteger d. It is shown that the solution always leads to convergent integrals in the right hand side of these equations. Although this solution cannot be extended directly to $\varepsilon = 2$, it illustrates the main idea of the paper, i.e. that the Hamiltonian unrenormalizable in the usual way can have a renormalizable finite solution.

The possibility of a construction of the "parquet" equations solution in the real 4-dimensional space is discussed in Sections 5,7. In Section 6 we show that the existence of the finite solution requires the number of fields to be more than 2, the presence of neutral currents in the Hamiltonian and SU(2) invariance of the weak interaction with $G > 0$.

Moreover, it exists only at the definite relation (Section 9) between the constants of the neutral currents of the types $(\bar{\nu}_\mu \nu_\mu)(\bar{e}e)$ and $(\bar{\mu}\mu)(\bar{e}e)$.

Values of the effective charges at the fixed point turn out to be small. This leads to the negligible contribution of the "nonparquet" terms omitted in the integral equations in the first approximation.

The equation for the fermion Green function should be solved together with the set of the "parquet" equations. It is examined in Section 8.

In conclusion (Section 9) we discuss briefly the physical consequences of our results.

2. "Parquet" Equations

Consider the SU(2) symmetric interaction

$$H = \frac{F_o}{4} J_o J_o + \frac{G_o}{4} \vec{J}\,\vec{J} \quad , \qquad (2)$$

where $J_o = (\sum_{i=1}^{N} \bar{\psi}_i \psi_i)$ is an "isoscalar" and $\vec{J} = (J_x, J_y, J_z) = (\sum_{i=1}^{N} \bar{\psi}_i \vec{\tau} \psi_i)$ is an "isovector" current. $\vec{\tau} = (\tau_x, \tau_y, \tau_z)$ are the Pauli matrices acting on the lepton doublets $\psi_i = \binom{\nu_i}{e_i}$; $i = 1, \ldots, N$. The spin matrices O_α are omitted. We do not consider the simpler interaction (e.g. with $G_o=0$ or with $N = 1$*)) since it is shown below that corresponding equations have no finite solutions.

The current J_o and J may be rewritten as $J_o = \bar{\psi}\psi$, $\vec{J} = \bar{\psi}\vec{\tau}\psi$, where $\psi = \binom{\nu}{e}$; $\nu = \begin{pmatrix} \nu_1 \\ \vdots \\ \nu_N \end{pmatrix}_1$ $e = \begin{pmatrix} e_1 \\ \\ e_N \end{pmatrix}$. This is the very form which we shall use below.

"Parquet" equations were obtained[6] by the following approach. Define the fermion-fermion scattering amplitude T to be the sum of all connected Feynman graphs with four external lines. Divide it (Fig. 1) into parts $R, \phi^s, \phi, \tilde{\phi}$:

$$T = R + \phi^s + \phi - \tilde{\phi} \quad , \qquad (3)$$

---

*) For N = 1, because of the Fierz identity $(\bar{e}\nu)(\bar{\nu}e) \equiv (\bar{\nu}\nu)(\bar{e}e)$ the "isoscalar" interaction coincides with the "isovector" interaction $(\bar{\psi}\psi)(\bar{\psi}\psi) = (\bar{\psi}\vec{\tau}\psi)(\bar{\psi}\vec{\tau}\psi)$.

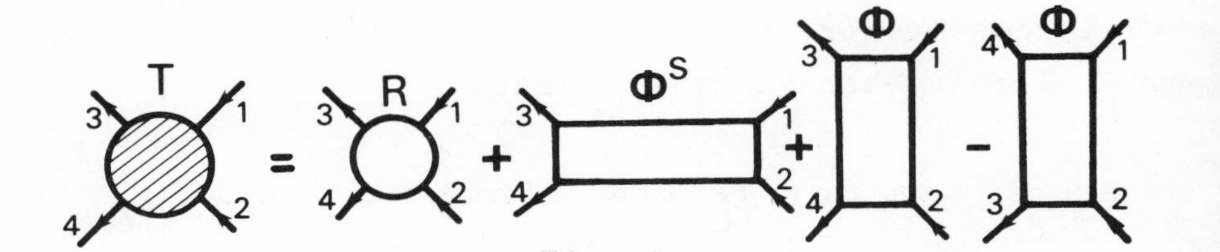

Fig. 1

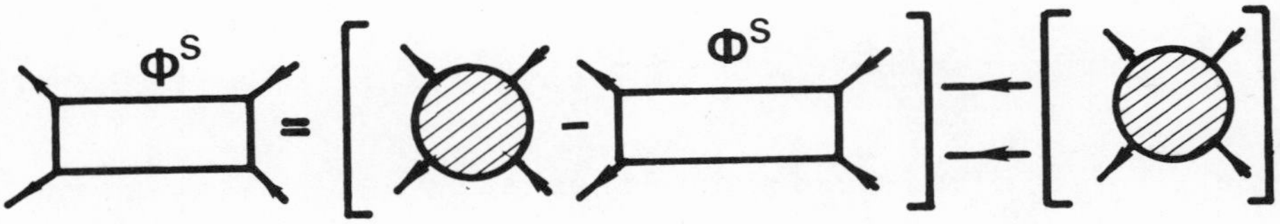

Fig. 2a

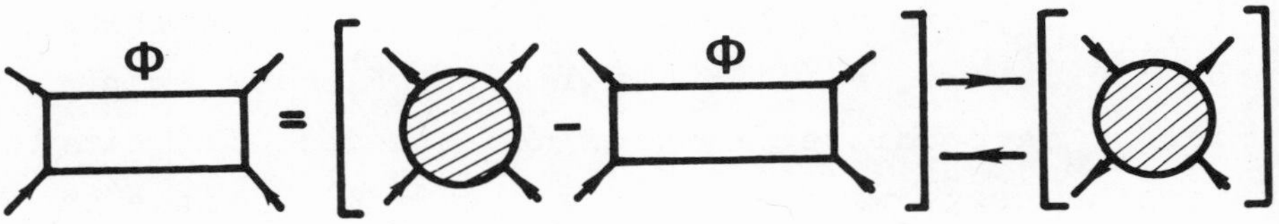

Fig. 2b

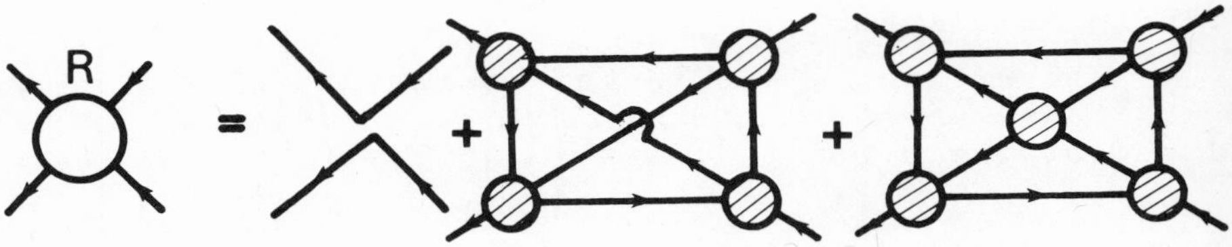

Fig. 3

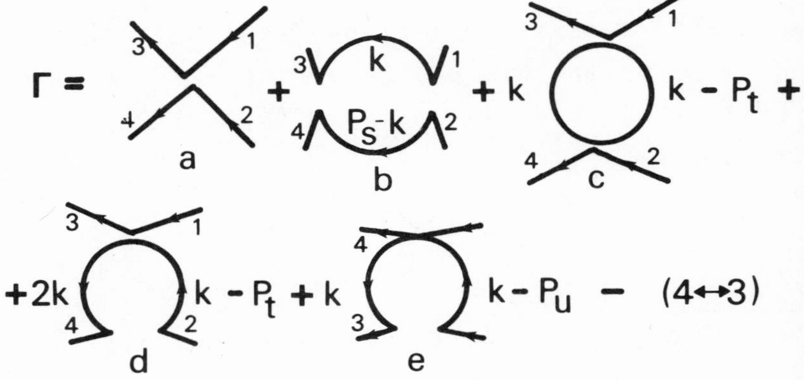

Fig. 4

where ϕ^s is the sum of all graphs consisting of two parts connected along the s-channel by only two fermion lines (with total momentum $p_s = p_1 + p_2$). It obeys an integral equation of the Bethe-Salpeter type (see Fig. 2)

$$\phi^s = \frac{1}{2} \int (T-\phi^s) S(k) S(p_s-k) T \frac{d^4k}{(2\pi)^4 i} \quad . \qquad (4)$$

Here $S(p) = S_\nu(p) = S_e(p) = \frac{\beta(p^2)}{\hat{p}}$ are the fermion Green's functions. (They are the same for electron and neutrino because the electron mass is neglected). The factor 1/2 appears as a result of the fermion identity.

Similarly denote $p_t = p_1 - p_3$ the total t-channel momentum. Then (see Fig. 2b)

$$\phi = -\int (T-\phi) S(k) T S(k-p_t) \frac{d^4k}{(2\pi)^4 i} \quad , \qquad (5)$$

and $\tilde{\phi} = (\phi)_{1\leftrightarrow 2}$ stands for the sum of diagrams which consists of two parts connected by two (fermion and anti-fermion) lines along the t or u channels respectively. R stands for the sum of diagrams which could not be separated in parts connected only by two particle lines. The simplest of them are given in Fig. 3.

The expression for any graph contribution may be simplified with the help of the identities

$$O_\alpha \gamma_\mu O_\beta = \chi_{\alpha\mu\beta\nu} O_\nu \quad ,$$

$$\chi_{\alpha\mu\beta\nu} = g_{\alpha\mu\beta\nu} - i\, e_{\alpha\mu\beta\nu} \quad ,$$

$$g_{\alpha\mu\beta\nu} = g_{\alpha\mu} g_{\beta\nu} + g_{\alpha\nu} g_{\beta\mu} - g_{\alpha\beta} g_{\mu\nu} \quad , \qquad (6)$$

$$\tau_i \tau_k = \delta_{ik} + i\, e_{ik\ell}\, \tau_\ell \quad ,$$

to the form*)

$$T = -\frac{1}{2}\Gamma_{f\alpha\beta}(O_\alpha)_{31}(O_\beta)_{42} - \frac{1}{2}\Gamma_{g\alpha\beta}(\vec{\tau}O_\alpha)_{31}(\vec{\tau}O_\beta)_{42} + (2\leftrightarrow 1),$$

$$\phi^s = -\frac{1}{2}\phi^s_{f\alpha\beta}(O_\alpha)_{31}(O_\beta)_{42} - \frac{1}{2}\phi^s_{g\alpha\beta}(\vec{\tau}O_\alpha)_{31}(\vec{\tau}O_\beta)_{42} + (2\leftrightarrow 1),$$

$$\phi = -\frac{1}{2}\phi^t_{f\alpha\beta}(O_\alpha)_{31}(O_\beta)_{42} + \frac{1}{2}\phi^u_{f\alpha\beta}(O_\alpha)_{32}(O_\beta)_{41} - \qquad (7)$$

$$- \frac{1}{2}\phi^t_{g\alpha\beta}(\vec{\tau}O_\alpha)_{31}(\vec{\tau}O_\beta)_{42} + \frac{1}{2}\phi^u_{g\alpha\beta}(\vec{\tau}O_\alpha)_{32}(\vec{\tau}O_\beta)_{41}.$$

Then (4) may be rewritten as

$$\phi^s_{f\alpha\beta} = -\frac{1}{2}\int\{[\Gamma_{f\gamma\delta}-\phi^s_{f\gamma\delta}]\Gamma_{f\sigma\rho}+3[\Gamma_{g\gamma\delta}-\phi^s_{g\gamma\delta}]\Gamma_{g\sigma\rho}\} \times$$

$$\times \chi_{\gamma\mu\sigma\alpha}\chi_{\delta\nu\rho\beta}\frac{k_\mu(p_s-k)_\nu}{k^2(p_s-k)^2}\beta(k^2)\beta((p_s-k)^2)\frac{d^4k}{(2\pi)^4 i},$$

$$\phi^s_{g\alpha\beta} = -\frac{1}{2}\int\{[\Gamma_{f\gamma\delta}-\phi^s_{f\gamma\delta}]\Gamma_{g\sigma\rho}+[\Gamma_{g\gamma\delta}-\phi^s_{g\gamma\delta}]\Gamma_{f\sigma\rho} - \qquad (8)$$

$$- 2[\Gamma_{g\gamma\delta}-\phi^s_{g\gamma\delta}]\Gamma_{g\sigma\rho}\}\chi_{\gamma\mu\sigma\alpha}\chi_{\delta\nu\rho\beta}\frac{k_\mu(p_s-k)_\nu}{k^2(p_s-k)^2} \times$$

$$\times \beta(k^2)\beta((p_s-k)^2)\frac{d^4k}{(2\pi)^4 i}.$$

In such a way we get 4 more equations for $\phi^t_f, \phi^u_f, \phi^t_g$ and ϕ^u_g.

Let us look for an approximate solution of these equations for which all integrals converge slowly so that large momenta of integration are essential. For this solution we may put $\Gamma_{f\alpha\beta} = g_{\alpha\beta}\,\Gamma_f$, where Γ_f is the function of scalar arguments only. The same is valid

---

*) If there is an isospin structure, the Fierz identity is not valid.

for $\Gamma_{g\alpha\beta}$, as well as for $\phi_{\alpha\beta}$, since integrals standing in front of factors $p_\alpha p_\beta$ converge rapidly and consequently are numerically small (see below).

For the scalar functions ϕ_f^s and ϕ_g^s the following equations hold:

$$\phi_f^s = 2\int\{[\Gamma_f-\phi_f^s]\Gamma_f+3[\Gamma_g-\phi_g^s]\Gamma_g\}\ \frac{\beta(k^2)\beta((p_s-k)^2)}{(p_s-k)^2}\ \frac{d^4k}{(2\pi)^4 i}\ ,$$

$$\phi_g^s = 2\int\{[\Gamma_g-\phi_g^s]\Gamma_f+[\Gamma_f-\phi_f^s]\Gamma_g-2[\Gamma_g-\phi_g^s]\Gamma_g\}\times \tag{9}$$

$$\times\ \frac{\beta(k^2)\beta((p_s-k)^2)}{(p_s-k)^2}\cdot\frac{d^4k}{(2\pi)^4 i}\ .$$

The entire system of equations is given in Appendix 1.

It will be shown that "nonparquet" graphs of Fig. 3 b), c) have small values and could be omitted in the first approximation. Thus, only the simplest graph 3a) contributes in R:

$$R = R_o = -\frac{F_o}{2}[(O_\alpha)_{31}(O_\alpha)_{42} - (O_\alpha)_{32}(O_\alpha)_{41}] -$$

$$-\frac{G_o}{2}[(\vec{\tau}O_\alpha)_{31}(\vec{\tau}O_\alpha)_{42} - (\vec{\tau}O_\alpha)_{32}(\vec{\tau}O_\alpha)_{41}]\ . \tag{10}$$

The system (10) with $R = R_o$ is called "parquet" equations. It is closed and allows one to find T (or Γ) at a given $\beta(\tilde{p}^2)$. To obtain the function β it is necessary to add to the system the corresponding Dyson-Schwinger equation (or use the unitarity equation for $S^{-1}(p)$; see Section 8).

3. THE SCALING SOLUTION

If the four-fermion vertex Γ decreases, ensuring the convergence of integrals in (4), (5) at large integration momenta, then the bare charges F_o and G_o vanish (see below) and general arguments of dimensional analysis yield the following asymptotic scaling solution[11-13]

$$\Gamma(p_1^2,p_2^2,p_3^2,p_4^2,p_s^2,p_t^2)=(\tilde{p}_1^2/\lambda^2)^{-\gamma}\ F\left(\frac{p_2^2}{p_1^2},\ldots,\frac{p_t^2}{p_1^2}\right) ,$$

$$\beta(p^2) = (\tilde{p}^2/\lambda^2)^{\Delta} \ ; \quad \tilde{p}^2 = -p^2 = \vec{p}^2 - p_o^2 \ , \tag{11}$$

where

$$\gamma = \frac{d}{2} - 1 + 2\Delta \ , \tag{12}$$

while F is some function and $d=4$ is the dimension of the space, Δ is a positive (due to Lehman theorem) number, λ^2 is the normalization momentum.

The condition (12) of strong coupling follows from the fact that each new isoscalar of isovector vertex in Feynman integrals yields the following additional factor in the integrand, where $\Gamma(\tilde{p}^2) = \Gamma(\tilde{p}^2,\ldots,\tilde{p}^2)$.

$$f(p^2) = \Gamma_f(\tilde{p}^2)\beta^2(\tilde{p}^2)(\tilde{p}^2)^{\frac{d}{2}-1} ,$$

$$g(p^2) = \Gamma_g(\tilde{p}^2)\beta^2(\tilde{p}^2)(\tilde{p}^2)^{\frac{d}{2}-1} . \tag{13}$$

It is called the "effective charge". The factor $(\tilde{p}^2)^{\frac{d}{2}-1}$ is due to the phase volume.

For the scaling form (11), (12) of the solution the effective charges have constant values f_1 and g_1, respectively.

4. FOUR-FERMION INTERACTION IN THE 2 + ε DIMENSIONAL SPACE

a) Analytic Continuation Method

The amplitude of the four-fermion interaction (2) to the second order in perturbation theory is determined by the graphs shown in Fig. 4 b), c), d), f) where it is assumed that γ matrices are multiplied along the solid lines. In the analytic continuation of γ-matrices to noninteger d a known difficulty arises, which is that the quantity $e_{\alpha\beta\gamma\delta}$ cannot be determined for noninteger d.

However, we remark that every graph contribution contains only products of an even number of the $e_{\alpha\beta\gamma\delta}$. These products are tensor (not pseudotensor as $e_{\alpha\beta\gamma\delta}$ for d=4) and could be continued to any noninteger d.

To show this let us use the equation which is true in 4-dimensional space

$$e_{\alpha_1\alpha_2\alpha_3\alpha_4}\, e_{\beta_1\beta_2\beta_3\beta_4} = -\det\|g_{\alpha_i\beta_j}\| \quad , \; i,j=1,..,4.$$

We take it as the definition of the symbol $e_{\alpha\beta\gamma\delta}$ for d≠4 assuming that $g_{\mu\mu}=d$. This analytic continuation method is not unique. However, the results of the calculations to the first order in ε will not depend on the continuation method. To make it clear note that due to the covariance and antisymmetry

$$e_{\alpha\mu\beta\rho}e_{\alpha\nu\beta\delta} = -f(d)(d-2)(d-3)(g_{\mu\nu}g_{\rho\delta}-g_{\rho\nu}g_{\mu\delta}), \qquad (14)$$

where f(2) ≠ ∞ and f(4) = 1 for any analytic continuation (for the above continuation method f(d) = 1).

To the first order in ε=d-2 graphs containing this quantity could be omitted entirely, since they yield the additional smallness of order ε.

All integrals will be normalized as follows:

$$\int \Psi(k^2)\frac{d^dk}{(2\pi)^{d/2}} = \frac{1}{(4\pi)^{d/2}\Gamma(d/2)} \int_0^\infty \Psi(-\tilde{k}^2)(\tilde{k}^2)^{\frac{d}{2}-1} d\tilde{k}^2 .$$

b) Gell-Mann - Low Equations

Consider the contribution of the graph of Fig. 4c),

$$\Gamma^{c)}_{f\alpha\beta} = F_o^2.N.Sp[O_\alpha\gamma_\mu O_\beta\gamma_\nu] \int \frac{k_\mu(k-p_t)_\nu}{k^2(k-p_t)^2}\frac{d^dk}{(2\pi)^d i} .$$

Calculating the integral in a usual way[14,15] we get:

$$\Gamma^{c)}_{f\alpha\beta} = F_o^2 NSp(1) g_{\alpha\mu\beta\nu}(\tilde{p}_t^2)^{\frac{d}{2}-1}\left[\frac{g_{\mu\nu}}{d-2} - \frac{p^t_\mu p^t_\nu}{p_t^2}\right] \times$$

$$\times \frac{\Gamma(2-\frac{d}{2})\Gamma^2(d/2)}{(4\pi)^{d/2}\Gamma(d)} = - F_o^2 NSp(1)\frac{2\Gamma(2-\frac{d}{2})\Gamma^2(\frac{d}{2})}{(4\pi)^{d/2}\Gamma(d)}(\tilde{p}_t^2)^{\frac{d}{2}-1}\left[g_{\alpha\beta}-\frac{p^t_\alpha p^t_\beta}{p_t^2}\right]$$

Here $SpO_\alpha\gamma_\beta = g_{\alpha\beta}.Sp(1)$. At $d=4$ $Sp(1) = 2$.

The singularity at $d=2$ was cancelled here and the matrix element has become transverse[14]. The same cancell tion takes place in graph 4d). Its value is

$$\Gamma^{d)}_{f\alpha\beta} = \frac{F_o+3G_o}{F_o}\frac{d-2}{2\,NSp(1)}\Gamma^{c)}_{f\alpha\beta} .$$

Contributions of diagrams 4b) and 4f) for $d=2$ differ only in signs. This leads to the cancellation of divergences in F amplitudes which is similar to that in the Thirring model*).

---

*) In the 2-dimensional case the matrices O_α coincide with the Thirring γ-matrices $O_o=\gamma_o=\sigma_x$, $O_1=\gamma_1=i\sigma_y$, the Fierz identity $(\bar{e}O_\alpha\nu)(\bar{\nu}O_\alpha e) = (\bar{e}O_\alpha e)(\bar{\nu}O_\alpha\nu)$ for them being invalid.

The corresponding two dimensional model (with G=0) is scale invariant for any value of bare charge F_o. The ε-expansion method[15,17] cannot be applied to it.

Let us develop the Wilson's ε-expansion method for the interaction (2) at $G_o \neq 0$. In two dimensions the corresponding model has the asymptotically free solution[18,19]. Just in this case we get in $d=2+\varepsilon$ dimensions the fixed point theory with numerically small effective charge $g_1 \sim \varepsilon$. As a result, the solution can be found in the form of the power series in ε.

To obtain the first term of this expansion let us calculate the scattering amplitude in the second order perturbation theory for the interaction (2). Calculating the contributions of graphs in Fig. 4b,f**) to order ε we get for the amplitudes F and G:

$$\Gamma_f = F_c - \frac{1}{4\pi}\,[F_c^2 + 3G_c^2]\;\frac{(\tilde{p}_u^2)^{\varepsilon/2} - (\tilde{p}_s^2)^{\varepsilon/2}}{\varepsilon/2}\,,$$

$$\Gamma_G = G_c - \frac{1}{2\pi}[G_c^2 - G_c F_c]\frac{(\tilde{p}_s^2)^{\varepsilon/2} - \lambda^{\varepsilon}}{\varepsilon/2} - \frac{1}{2\pi}[G_c^2 + F_c G_c]\frac{(\tilde{p}_u^2)^{\varepsilon/2} - \lambda^{\varepsilon}}{\varepsilon/2}\,. \quad (15)$$

Here $F_c = \Gamma_f(\lambda^2)$ and $G_c = \Gamma_g(\lambda^2)$ are the physical values of the coupling constants. Assume now that $F_c\lambda^{\varepsilon}$, $G_c\lambda^{\varepsilon} \ll 1$ but F_c, $G_c \dfrac{(\tilde{p}^2)^{\varepsilon/2} - \lambda^{\varepsilon}}{\varepsilon/2} \sim 1$ and sum leading graphs contributions by the renormalization group approach[16].

Note that effective charges in the case considered have the form (13). They coincide at d=2 with vertices as in the approximation in question, $\beta = 1$, i.e. the wave function renormalization is absent (it is expressed

---

**) Graphs in Fig. 4c,d include products of quantities $e_{\alpha\mu\beta\gamma}$ mentioned above (see (14)) and to order ε do not contribute at all.

by the graphs in Fig. 5 of order ε^2 and by graphs of higher orders).

The Gell-Mann-Low equations can be obtained from (15) by the standard methods

$$f' = \frac{\varepsilon}{2} f + \frac{3}{8}\frac{\varepsilon}{\pi} g^2 \ ; \quad g' = \frac{\varepsilon}{2} g - \frac{g^2}{\pi} \ , \tag{16}$$

where f' means $\frac{df}{d\ell n\tilde{p}^2}$. The equation for g has the solution

$$g(\tilde{p}^2) = \frac{G_c(\tilde{p}^2)^{\varepsilon/2}}{1 + \frac{G_c}{g_1}[(\tilde{p}^2)^{\varepsilon/2} - \lambda^{\varepsilon}]} \ , \tag{17}$$

with $g_1 = \frac{\pi}{2}\varepsilon$. At $G_c > 0$ and $\tilde{p}^2 \to \infty$ we get $g(\tilde{p}^2) \to g_1$. Thus, the renormalization group equations have the scaling power form solution only at very large $\tilde{p}^2$ when $G_c(\tilde{p}^2)^{\varepsilon/2} >> 1$.

On the right hand side of eq. (16) for f, terms of order ε^3 were taken into account. Substitution of $g(\tilde{p}^2)$ from (17) yields its solution in the form

$$f = \frac{3}{8}\varepsilon g + [F_c - \frac{3}{8}\varepsilon G_c](\tilde{p}^2)^{\varepsilon/2} \ .$$

At $\tilde{p}^2 \to \infty$ it has the stable (not rising at $\tilde{p}^2 \to \infty$) value $f \to \frac{3}{8}\varepsilon g_1 = \mathrm{Const}$ only if the second term vanishes, i.e. at $F_c = \frac{3}{8}\varepsilon G_c$. We shall discuss this relation between physical charges in Section 6.

c) Anomalous Dimensions

Calculate now the dimension of ψ-field Δ in the scale invariant theory. Graphs of the second order for $\Sigma(p) = (1 - \beta^{-1}(p^2))\hat{p}$ are shown in Fig. 5. Calculating their contribution by the above explained method we find that graphs of Fig. 5b contribution do not contain terms of order ε, but the graph of Fig. 5a yields (see Appendix 4)

$$\Sigma = \frac{F_c^2+3G_c^2}{2(2\pi)^2} \; \frac{(\tilde{p}^2)^{\varepsilon}-\lambda^{2\varepsilon}}{\varepsilon} \; .n.\hat{p}$$

where*)

$$n = N \frac{Sp(1)}{2}$$

and N is the number of fermion doublets $\psi_i = \begin{pmatrix} \nu_i \\ e_i \end{pmatrix}$.

The renormalization group equation for β gives[16]:

$$\frac{d\ell n\beta}{d\ell n\tilde{p}^2} = \frac{f^2+3g^2}{2(2\pi)^2} \; .n$$

As we have seen, at large $\tilde{p}^2$ we can put $g^2=g_1^2$, $f^2=f_1^2$. Then

$$\beta = (\tilde{p}^2/\lambda^2)^{\frac{3}{32}\varepsilon^2 n} \quad ,$$

which means that $\Delta = \frac{3}{32}\varepsilon^2 n$, i.e. it is positive in accordance with the Lehmann theorem. The ψ-field dimension is

$$\psi = [x]^{-\frac{1}{2} - \frac{\varepsilon}{2} - \frac{3}{32} n\varepsilon^2} \quad .$$

The index γ in (11) can be found from the condition (12) of strong coupling: $\gamma = \frac{\varepsilon}{2} + \frac{3}{16} n\varepsilon^2$ to order ε^2.

---

*) Value of Sp(1) cannot be defined uniquely in our analytic continuation method.

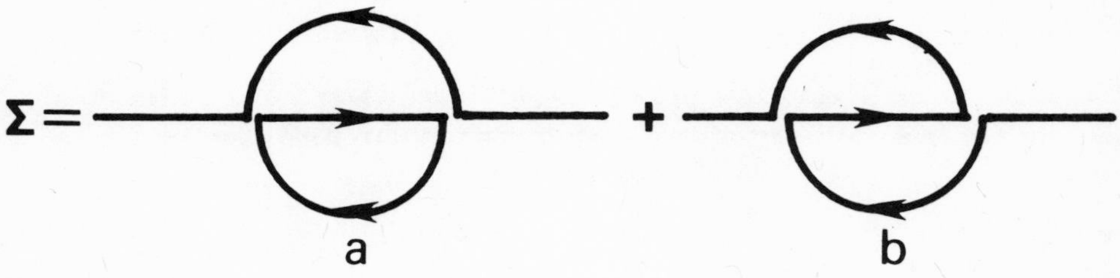

Fig. 5

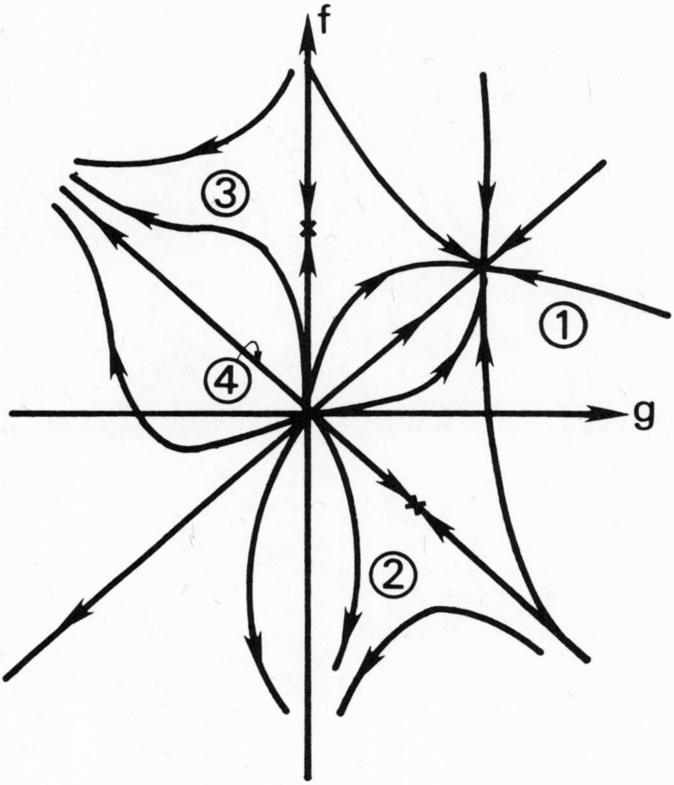

Fig. 6

d) Solution of the "Parquet" Equations

Let us consider the solution of the "parquet" equations system in $d=2+\varepsilon$ dimension space. To the first order in ε it reduces to a very simple form and can be solved by methods used in the logarithmic case[7,9,20]. The first simplification is that ϕ^s depends only on three variables

$$\xi = \max\{\tilde{p}_1^2, \tilde{p}_2^2, \tilde{p}_s^2\} \quad ,$$
$$\zeta = \max\{\tilde{p}_3^2, \tilde{p}_4^2, \tilde{p}_s^2\} \quad , \tag{18}$$
$$\eta = \tilde{p}_s^2 \quad .$$

One may admit, without losing generality, that momenta p_1 and p_2 are large, i.e. $\xi > \zeta > \eta$. If p_1 and p_3 (or p_1 and p_4) are large then $\xi \sim \zeta \sim \eta$ and ϕ^s will depend on a single variable only. This is confirmed by the perturbation theory calculations.

The system of equations for the space-like momenta is closed and does not require for its solution the knowledge of vertex parts in other regions of momenta space. Let us make the Wick rotation[8] $k_o \to i\,k_o$ and deal only with the Euclidean matrix. We shall consider the vertex for $\xi \gtrsim \zeta \sim \eta$, since just that very **region** will be sufficient[7] for the following. Let $x=\xi^{\varepsilon/2}$; $y=\eta^{\varepsilon/2}$; $z=(k^2)^{\varepsilon/2}$. Then the system of "parquet" equation takes the form

$$\phi_f^s(x,y) = -\frac{1}{2\pi\varepsilon}\int_y^x \{[\Gamma_f(x)-\phi_f^s(x)]\Gamma_f^s(z,y) + 3[\Gamma_g(x)-\phi_g^s(x)] \times \Gamma_g^s(z,y)\}dz - \frac{1}{2\pi\varepsilon}\int_x^\infty \{[\Gamma_f(z)-\phi_f^s(z)]\Gamma_f^s(z,y) + 3[\Gamma_g(z)-\phi_g^s(z)]\Gamma_g^s(z,y)\}dz \; . \tag{19}$$

Only the integration region indicated here with the lower limit of integration given in (19) yields the contribution in the first order in ε.

Equations for ϕ_g^s, ϕ_f^u and ϕ_g^u can be written in a similar form.

The function $\phi(x) = \phi(x,x)$ could be determined only if the value of $\phi(x,y)$ depending on two variables is known. It makes it difficult to solve eq. (19). The situation becomes much simpler if "parquet" equations (19) are rewritten in a similar form due to Sudakov[7,20,13]. It is obtained by cutting any "reducible" graph for T (i.e. the graph consisting of two parts connected by lines only of two particles; see Section 2) across the pair of lines carrying the least momentum of integration in the Feynman integrals. In our case these equations have the form

$$\phi_f^s(x,y) = -\frac{3}{2\pi\varepsilon}\int_y^x \Gamma_g^s(x,z)\Gamma_g(z)dz - \frac{3}{2\pi\varepsilon}\int_x^\infty \Gamma_g^2(z)dz ,$$

$$\phi_g^s(x,y)=-\frac{1}{2\pi\varepsilon}\int_y^x[\Gamma_f^s(x,z)-2\Gamma_g^s(x,z)]\Gamma_g(z)dz+\frac{1}{\pi\varepsilon}\int_x^\infty \Gamma_g^2(z)dz ,$$

$$\phi_f^u(x^u,y^u)=-\frac{3}{2\pi\varepsilon}\int_{y^u}^{x^u} \Gamma_g^u(x^u,z^u)\Gamma_g(z^u)dz^u-\frac{3}{2\pi\varepsilon}\int_{x^u}^\infty \Gamma_g^2(z^u)dz^u, \quad (20)$$

$$\phi_g^u(x^u,y^u)=-\frac{1}{2\pi\varepsilon}\int_{y^u}^{x^u} [\Gamma_f^u(x^u,z^u)+2\Gamma_g^u(x^u,z^u)]\Gamma_g(z^u)dz^u - \frac{1}{\pi\varepsilon}\int_{x^u}^\infty \Gamma_g^2(z)dz .$$

Here $\Gamma_g^s(x,y)$ or $\Gamma_g^u(x^u,y^u)$ stands for the same four-fermion vertex $\Gamma_g(p_1,p_2,p_3,p_4)$ in the cases when momenta p_1 and p_2 or p_1 and p_4 are very large, respectively. They are determined by the following equations (see Appendix 1 for more details):

$$\Gamma_f^s(x,y)=\phi_f^s(x,y)+\phi_f^u(x) \; ; \; \Gamma_g^s(x,y)=G_o+\phi_g^s(x,y)+\phi_g^u(x) \; ;$$

$$\Gamma_f^u(x^u,y^u)=\phi_f^s(x^u)+\phi_f^u(x^u,y^u) \; ; \; \Gamma_g^u(x^u,y^u)=G_o+\phi_g^s(x^u)+\phi_g^u(x^u,y^u),$$

where $x^u = (\xi^u)^{\varepsilon/2}$, $\xi^u = \max\{\tilde{p}_1^2, \tilde{p}_4^2, \tilde{p}_u^2\}$, $y^u = (\tilde{p}_u^2)^{\varepsilon/2}$ are the variables determined similarly to (18) for u-channel. We have put $f_o = 0$, since it was shown that $f_o = O(\varepsilon^2)$. Summing up the equations of the type (19) we obtain the following equations for $\Gamma_f^s(x,y)$ and $\Gamma_g^s(x,y)$:

$$\Gamma_f^s(x,y) = -\frac{3}{2\pi\varepsilon}\int_y^x \Gamma_g^s(x,z)\Gamma_g(z)dz \quad ,$$

$$\Gamma_g^s(x,y)=G_1-\frac{1}{2\pi\varepsilon}\int_y^x [\Gamma_f^s(x,z)-2\Gamma_g^s(x,z)]\Gamma_g(z)dz+\frac{2}{\pi\varepsilon}\int_x^\infty \Gamma_g^2(z)dz.$$

Let us find the solution for $x = y$. Differentiating both parts of the equations with respect to x one obtains: $\Gamma_g(x) = \frac{g_1}{x}$, $\Gamma_f(x) = 0$. This solution coincides, of course, with the result of renormalization group calculations (17). Differentiating with respect to y we get:

$$\Gamma_f = \frac{3}{4} g_1 \left[\frac{y^{1/4}}{x^{5/4}} - \frac{1}{x^{1/4}y^{3/4}}\right] \quad ,$$

$$\Gamma_g = \frac{3}{4} g_1 \frac{1}{x^{1/4}y^{3/4}} + \frac{1}{4} g_1 \frac{y^{1/4}}{x^{5/4}} \quad . \qquad (22)$$

In the other momentum region where p_1 and p_4 are large the vertices can be determined similarly:

$$\Gamma_f^u = -\Gamma_f^s(x^u,y^u); \; \Gamma_g^u = \Gamma_g^s(x^u,y^u) \text{ (see (22))}.$$

Finally let us determine the functions $\phi(x)$ which

will be also used below. Substituting $\Gamma_g(x) = \frac{g_1}{x}$ in eq. (20) we obtain:

$$\phi_f^s(x) = -\frac{3}{4}\frac{g_1}{x} \; ; \; \phi_g^s(x) = \frac{g_1}{2x} \quad . \tag{23}$$

e) Convergence of Integrals in the "Parquet" Equations for $\varepsilon > 0$.

Let us substitute the solution (22), (23) obtained from Sudakov's equations in the system of "parquet" equations (19) and show that it ensures the convergence of all integrals. Eqs. (19), unlike Sudakov's equations which hold to the first order in ε only, can serve a good model for the 4-dimensional case. Let us show that all the integrals in them are convergent if the vertex decreases as a function of the largest momentum (like (22)).

Substituting solution (22), (23) in (19) we obtain the following convergent integral

$$\phi_f^s(x) = -\frac{3g_1}{2^6}\int_x^\infty \left\{\frac{3}{z^{1/4}x^{3/4}} + \frac{5x^{1/4}}{z^{5/4}}\right\}\frac{dz}{z} = -\frac{3}{4}\frac{g_1}{x} \; . \tag{24}$$

This is identical to $\phi_f^s(x)$ in (23). The other equations of the system (19) may be analyzed in the same way.

A similar discussion for the 4-dimensional space is given in Section 7.

5. SOLUTION OF 4-DIMENSIONAL EQUATIONS BY THE ε-EXPANSION APPROACH

In this Section the approximate solution of the 4-dimensional equation system (9) will be obtained. As was noted in Section 2, we shall search for the solution leading to the slowly convergent integrals with large effective integration momenta region. If this type of solution exists, it can be found by the ε-expansion method, as in $2 + \varepsilon$ dimension space, only the large momentum region contributes to the first order in ε. This solution may be found as follows.

The system of the "parquet" equations for the $d = 2 + \varepsilon$ - dimensional space is given in Appendix 1 (see (A.1)). In this system both the spinor structure and the scalar integrals are transformed to noninteger d, the factors $a(\varepsilon)$, $b(\varepsilon)$ being due to the elimination of the spinor structure. At $d = 2$ (when $a(0) = 1$, $b(0) = 1$) the system (A.1) coincides with 2-dimensional system (19). At $d=4$ ($a(2) = 2$, $b(2) = 1/2$) it coincides with the system (9) of the 4-dimensional equations.

We shall find in this Section a solution of the system (A.1) with $a = 2$, $b = 1/2$, i.e. taking the spinor structure to be the same as in the 4-dimensional case and transforming to noninteger d only the scalar integrals. We shall find such a solution to the first order in $\varepsilon = d - 2$ and use it at $\varepsilon = 2$. We shall show (Sect.7) that, due to the small numerical factors, the region of the large integration momenta is essential even at $\varepsilon = 2$. This is the reason for this solution approximately satisfying the system also in 4 dimensions.

Introduce dimensionless charges f_c and g_c, including in their definition the factors $(4\pi)^{d/2}f_c = F_c\lambda^{\varepsilon}/(4\pi)^{d/2}$, $g_c = G_c\,\lambda^{\varepsilon}/(4\pi)^{d/2}$.

Thus, let us find a solution of system (A.1) with $a=2$, $b = 1/2$*). To solve it, to the first order in ε

*) This system does have symmetries of the 4 dimensional space (e.g. the Fierz identity) in contrast with system (19) and its solution differs from (22), (23).

it is also convenient to use the corresponding Sudakov's equations (see (21)):

$$\Gamma^s_f(x,y)_\infty = \frac{f_1}{\lambda^\varepsilon} - \frac{4}{\varepsilon}\int_y^x [\Gamma^s_f(x,z)\Gamma_f(z)+3\Gamma^s_g(x,z)\Gamma_g(z)]dz +$$
$$+ \frac{1}{\varepsilon}\int_x^\infty [(2N-1)\Gamma^2_f(z) + 6\,\Gamma_f(z)\Gamma_g(z) - g\Gamma^2_g(z)]dz ,$$
$$\Gamma^s_g(x,y) = \frac{g_1}{\lambda^\varepsilon} - \frac{4}{\varepsilon}\int_y^x [\Gamma^s_f(x,z)\Gamma_g(z)+\Gamma^s_g(x,z)\Gamma_f(z) -$$
$$- 2\Gamma^s_g(x,z)\Gamma_g(z)]dz + \frac{2}{\varepsilon}\int_x^\infty [(N+4)\Gamma^2_g(z)-2\Gamma_f(z)\Gamma_g(z)]dz . \qquad (25)$$

Differentiation with respect to x=y results in the following Gell-Mann - Low equation system for the charges

$$f(x) = x\Gamma_f(x) \text{ and } g(x) = x\,\Gamma_g(x) ,$$

$$\frac{d}{d\ell n\xi} f' = \frac{\varepsilon}{2} f - f^2(N-\tfrac{1}{2}) + \frac{9}{2} g^2 - 3fg ,$$
$$\frac{d}{d\ell n\xi} g' = \frac{\varepsilon}{2} g - g^2(N+4) + 2fg . \qquad (26)$$

The solution of this system will be discussed in the next Section.

6. STABILITY AND SYMMETRIES

The phase plane of the solutions of the Gell-Mann - Low equations (26) is shown in Fig. 6. The arrows show the direction of "motion" of the effective charges with $\tilde{p}^2$ increasing. The system has four fixed points, lying on the special solutions $f = a^{(1,2)}g$, $g = 0$, where

$$a^{(1,2)} = (N+1\pm\sqrt{N^2+20N+28})(2N+3)^{-1} .$$

The positions of the fixed points are:

$$g_1^{(1,2)} = \frac{\varepsilon}{2(N+4-2a^{(1,2)})} \; ; \; f_1^{(1,2)} = a^{(1,2)} g_1^{(1,2)} \; ;$$

$$g_1^{(3)} = 0 \; ; \quad f_1^{(3)} = \frac{\varepsilon}{2N-1} \; ; \quad g_1^{(4)} = f_1^{(4)} = 0 \; .$$

The first and the fourth points are stable and unstable nodes respectively. Two others are "saddles". The corresponding solutions will be stable only if one takes the physical values of charges lying on special solutions at $g > 0$ for the second point and $f > 0$ for the third one. Then the effective charges have the constant limit at $\tilde{p}^2 \to \infty$. The presence of the saddle points on the phase plane enables one to fix some features of the Hamiltonian. E.g. if the SU(2) symmetry is broken and

$$H = \frac{F_o}{4} J_o J_o + \frac{G_{oz}}{4} J_z J_z + G_o J_+ J_- \quad ,$$

then in the (G, G_z) plane the saddle arises and the special solution $G = G_z$ will correspond to SU(2) symmetrical form. Similarly, for the same reasons the isoscalar current cannot have a nonuniversal form (e.g. of the type $J_o = \sum_{i=1}^{W} C_i \bar{\Psi}_i \psi_i$ with $C_i \neq 1$).

These questions will be discussed in Appendix 3.

7. SOLUTION IN THE 4-DIMENSIONAL SPACE FOR MOMENTA OF DIFFERENT ORDER OF MAGNITUDE

Let us find $\Gamma^s(x,y)$ from (25). Having $\Gamma_f(x) = \frac{f_1}{x}$, $\Gamma_g(x) = \frac{g_1}{x}$ where f_1 and g_1 are the asymptotic values of effective charges one obtains

$$\Gamma_f^s(x,y) = -\frac{\varepsilon}{16}\left[(1-\delta_o)y^{-1+\delta_o}x^{-\delta_o} + 3(1-\delta_1)y^{-1+\delta_1}x^{-\delta_1}\right] ,$$

$$\Gamma_g^s(x,y) = \frac{\varepsilon}{16}\left[(1-\delta_o)y^{-1+\delta_o}x^{-\delta_o} - (1-\delta_1)y^{-1+\delta_1}x^{-\delta_1}\right] , \tag{27}$$

where $\delta_o = 1 + \frac{4}{\varepsilon}(f_1-3g_1)$, $\delta_1 = 1 + \frac{4}{\varepsilon}(f_1+g_1)$. After some transformation these numbers can be reduced to the form:

$$\delta_o = \frac{N-2}{N+4-2a} \; ; \; \delta_1 = \frac{N+6}{N+4-2a} \quad , \tag{28}$$

where, for example, in the second fixed point $a = \frac{N+1-\sqrt{N^2+20N+28}}{2N+3}$. Note that at $N > 2$ both indices δ_o, δ_1 turn out to be positive.

In Appendix 2 the same vertices are given for large values of the momenta p_1 and p_3, p_1 and p_4.

For $\phi^s(x)$ one gets:

$$\phi^s_f(x) = -\frac{4}{\varepsilon x}(f_1^2+3g_1^2) \; ; \; \phi^s_g(x) = \frac{8}{\varepsilon x}(g_1^2-f_1g_1) \; . \tag{29}$$

Now we substitute the functions (27), (29) into the "parquet" equation system as it was done in Section 4e to show that the integrals are convergent and that the solution may be extended to $\varepsilon = 2$, i.e. to the 4-dimensional space. Consider, for example, the equation for ϕ^s_g. After some transformations it results in

$$\phi^s_g(x) = \frac{\varepsilon}{16x}\int_x^\infty \{(1-\delta_o)^2\delta_o \frac{1}{1-\delta_o \; \delta_o}{}_{x \quad\; z} - (1-\delta_1)^2\delta_1 \frac{1}{1-\delta_1 \; \delta_1}{}_{x \quad\; z}\}\frac{dz}{z} \; . \tag{30}$$

Integrals are convergent since $\delta_o, \delta_1 > 0$. Integrating one gets:

$$\phi^s_g(x) = \frac{\varepsilon}{16x}[(1-\delta_o)^2-(1-\delta_1)^2] = \frac{8g_1(g_1-f_1)}{\varepsilon x} \quad .$$

In other equations the integrals are convergent, too. The equation for ϕ^t_f needs a special discussion since

$$\delta_o^t = \frac{(2N+5)(1-a)}{2(N+4-2a)}$$

is positive in the second fixed point but it is negative in others since in them $a > 1$.

So we see that the finite scale invariant solution corresponds only to the second fixed point and consequently exists for the physical charges obeying the relation $F_c = a\, G_c$.

The smallness of δ_o (e.g. in (28) $\delta_o = 0.12;\ 0.22$ when $N = 3;\ 4$) enables one to extend the solution (27), (28) to the case $\varepsilon = 2$, i.e. to the real 4-dimensional space. For this purpose note that the contribution of the first term in (30) from the integration region of not large $\tilde{p}^2 \sim \lambda^2$ includes the smallness of δ_o (as a fact this region gives contribution ~ 1 but the contribution of region of large $\tilde{p}^2$ is much larger $\sim \frac{1}{\delta_o}$). The contribution of the second term from the same region $\tilde{p}^2 \sim \lambda^2$ formally is not small, but this term is very small compared to the first one since it includes the small factor $(1-\delta_1)^2$ (for all N the index δ_1, turns out to be near unity).

Similar arguments are valid for ϕ_f^s, ϕ_g^t too. Although there is no sufficient "stretching" of the integration region ($\delta_a^u = 0{,}86$) for ϕ_f^t, ϕ_f^u, ϕ_g^u the proximity of δ_o^u to unity makes the contribution of these quantities to zero order approximation negligible. Since in this case $\delta_o^u < 1$ the iterations of these equations seem to be convergent, as well as in the case of ϕ_f^s, ϕ_g^s, ϕ_g^t.

At small ε the "nonparquet" graphs included the extra factor g_1^2 and were small with respect to "parquet" ones. They do not contribute to the zero order approximation when $\varepsilon = 2$ too, since g_1 remains numerically small even at $\varepsilon = 2$: $\frac{g_1}{2} \sim 0.05$, $\frac{f_1}{2} \sim -0.03$. Because of the fast convergence of all the integrations except the last one in the contribution of these graphs the factor $\frac{1}{(k-1)!}$ appears (k is the order of "nonparquet" graph) which, perhaps, compensates for the number of graphs increasing, at least if k is not too large. The structure of types $p_\alpha\ p_\beta$ do not contribute to the zero approximation for similar reasons.

8. EQUATION FOR THE PROPAGATOR

The solution of the "parquet" equations was found taking $\beta(p^2) = 1$. Substitute now β of the form (11). It is easy to note that this leads only to the change $\varepsilon \to \varepsilon + 4\Delta, (x=(\xi)^{\frac{\varepsilon}{2}+2\Delta}; \text{etc.})$ in the relations (27), (29) and in the expressions for f_1 and g_1. To determine Δ one should supplement the "parquet" equation system with an equation for the propagator.

The Dyson-Schwinger equation is inconvenient for calculations since for its solution much more accurate knowledge of the vertex part behavior, than it was found above, is required[19]. We shall use then the "stream unitarity" expansion for $S^{-1}(p)$ suggested by Polyakov[21]. Due to the numerical smallness of the limiting fixed point values of charges f_1 and g_1 we can consider only the first "three streams" term of this expansion which dominates in our case. Than we get

$$I_m S^{-1}(p) = \int \Gamma_{\alpha\beta}(p,k,q-k,p-q) \mathrm{Sp}[O_\alpha I_m S(k) O_\alpha I_m S(k-q)] \times \\ \times O_\beta I_m S(p-q) O_{\beta'}] \Gamma_{\alpha'\beta'}(k,q-k,p-q,p) \frac{d^4k\, d^4q}{\pi^4} \quad . \tag{31}$$

For the scaling form (11) of the propagator the following expression holds: $\Theta(p_0) I_m \beta(p) = \sin\pi\Delta (p^2/\lambda^2)^\Delta \Theta(p)$ where

$\Theta(p) = \Theta(p^2)\Theta(p_0)$ and $p^2 = p_0^2 - \vec{p}^2$. We estimate the right-hand side integral in (31) with the assumptions that the vertices $\Gamma_{\alpha\beta}$ are slowly varying functions in the integration region where they are approximately equal to $\Gamma_{G\alpha\beta} \sim g_{\alpha\beta} \frac{g_1\lambda^{4\Delta}}{(p^2)^{1+2\Delta}}$. Then (31) takes the form

$$I_m\ S^{-1}(p) = [\frac{f_1^2}{2}(N+\frac{1}{2}) + \frac{3g_1^2}{2}(N-\frac{1}{2}) + \frac{3}{2}\ f_1 g_1]\lambda^{2\Delta} \times$$

$$\times \int \frac{S_p[O_\alpha \hat{k} O_\beta(\hat{k}-\hat{q})]O_\alpha(\hat{p}-\hat{q})O_\beta}{(p^2)^{2+2\Delta}[k^2(k-q)^2(p-q)^2]^{1-\Delta}} \Theta(k)\Theta(q-k)\Theta(p-q)\frac{d^4kd^4q}{\pi^4} .$$

Simple but tedious calculations (see Appendix 4) yield

$$I_m\ S^{-1}(p) = \hat{p}\ (\frac{p^2}{\lambda^2})^{-\Delta} \frac{8\pi}{\Gamma(3+3\Delta)\Gamma(5+3\Delta)} [\frac{\Gamma(2+\Delta)}{\Gamma(1-\Delta)}]^3 \times$$

$$\times\ [\frac{f_1^2}{2}(N+\frac{1}{2}) + \frac{3g_1^2}{2}\ (N-\frac{1}{2}) + \frac{3}{2}\ f_1 g_1] \quad . \tag{32}$$

This equation has the approximate solution,

$$\Delta = \frac{1}{6}\ [\frac{f_1^2}{2}(N+\frac{1}{2}) + \frac{3g_1^2}{2}\ (N-\frac{1}{2}) + \frac{3}{2}\ f_1 g_1] \quad .$$

9. CONCLUSION

The four-fermion interaction (2) of N lepton doublets of the types $\binom{\nu_e}{e}$, $\binom{\nu_\mu}{\mu}$ was considered above. Such doublets are used in unified theories[2] of weak and electromagnetic interactions. To include strong interacting particles the quark doublets $\binom{p}{n_\theta}$, $\binom{p'}{\lambda_\theta}$ with

$$n_\theta = n\cos\theta+\lambda\sin\theta \ ; \ \lambda_\theta = -n\sin\theta+\lambda\cos\theta \quad ,$$

should be introduced where the 4-th (charmed) quark was suggested[22], as it is well known, to exclude strange neutral currents. It is clear that until the strong interaction between quarks is taken into account the introduction of them into (2) leads only to the increase of the number N of doublets. For real weak interaction N seems to be, thus, equal to 4 or is larger[23].

Consider the isospin structure of the Hamiltonian (2) and rewrite it in the form

$$H=\frac{F_o}{4}\Big(\sum_{i=1}^{N}(\bar{\nu}_i\nu_i+\bar{e}_ie)\Big)^2+\frac{G_o}{4}\Big(\sum_{i=1}^{N}(\bar{\nu}_i\nu_i-\bar{e}_ie_i)\Big)^2+G_o\Big(\sum_{i=1}^{N}\bar{\nu}_ie_i\Big)\times$$

$$\times\Big(\sum_{i=1}^{N}\bar{e}_i\nu_i\Big).$$

Here the charged current interaction is represented by the last term of the type $G_o(\bar{\nu}_\mu\mu)(\bar{e}\nu_e)$, the neutral by $\frac{1}{2}(F_o+G_o)(\bar{\mu}\mu)(\bar{e}e)$ and by the terms of the type $\frac{1}{2}(F_o-G_o)(\bar{\nu}_\mu\nu_\mu)(\bar{e}e)$. Thus in the neutral current we meet two kinds of terms: those describing down-down or up-up doublets components interaction and having $G_n=\frac{1}{2}(G_c+F_c)$ as a coupling constant, and the up-down components interaction with coupling constant $G_n' = \frac{1}{2}(F_c-G_c)$.

The relation $F_c = a\ G_c$, with $a = \frac{N+1-\sqrt{N^2+20N+28}}{2N+3}$ (where a is negative, e.g. for N = 4 a = - 0.56 and a = - 0.36 for N = 8), between the constants represents, as we have seen above, the condition of the existence of the scaling solution. If the value of the factor "a" found here for large energies, $G_cp^2>>1$, turns out to be approximately valid for $G_cp^2<1$ too, then the inequality $|G_n'|>G_n|$ should be expected. At small "a" the cross sections of charged current processes should be about four times larger than the corresponding one caused by neutral currents.

The scaling solution obtained above is valid only

at $G_c\tilde{p}^2 \gg 1$. It will be very interesting to use it at $G_c p^2 < 1$ instead of the usual Born term as a zero approximation of the iteration method in "parquet" equations.

The unique prediction for the sign of the constant G ($G>0$) was made above. It remains still valid at low energies (where $G = G_c$) as the line $G = 0$ corresponds to the new special solution. The sign of G_c can be found for example from the asymmetry of the scattering of the polarized charged leptons and anti-leptons with the energy larger than 10^2 GeV on the nucleons[24].

APPENDIX 1

Let us write the entire system of the "parquet" equations (9). Using the notation of Section 2 we obtain

$$\phi_f^s=a(\varepsilon)\int\{[\Gamma_f-\phi_f^s]\Gamma_f^s+3[\Gamma_g-\phi_g^s]\Gamma_g^s\}\ \frac{\beta(k^2)\beta((p_s-k)^2)}{(p_s-k)^2}\ \frac{d^dk}{(2\pi)^{d_i}},$$

$$\phi_g^s=a(\varepsilon)\int\{[\Gamma_g-\phi_g^s]\Gamma_f^s+[\Gamma_f-\phi_f^s]\Gamma_g^s-2[\Gamma_g-\phi_g^s]\Gamma_g^s\}\times$$
$$\times\frac{\beta(k^2)\beta((p_s-k)^2)}{(p_s-k)^2}\ \frac{d^dk}{(2\pi)^{d_i}},$$

$$\phi_f^u=-b(\varepsilon)\int\{[\Gamma_f-\phi_f^u]\Gamma_f^u+3[\Gamma_g-\phi_g^u]\Gamma_g^u\}\ \frac{\beta(k^2)\beta((p_t-k)^2)}{(p_u-k)^2}\times \qquad (A.1)$$
$$\times\frac{d^dk}{(2\pi)^{d_i}},$$

$$\phi_g^u=-b(\varepsilon)\int\{[\Gamma_f-\phi_f^u]\Gamma_g^u+[\Gamma_g-\phi_g^u]\Gamma_f^u+2[\Gamma_g-\phi_g^u]\Gamma_g^u\}\times$$
$$\times\frac{\beta(k^2)\beta((p_t-k)^2)}{(p_u-k)^2}\ \frac{d^dk}{(2\pi)^{d_i}},$$

$$\phi_f^t=-\frac{\varepsilon}{4}\int\{2N[\Gamma_f-\phi_f^t]\Gamma_f^t+\frac{\varepsilon}{2}[\Gamma_f-\phi_f^t]\Gamma_f^u+\frac{\varepsilon}{2}[\Gamma_f-\phi_f^u]\Gamma_f^t+$$
$$+\frac{3\varepsilon}{2}[\Gamma_f-\phi_f^t]\Gamma_g^u+\frac{3\varepsilon}{2}[\Gamma_g-\phi_g^u]\Gamma_f^t\}\ \frac{\beta(k^2)\beta((p_t-k)^2)}{(p^t-k)^2}\ \frac{d^dk}{(2\pi)^{d_i}},$$

$$\phi_g^t=-\frac{\varepsilon}{4}\int\{2N[\Gamma_g-\phi_g^t]\Gamma_g^t+\frac{\varepsilon}{2}[\Gamma_g-\phi_g^t]\Gamma_f^u+\frac{\varepsilon}{2}[\Gamma_f-\phi_f^u]\Gamma_g^t-$$
$$-\frac{\varepsilon}{2}[\Gamma_g-\phi_g^t]\Gamma_g^u-\frac{\varepsilon}{2}[\Gamma_g-\phi_g^u]\Gamma_g^t\}\ \frac{\beta(k^2)\beta((p_t-k)^2)}{(p_t-k)^2}\ \frac{d^dk}{(2\pi)^{d_i}},$$

where $a(\varepsilon) = 1 + \frac{\varepsilon(\varepsilon-1)}{2} - \frac{\varepsilon(\varepsilon-2)}{2(\varepsilon+2)}$ and

$b(\varepsilon) = 1 - \frac{\varepsilon(\varepsilon-1)}{2} + \frac{\varepsilon^2}{2(\varepsilon+2)}$. The following notation for the vertex T in the cases of large momenta p_1 and p_3 are used:

$$T_f = -\frac{1}{2}\Gamma_f^t(O_\alpha)_{31}(O_\alpha)_{42} + \frac{1}{2}\Gamma_f^u(O_\alpha)_{32}(O_\alpha)_{41} ,$$

$$T_g = -\frac{1}{2}\Gamma_g^t(\vec{\tau}O_\alpha)_{31}(\vec{\tau}O_\alpha)_{42} + \frac{1}{2}\Gamma_g^u(\vec{\tau}O_\alpha)_{32}(\vec{\tau}O_\alpha)_{41} .$$

For small ε

$$\Gamma_f^t = F + \phi_f^s(\xi^t) + \phi_f^t(\xi^t,\eta^t) + \phi_f^u(\xi^t) ,$$

$$\Gamma_f^u = F + \phi_f^s(\xi^t) + \phi_f^t(\xi^t) + \phi_f^u(\xi^t,\eta^t) ,$$

$$\Gamma_g^t = G + \phi_g^s(\xi^t) + \phi_g^t(\xi^t,\eta^t) + \phi_g^u(\xi^t) ,$$

$$\Gamma_g^u = G + \phi_g^s(\xi^t) + \phi_g^t(\xi^t) + \phi_g^u(\xi^t,\eta^t) ,$$

where t-channel variables, ξ^t, η^t, are defined quite similarly to the s-channel ones ξ, η (18)

$$\xi^t = \max\{\tilde{p}_1^2,\tilde{p}_3^2,\tilde{p}_t^2\} ,$$

$$\eta^t = \tilde{p}_t^2 .$$

APPENDIX 2

We give expression for Γ^t and Γ^u determined similarly to (27). Here $x = (\xi^t)^{\varepsilon/2}$, $y = (\eta^t)^{\varepsilon/2}$.

$$\Gamma_f^u = \frac{\varepsilon}{4}(1-\delta_o^u)y^{-1+\delta_o^u}x^{-\delta_o^u} + \frac{3\varepsilon}{4}(1-\delta_1^u)y^{-1+\delta_1^u}x^{-\delta_1^u} ,$$

$$\Gamma_g^u = \frac{\varepsilon}{4}(1-\delta_o^u)y^{-1+\delta_o^u}x^{-\delta_o^u} - \frac{\varepsilon}{4}(1-\delta_2^u)y^{-1+\delta_1^u}x^{-\delta_1^u} ,$$

$$\Gamma_f^t = -\frac{\varepsilon}{2N}(1-\delta_o^u)y^{-1+\delta_o^u}x^{-\delta_o^u} + \frac{\varepsilon}{2N}(1-\delta_o^t)y^{-1+\delta_o^t}x^{-\delta_o^t} ,$$

$$\Gamma_g^t = -\frac{\varepsilon}{2N}(1-\delta_1^u)y^{-1+\delta_1^u}x^{-\delta_1^u} + \frac{\varepsilon}{2N}(1-\delta_1^t)y^{-1+\delta_1^t}x^{-\delta_1^t} ,$$

where

$$\delta_o^u = \frac{2N+5-5a}{2(N+4-2a)} ; \quad \delta_o^t = \frac{(2N+5)(1-a)}{2(N+4-2a)} ;$$

$$\delta_1^u = \frac{2N+9-5a}{2(N+4-2a)} ; \quad \delta_1^t = \frac{-5a+9}{2(N+4-2a)} .$$

Similar to (29) one obtains

$$\phi_f^u = -\frac{1}{\varepsilon x}(f_1^2+3g_1^2) ; \quad \phi_f^t = \frac{2f_1}{\varepsilon x}[(N+1)f_1+3g_1] ;$$

$$\phi_g^u = -\frac{2}{\varepsilon x}g_1(f_1+g_1) ; \quad \phi_g^t = \frac{2g_1}{\varepsilon x}[(N-1)g_1+f_1] .$$

APPENDIX 3

Let us show how the presence of the saddle fixed points allows one to fix a type of the interaction. Consider the interaction Hamiltonian of the type (2) with a broken SU(2) symmetry:

$$H = \frac{F_o}{4} J_o J_o + \frac{G_{zo}}{4} J_z J_z + G_o J_+ J_- .$$

The system of Gell-Mann - Low equations for effective charges (see Section 5) is as follows:

$$\begin{aligned} f' &= \frac{\varepsilon}{2} f - f^2(N-\tfrac{1}{2}) + \frac{3}{2} g_z^2 - fg_z - 2fg , \\ g_z' &= \frac{\varepsilon}{2} g_z - g_z^2(N+1) - 5g^2 + 2fg_z + 2gg_z , \\ g' &= \frac{\varepsilon}{2} g - Ng^2 - 4gg_z + 2fg . \end{aligned} \qquad (A.3)$$

A straight line $g = g_z = \frac{f}{a}$ represents the special solution having the fixed point

$$g_1 = g_{z1} = \frac{f_1}{a} = \frac{\varepsilon}{2(N+4-2a)}$$

At $N < 8$ it is just the saddle point in both fg (see Section 6) and $g_z g$ planes (the last saddle corresponds to the eigenvalue $\frac{g_1}{8-N}$ of the linearized system (A.3); it is positive and given a saddle at $N < 8$ only). Thus, at $N < 8$ the stable scaling solution exists only for initial (i.e. physical) values of charges lying on the line $g = g_z = \frac{f}{a}$. Physically this means that only the SU(2) symmetric interaction exists.

There are other special solutions in phase space with other points. It will be very interesting to show that a finite physically acceptable solution corresponds only to an SU(2) symmetric point $g_1 = g_{z1}$ (like the case of fg plane where only one fixed point is physically acceptable, see Section 6).

APPENDIX 4

Let us calculate the integral

$$I_{\mu\nu} = \int Sp[O_\mu I_m S(k) O_\nu S(q-k)] \frac{d^4k}{(2\pi)^4} =$$

$$= \sin^2\pi\Delta \int \frac{Sp[O_\mu \hat{k} O_\nu (\hat{q}-\hat{k})]}{[k^2(q-k)^2]^{1-\Delta}} \Theta(k)\Theta(q-k) \frac{d^4k}{(2\pi)^4} .$$

It is equal to the discontinuity on the cut of the quantity

$$I'_{\mu\nu} = \int \frac{Sp[O_\mu \hat{k} O_\nu (\hat{q}-\hat{k})]}{[(-k^2-i0)(-(q-k)^2-i0)]^{1-\Delta}} \frac{d^4k}{(2\pi)^4 i} . \qquad (A.4)$$

To show this we note that the discontinuity of the $I'_{\mu\nu}$ appears only when singularities of the denominator pinch the contour of integration.

The divergent part of (A.4) has no discontinuity and is not interesting. The calculations of the convergent part may be carried out using the usual Feynman parametrization of the denominator

$$\frac{1}{(k^2)^{\alpha}((q-k)^2)^{\beta}} = \frac{\Gamma(\alpha+\beta)}{\Gamma(\alpha)\Gamma(\beta)} \int_0^1 \frac{x^{\beta-1}(1-x)^{\alpha-1}}{(k^2-2qkx+q^2x)^{\alpha+\beta}} .$$

Now it is easy to obtain

$$I_{\mu\nu} = \text{disc}\, I'_{\mu\nu}\, [(\frac{1+\Delta}{1+2\Delta})g_{\mu\nu} - \frac{q_\mu q_\nu}{q^2}]$$

$$\frac{i\pi^2 4\sin(2\pi\Delta)\Gamma^2(2+\Delta)\Gamma(-2\Delta)(q^2)^{1+2\Delta}}{\Gamma(4+2\Delta)\Gamma^2(1-\Delta)(2\pi)^4} .$$

Similar calculations for the second integral

$$I_m S^{-1}(p) = I_m \int \frac{O_\mu(\hat{p}-\hat{q})O_\nu}{[-(p-q)^2-i0]^{1-\Delta}}\, I'_{\mu\nu}(q)\, \frac{d^4q}{(2\pi)^4}$$

lead to the formula (32).

REFERENCES

1. W. Heisenberg, Zs. Phys., 101, 533, 1936.
2. S. Weinberg, Phys. Rev. Lett., 19, 1264 (1967)
   A. Salam. Proceedings of the 8-th Nobel Symposium, Stockholm, 1968, p. 367.
3. L.D. Landau, I.Ya. Pomeranchuk, Dokl. Akad. Nauk SSSR, 102, 489, 1955.
4. A.D. Dolgov, L.B. Okun, V.I. Zakharov, Yad. Fiz., 14, 1044, 1247, 1971; Nucl. Phys., B37, 493, 1972,
   T. Appelquist, J. Bjorken, Phys. Rev., D4, 3726, 1971.
5. I.Ya. Pomeranchuk, Yad. Fiz., 11, 852, 1970,
   A.D. Dolgov, L.B. Okun, V.I. Zakharov, Yad. Fiz., 15, 808, 1972,
   A.D. Dolgov, V.N. Gribov, L.B. Okun, V.I. Zakharov, Nucl. Phys., B59, 611, 1973.
6. K.A. Ter-Martirosyan, Phys. Rev., 111, 948, 1958.
7. I.T. Dyatlov, V.V. Sudakov, K.A. Ter-Martirosyan, ZhETF 32, 767, 1957,
   I.Ya. Pomeranchuk, V.V. Sudakov, K.A. Ter-Martirosyan, Phys. Rev., 103, 784, 1956.
8. S.F. Edwards, Phys. Rev., 90, 824, 1953.
9. L.D. Landau, A.A. Abrikosov, I.M. Khalatnikov, Dokl. Akad. Nauk SSSR, 95, 497, 773, 1177, 1954.
10. A.A. Abrikosov, A.D. Galanin, L.P. Gorkov, L.D. Landau, I.Ya. Pomeranchuk, K.A. Ter-Martirosyan, Phys. Rev., 111, 321, 1958.
11. V.N. Gribov, A.A. Migdal, ZhETF, 55, 1498, 1968.
12. A.A. Migdal, ZhETF, 55, 1964, 1968.
13. A.M. Polyakov, ZhETF, 55, 1026, 1968; 57, 271, 1969.
14. G. t'Hooft, M. Veltman, Nucl. Phys., B44, 189, 1972.
15. K.G. Wilson, Phys. Rev., D7, 2911, 1973.

16. M. Gell-Mann, F. Low, Phys. Rev., 95, 1300, 1954, N.N. Bogolyubov, D.V. Shirkov, Introduction to the Theory of Quantized Fields, Interscience, New York, 1959, Sect. 8.
17. K.G. Wilson, M.F. Fisher, Phys. Rev. Lett., 28, 240, 1972; K.G. Wilson, S. Kogut, Phys. Rep., 12c, No. 2, 1974.
18. A.A. Anselm, ZhETF, 36, 863, 1959.
19. V.G. Vaks, A.I. Larkin, ZhETF 40, 1392, 1961.
20. V.V. Sudakov, Dokl. Acad. Nauk SSSR, 111, 338, 1956.
21. A.M. Polyakov, ZhETF 59, 542, 1970.
22. S.L. Glashow, S. Iliopuolus, L. Maiani, Phys. Rev., D2, 1285, 1970.
23. M.Y. Han, Y. Nambu, Phys. Rev., 139B, 1006, 1965.
24. S.M. Bilenky, N.A. Dadayan, E.Kh. Khristova, Yad. Fiz., 21, 360, 1975.

POSSIBLE STRUCTURE OF WEAK INTERACTION*

A. Zee†

Princeton University

Princeton, New Jersey 08540

There has been a great deal of theoretical activity attempting to endow the weak interaction with additional structure, beyond the "standard" four quarks coupled via left-handed currents. In particular, within this past year the question of right-handed currents received much attention.

The question of right-handed currents was initiated last spring by de Rujula, Georgi, and Glashow[1], who proposed adding the current $\bar{c}_R\gamma_\mu n_R$ to weak interaction. Subsequently, a group at Princeton (Kingsley, Treiman, Wilczek, and Zee[2]) and a group at Caltech (Fritsch, Gell-Mann, and Minkowski[3]) proposed, as an alternative, the current $\bar{c}_R\gamma_\mu\lambda_R$. The questions were many. Firstly, whether right-handed currents occur, and if they do, which of the two forms just mentioned is relevant. There were also attempts to incorporate the right handed currents in a symmetrical fashion, in the so-called vector-like

*Supported by the National Science Foundation under grant number MPS73-4997.

†A. P. Sloan Fellow

theories which involves six (or more) quarks. Since the situation has not changed substantially since last summer, I won't discuss it in detail here. For a summary of the issues involved, I refer you to the excellent review given by B. Lee[4] at the Stanford Conference last summer and to a paper by Kingsley, Wilczek, and myself[5]. I will merely list some of the salient points here.

(a) Considerations of the neutral kaon mass difference and current algebraic theorems[6] for non-leptonic decay appear to rule out the presence of a $\bar{c}_R\gamma_\mu n_R$ current.

(b) $\Delta I=\frac{1}{2}$ rule for non-leptonic decay[1]: The situation is murky. All the operators with canonical dimension < 6 have been listed and studied[2]; none of them appears to account for the $\Delta I=\frac{1}{2}$ rule in a decisive way. Right handed current helps to the extent that the anomalous dimensions associated with operators formed out of products of left and right handed currents are somewhat larger. More quark flavors also help (up to a point). Finally, the origin of the $\Delta I=\frac{1}{2}$ rule may well lie in the low frequency region.

(c) Phenomenology[1,2,3,4,5]

(i) Productions of dileptons $\mu^+\mu^-$, $\mu^-\mu^-$ and $e^+\mu^-$ in neutrino reactions.

(ii) x and y distribution in neutrino deep inelastic scattering,

(iii) The structure of the neutral current.

(d) Some of the motivations for introducing right-handed currents are based on aesthetics and theoretical speculation. I refer you to Professor Gell-Mann's talk for an eloquent summary of some of these points

We will also mention some of these in (II) below.

I will limit myself in this talk to three topics:

(I) Radiative weak decay of hyperons

(II) The orientation of weak interaction relative to strong interaction

(III) Radiative decays of heavy leptons

I. RADIATIVE WEAK DECAY OF HYPERONS

These rare decay modes of hyperons have been recently re-examined by a graduate student of mine, N. Vasanti.[7] It is interesting to study these decays because there appears to be some discrepancy between the (rather uncertain) experimental data and theory which we will now explain.

Let us first recall the available experimental information. Until recently the only measured asymmetry parameter is for $\Sigma^+ \to p\gamma$ given by Tripp, et al.,[8] in 1969 to have the value

$$\alpha(\Sigma^+ \to p\gamma) = -1.03^{+0.52}_{-0.42} \quad .$$

The branching ratio for the decay,

$$\frac{\Gamma(\Sigma^+ \to p\gamma)}{\Gamma(\Sigma^+ \to p\pi^0)} = (2.76 \pm .51) \times 10^{-3}$$

is consistent with simple order-of-magnitude estimates. The asymmetry parameter α is defined by

$$\frac{d\Gamma}{d\Omega} \propto (1 + \alpha \hat{p} . \vec{s}),$$

where $\hat{p}$ denotes the unit vector in the direction of the momentum of the outgoing baryon (in the rest frame of the decaying baryon) and $\vec{s}$ is the unit spin vector of the decaying baryon. The sign convention is most easily remembered by noting that if the daughter baryon is right handed she will tend to emerge in a direction opposite to the spin of the parent, a situation corresponding to $\alpha=-1$.

What can we say about these decays theoretically? We will, to begin with, discuss the situation in the "standard" theory with left-handed currents and then consider the effect of introducing right-handed currents.

Firstly, there is an old result[9] that implies that $\alpha(\Sigma^+\to p\gamma)$ vanishes, contradicting the experimental result of $\alpha\sim-1$. The result can be deduced by remarking that the effective weak Hamiltonian

$$H_{LL}\sim G\bar{n}\gamma_\mu(1-\gamma_5)p\bar{p}\gamma^\mu(1-\gamma_5)\lambda+\text{h.c.}$$

transforms like the 1st component of a U-spin vector (i.e. like λ_6) and by applying P and T. Note that the introduction of charm does not invalidate this result. It is also important to note that the prediction of a vanishing α does not apply to other hyperon decays such as $\Xi^0\to\Lambda+\gamma$ because Ξ^0 and Λ do not form a U-spin doublet.

One may argue that SU(3) breaking effects are unexpectedly large. However, simple pole model calculation typically obtain[10] $\alpha\sim0$. There is a somewhat more sophisticated calculation by Farrar[11] who included more intermediate states. She obtained an $\alpha\sim.8^{+.2}_{-1.1}$ and a rate $\sim1/3$ the measured rate.

What happens when right-handed currents are introduced? Consider the right-handed current $\bar{c}_R\gamma_\mu n_R$. The effective weak Hamiltonian is then

$$H_{LR} \sim G\bar{n}\gamma_\mu(1+\gamma_5)c\bar{c}\gamma^\mu(1-\gamma_5)\lambda + h.c.$$

Now the parity-conserving part of H_{LR} transforms like $\bar{n}\lambda+\bar{\lambda}n \sim \lambda_6$ while the parity-violating part of H_{LR} transforms like $\bar{n}\lambda-\bar{\lambda}n \sim \lambda_7$. Hence, with the introduction of the current $\bar{c}_R\gamma_\mu n_R$, $\alpha(\Sigma^+\to p\gamma)$ no longer necessarily vanishes (and $K_S\to 2\pi$ is no longer forbidden[12] in the limit of exact SU(3) symmetry). The current $\bar{c}_R\gamma_\mu\lambda_R$ leads to exactly the same transformation properties and hence the same conclusion.

So much for group-theoretic considerations. The pole model calculations are, however, not at all affected by the introduction of the current $\bar{c}_R\gamma_\mu\lambda_R$. The reason is that these calculations involve the quantity $\langle p|H_w|\Sigma^+\rangle$ which is determined via current algebra from the decay $\Sigma\to N\pi$. The derivation of this current algebraic relation makes use only of the chiral nature of p and n quarks in weak interaction and so is unaffected by the current $\bar{c}_R\gamma_\mu\lambda_R$. (The current $\bar{c}_R\gamma_\mu n_R$ invalidates the current algebra relation and so the above remarks do not apply to it.)

How about the high frequency region[13]? (For the sake of definitness I shall refer to the pole model calculation as giving the contribution of the low frequency region. By high frequency contribution I mean the estimates of the contribution of the short distance region using asymptotic freedom to the effective $\bar{n}\lambda\gamma$ interaction. The division into these two regions is useful but not well-defined. There is some double-counting involved but a complete disentangling of these effects would require a dynamical understanding which we do not have.) Let us first consider the contribution of left-handed currents alone. This contribution is negligible because of two reasons: (a) it is proportional to m_n and m_λ and (b) it is further suppressed by the Glashow-Ilioupoulous-Maiani[14] mechanism. With the introduction of right-handed currents however, the high frequency region may no longer be negligible. Firstly, it is now proportional to the charmed quark mass m_c and secondly, no GIM mechanism is operative here. The current $\bar{c}_R\gamma_\mu\lambda_R$ leads to the effective interaction

$$H_{eff}\sim\bar{n}(1+\gamma_5)\sigma_{\mu\nu}\lambda F^{\mu\nu} \quad .$$

However, we certainly do not know how to estimate the matrix element of this operator between $|\Sigma^+>$ and $|p>$ with any reliability.Rough estimates, however, indicate that it leads to an amplitude of the same order of magnitude as that given by the simple pole model so it may be quite important. Unfortunately, it tends to push α towards +1, thus worsening the discrepancy between theory and experiment. (The $\bar{c}_R\gamma_\mu n_R$ current leads to α=-1.

In summary the situation is as follows:

(a) Left-left interaction and SU(3) imply $\alpha(\Sigma^+p)=0$.

(b) Left-right interaction invalidates the conclusion in (a).

(c) Pole model calculations give $\alpha\sim0$. This result is unaffected by the introduction of a $(\bar{c}\lambda)_R$ current.

(d) "High frequency" contributions:
-left-left is unimportant
-left-right may be important but gives α=+1 for the $(\bar{c}\lambda)_R$ current.

(e) Experiment, on the other hand, indicates that $\alpha=-1.03^{+0.52}_{-0.42}$.

We should caution the reader that there may not be a real crisis in the brewing. After all, the experimental value for α is only two standard deviations away from zero, and a value of say $\alpha \sim -0.3$ or so might well be consistent with the (theoretically rather crude) pole model calculations. But in any case it would be desirable to perform the relevant experiment again and to try to reduce the errors. It is my understanding that this will in fact be done in the near future.

Recently, there has been another important measurement of hyperon radiative decay. Overseth, et al.[15] have analyzed the process $\Xi^0 \to \Lambda\gamma$ obtaining

$$\frac{\Gamma(\Xi^0 \to \Lambda\gamma)}{\Gamma(\Xi^0 \to \Lambda\pi^0)} = (1.4 \pm 0.4) \times 10^{-3}$$

and

$$\alpha(\Xi^0 \to \Lambda\gamma) = 0.22 \pm 0.34 \quad .$$

The measured value of $\alpha(\Xi^0 \to \Lambda\gamma)$ is quite important because some theorists were tempted to argue that for some unknown dynamical reasons the matrix elements of the operator $\bar{n}\sigma_{\mu\nu}(1 \pm \lambda_5)\lambda$ between baryon states are abnormally large and dominate the low frequency contribution. This would give a value for $\alpha(\Xi^0 \to \Lambda\gamma)$ of $\alpha = \pm 1$ depending on the theory, in contradiction with experiment.

II. ORIENTATION PROBLEM

I would like to discuss next the problem of understanding the orientation of weak interaction relative to the strong interaction. With the introduction of more than four quarks a number of angles have to be introduced as well. How could we understand the systematics of these angles?

The orientation problem appears to be closely linked with quark mass spectrum. To begin with,the Cabibbo angle is obviously undefined if the n and λ quarks are degenerate. Whatever lifts this degeneracy also determines the Cabibbo angle. Note that if we assume, as we will, that strong interaction is described by quantum chromodynamics,then its orientation is solely defined by its mass matrix.

Chiral symmetry informs us that the p and n quarks are almost massless while the λ quark is somewhat more massive. Recent experiments appear to indicate the existence of a considerably more massive quark, the charmed quark c, and possibly even more massive ones. It thus appears that there are at least two mass scales. On the other hand one would like to believe that Higgs particles are composite and that dynamical symmetry breaking is responsible for quark masses. In this case one expects to have only one mass scale (or a hierarchy of mass scales if symmetry is dynamically broken in a hierarchial manner).

In any case, we will assume as our working premise that the light quarks p, n, and λ are massless when weak interaction is switched off. When the weak interaction is switched on the light quarks manage to acquire masses by undergoing weak radiative processes involving heavy quarks, thus lifting the mass degeneracy. By identifying appropriate linear combinations of the n and λ fields as the physical n and λ states one in effect determines the Cabibbo angle. In this view, the orientation between the strong and weak interactions is controlled by the structure of weak interaction. The implementation of this program leads us quite naturally to a vector-like weak interaction theory which involves (at least) six quarks and right handed currents. The general ideas outlined here have been explored by myself in two publications. The reader is referred to Ref. 16 for details.

It is worth remarking on one feature of our point of view. Suppose p_L and n_L form a doublet under the weak gauge group SU(2) × U(1). Then when weak interaction is switched on the p and n quarks will remain massless while the λ quark acquires a mass. Thus the degeneracy between n and λ will be lifted but in a way such that the Cabibbo angle remains zero and such that chiral SU(2) × SU(2) is an exact symmetry. That an intimate connection exists between zero Cabibbo angle and exact chiral SU(2) × SU(2) has been suspected for a long time.

When one carries out the program outlined here one obtains an equation with the schematic form

$$\underset{\sim}{m}_{\ell} = \xi \, \underset{\sim}{g}^{L}_{\ell h} \, \underset{\sim}{m}_{h} \, \underset{\sim}{g}^{R}_{h\ell} \quad .$$

Here $\underset{\sim}{m}_{\ell}$ and $\underset{\sim}{m}_{h}$ denote the mass matrices of the light and heavy quarks respectively; g^L and g^R denote the couplings between light and heavy quarks. The unknown overall constant ξ expresses our ignorance of weak interaction at high frequency. Thus (in a six quark model, to be specific) the mass spectrum and angles of the light quarks p, n, and λ are related to the mass spectrum and angles of the heavy quarks c, t, and b.[17]

Since we do not know much about the other parameters in the theory we certainly can not calculate the Cabibbo angle θ. Nevertheless the scheme does have predictive power to the extent that we know something about m_p/m_n, m_n/m_λ, and θ. The reader is referred to Ref. 16

for a detailed analysis. Here we will merely state some of the results of such an analysis. With a "reasonable" choice of parameters as discussed in Ref. 16 the masses of the t and b quarks and all the angles in the theory are fixed. In particular, a $(\bar{b}b)$ resonance with mass around 5 or 6 GeV is expected.

The actual implementation of the program outlined here is hampered by our present ignorance of dynamical symmetry breaking. Nevertheless, we believe that the basic point of view as expressed in Ref. 16 is correct and should survive in one form or another.

III. RADIATIVE DECAY OF LEPTONS

Recent experiments[18] have hinted at the possible existence of a new charmed heavy lepton U with a mass of approximately 1.8 GeV. It will presumably be a long time before a detailed theory of leptons is worked out. In the meantime, it is obviously important to ask whether U is related to the electron and/or the muon. I would like to talk a little about possible radiative decay modes $U \to e + \gamma$ and/or $U \to \mu + \gamma$ which could shed light on this question. The relevant experiments can in fact be done in the immediate future. The detection or non-detection of these processes is comparable in historical importance to the non-detection of $\mu \to e\gamma$.

Frank Wilczek and I[19] had classified and catalogued a number of possible models of leptons in order to have available a framework in which to discuss these radiative decays. In certain models these decays are forbidden, while in others they may even dominate. For details of these remarks and for discussion of experimental signatures the reader is referred to Ref. 19.

REFERENCES

1. A. de Rujula, H. Georgi, and S. L. Glashow, Phys. Rev. Letters 35, 69 (1975) and Harvard preprint (1975) to be published.
2. F. Wilczek, A. Zee, R. L. Kingsley, and S. Treiman, "Weak Interaction of Heavy Quarks": Phys. Rev. D, to be published.
3. H. Fritsch, M. Gell-Mann, and P. Minkowski, to be published.
4. B. Lee, Fermilab - Conf. 75/72 - THY.
5. R. L. Kingsley, F. Wilczek, and A. Zee, Princeton preprint (1975).
6. This has been remarked by a number of authors, including those of Ref. (2) and M. A. B. Beg and A. Zee, Phys. Rev. D8, 1480 (1973). A detailed analysis was given by E. Golowich and B. Holstein, Phys. Rev. Letters 35, 831 (1975).
7. N. Vasanti, Phys. Rev. D, to be published. See also M. Ahmed and G. Ross, CERN preprint (1975).
8. L. K. Gershwin, et al., Phys. Rev. 188, 2077 (1969).
9. Y. Hara, Phys. Rev. Letters 12, 378 (1964).
10. For example, R. H. Graham and S. Pakvasa (Phys. Rev. 140B, 1144 (1965) obtained roughly the correct rate and $\alpha \sim .06$, while B. Holstein (Nuovo Cimento 2A, 561 (1971)) obtained a rate a factor ~ 1/6 too small and $\alpha \sim 0$.
11. G. Farrar, Phys. Rev. D4, 212 (1971).
12. N. Cabibbo, Phys. Rev. Letters 12, 62 (1964). M. Gell-Mann, Phys. Rev. Letters 12, 155 (1964).
13. For a detailed discussion, see N. Vasanti, Ref. 7.
14. S. Glashow, J. Ilioupoulous, and L. Maiani, Phys. Rev. D2, 1285 (1970).
15. Overseth et al., University of Michigan preprint

(Nov. 1975).

16. A. Zee, Princeton preprint (September 1975) Phys. Rev. D, to be published. For an earlier reference, see A. Zee, Phys. Rev. D9, 1772 (1974).

17. In Ref. 16 the t quark is called r and the b quark is called s. We switch notation here in order to alleviate somewhat the general confusion about quark names.

18. M. Perl, et al., SLAC Pub. 1626., LBL-4228 (1975).;

19. F. Wilczek and A. Zee, Princeton preprint (to be published).

ARE SLAC (μe)-EVENTS DECAY PRODUCTS OF INTEGER CHARGE QUARKS?

Jogesh C. Pati* (Presented by Jogesh C. Pati)
University of Maryland
College Park, Maryland 20742

Abdus Salam
International Centre for Theoretical Physics
Trieste, Italy
and
Imperial College, London
and
S. Sakakibara
City College of CUNY
New York, New York 10031

ABSTRACT

It is remarked that the (μe)-events as well as the jet-structure observed at SPEAR may have their origin in the pair-production and decays of integer-charge quarks. Several experimental distinctions between quark versus

*Supported in part by NSF Grant No. GP 43662X.

heavy lepton-hypotheses for the (μe)-events are noted.

Some years back, two of us suggested[1] that conservation of baryon and lepton numbers may not be absolute within the hypothesis of a unified gauge theory for all interactions with integer-charge quarks and leptons as members of one fermion multiplet. We estimated[1,2] that even though quarks could decay into leptons relatively rapidly ($\tau_q \approx 10^{-11}$ to 10^{-12} secs even for light quarks, $m_q \approx$ 2-3 GeV), the proton--a three quark composite--would remain comfortably stable within its present lifetime estimates. The hypothesis of quark-decay into leptons provides a simple resolution of the missing quark-mystery without having to assume quark-confinement.

We have no compelling theoretical reason to believe that quarks are as light as 1.7 ~ 2 GeV; however, if we accept this, and if we make one further assumption that colored vector gluons lighter than quarks exist,[3] then it appears that the quark-lepton decays could provide an interesting explanation of the (μe) events[4] as well as of the jet structure[5] recently observed at SLAC. In this note we sketch the chain of possible decays and the characteristic signatures which would distinguish between the alternatives of quark versus heavy-lepton parents for these events.

To illustrate our remarks, we follow here closely the restrictions of the basic model[1], based on the minimal local symmetry $G=SU(2)_L \times SU(2)_R \times SU(4)'_{L+R}$, and of the assumption that quark-charges are integral. We recall that the model requires a set of 16 fundamental four-component fermions consisting of twelve quarks and four leptons (ν_e, e^-, μ^-, ν_μ); the quarks possess four flavors[6] (p, n, λ, c = charm) and three colors (red, yellow, blue) = (r,y,b). The effective strong interactions in the model are generated by the octet of spin-1 color-gluons ($V_\rho^\pm$, $V_{K*}^\pm$, V_{K*}^0, $\bar{V}_{K*}^0$, $\tilde{U}$ and $\tilde{V}$) which are relatively light ($\lesssim$ Few GeV, see later), but not massless. We expect quarks and charged color-gluons to be produced in pairs by e^-e^+-annihilation,

$$e^-+e^+ \rightarrow (q+\bar{q});\ (V_\rho^+ + V_\rho^-);\ (V_{K*}^+ + V_{K*}^-) \quad . \qquad (1)$$

It has been shown[7] that within the gauge-theory-framework, the color-octet part of quark-charges do not asymptotically contribute to R. Taking this into account, the contributions to the R-parameter from quarks and color-gluons (treated as <u>partons</u>) are given by:

$$R(p_r^0\bar{p}_r^0) = R(p_y^+p_y^-) = R(p_b^+p_b^-) = 4/9,$$

$$R(n^-n_r^+) = R(n_y^0\bar{n}_y^0) = R(n_b^0\bar{n}_b^0) = 1/9,$$

$$R(\lambda_r^-\lambda_r^+) = R(\lambda_y^0\bar{\lambda}_y^0) = R(\lambda_b^0\bar{\lambda}_b^0) = 1/9,$$

$$R(c_r^0\bar{c}_r^0) = R(c_y^+c_y^-) = R(c_b^+c_b^-) = 4/9,$$

$$R(V_\rho^+V_\rho^-) = R(V_{K*}^+V_{K*}^-) = 1/16. \qquad (2)$$

Note the intriguing feature that asymptotically the neutral pair ($p_r^0\bar{p}_r^0$) contributes the same amount to R as the charged pair ($p_{y,b}^+p_{y,b}^-$). With nonconfinement, we expect in general some of the parton-pairs to "recombine" to form known hadrons. However, a certain fraction of these partons will survive in the final state - as quarks and gluons. The contributions to R listed above (presumably) represent the inclusive hadronic cross section, which is a sum over all channels. The signatures of quarks and charged gluons depend upon their allowed decay modes which we list below:

Charged Color Gluon-Decays: Assuming that the gluons are the lightest color-octet states, the charged members ($V_\rho^\pm$ and $V_{K*}^\pm$) decay solely due to their mixing with the weak $W_L^\pm$ gauge mesons. Their effective decay constant is $\approx (G_F m_V^2/f)$, where m_V denotes the mass of the color gluons and f the effective strong gauge-coupling constant ($f^2/4\pi \approx 1$ to 10). This leads to a life-time $\tau(V_\rho^\pm) \sim \tau(V_{K*}^\pm) \approx 10^{-13}$ to 10^{-15} sec. for $m_V \approx 1$ to 3 GeV. Some of the allowed and forbidden[8] decay modes of the charged gluons are

$$\begin{aligned} (V_\rho^\pm, V_{K*}^\pm) &\to e\nu,\ \mu\nu\ , \\ &\to \pi\pi,\ \pi\pi\pi,\ K\bar{K},\ 5\pi\ , \\ &\to \pi\pi e\nu,\ K\bar{K}e\nu,\ \eta\eta e\nu, \\ &\not\to \pi e\nu,\ Ke\nu\ . \end{aligned} \tag{3}$$

We estimate (i) $\Gamma(V_\rho^+\to e\nu) \approx \Gamma(V_\rho^+\to\mu\nu)$; (ii) $\Gamma(V_\rho^+\to\pi\pi e\nu) \approx$ (1/40 to 1/10) $\Gamma(V_\rho^+\to e\nu)$ and (iii) $\Sigma\Gamma(V_\rho^+\to \text{Hadrons}) \approx$ (1 to 3)$\Gamma(V_\rho^+\to e\nu)$; the higher value for the inclusive nonleptonic rate being applicable[9] only for higher values of the gluon masses ($m_V \gtrsim 3$ GeV). Thus adding various modes, we expect each of the leptonic modes ($V_\rho^+\to e\nu$) and ($V_\rho^+\to\mu\nu$) to have a branching ratio $\approx$ (20 to 30) % if gluons are relatively light ($m_V \approx 1$ to 1.5 GeV) and $\approx$ (15 to 20)% if gluons are relatively heavy ($m_V \gtrsim 3$GeV).

Quark-Decays: First a few general remarks: (i)

In the basic model there are three gauge bosons (X^o,X^+ and X'^+), which (respectively) couple (red, yellow and blue) quarks of a given flavor to the lepton of the same flavor. Baryon and lepton-number-nonconserving quark-decays occur only because of the spontaneously induced mixing of the X-bosons with the weak gauge bosons W_L's. It is a property of the gauge structure of the basic model[1] that only $X^\pm$ and $X'^\pm$ can mix with $W_L^\pm$, but X^o can not mix with $(W_3)_{L,R}$. Thus the mechanism for yellow and blue-quark decays within the basic model is characteristically different from that for the red quarks. (ii) The dominant decay mechanism for yellow and blue quarks (involving baryon number violation) is given by the convergent loop diagram (Fig. 1), which is of order (Δ^2/m_x^2), Δ^2 being the W-X mixing $(\text{mass})^2$. Note that subject to electric charge and fermion number conservations, Fig. 1 may induce semi-leptonic decays ($q_{y,b}\to\nu$+mesons) involving emission of only neutral leptons (neutrinos); it cannot lead to charged lepton emission. Charged leptons may be emitted either via loop diagrams as in Fig. 2, where a $q\bar{q}$-composite (e.g. pion) is emitted from inside the loop, or alternatively via tree-diagrams as in Fig. 3. It is possible to see[10] that both Fig. 2 and Fig. 3 are suppressed by two large masses (m_X^2 as well as $m_{W_L}^2$); correspondingly the rates of processes induced via Figs. 2 and 3 are suppressed by a factor $\sim(m/m_{W_L})^4$ compared to those induced via Fig. 1, where $m\approx m_q$ or m_V. With these considerations, the dominant decay modes of the yellow and blue quarks are:

$$
\begin{aligned}
p^+_{yel,blue} &\to (\nu_e + \text{pions}); (\nu_\mu + K^o + \text{pions}),\\
n^o_{yel,blue} &\to \nu_e + \text{pions}; (\nu_\mu + K^o + \text{pions}),\\
\lambda_{yel,blue} &\to \nu_\mu + \eta; (\nu_e + K^o + \text{pions}),\\
c^+_{yel,blue} &\to p^+_{y,b} + \pi^o; \lambda^o_{y,b} + \pi^+,\\
&\to \nu_e + D^+; \nu_\mu + F^+ \ (?)
\end{aligned}
\tag{4}
$$

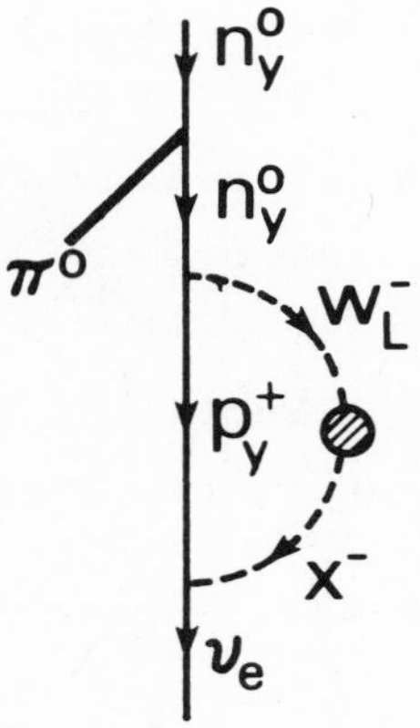

Fig. 1

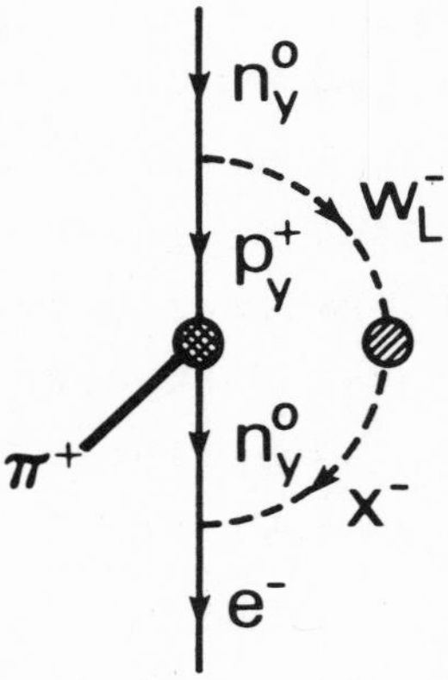

Fig. 2

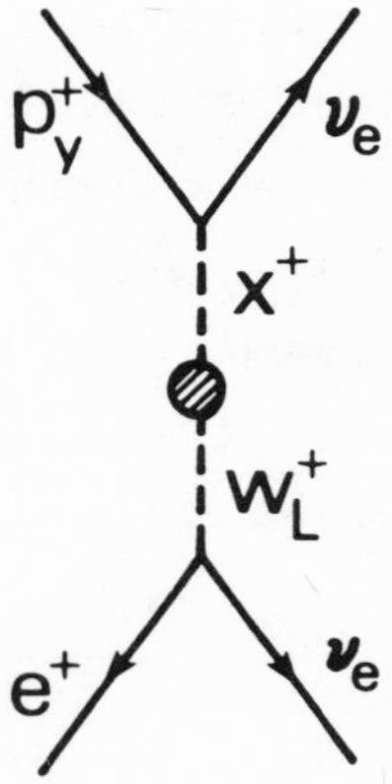

Fig. 3

Mechanisms for Yellow and Blue Quark-Decays

We obtain τ(yellow and blue quarks) ~ 10^{-11} to 10^{-12} secs. for $m_q \simeq 2$ GeV. The quark-number conserving ordinary weak-decays (e.g. $c^+_{y,b} \rightarrow \lambda_{y,b} + \pi^+$) are expected to be important only for the charmed quarks (due to mass considerations). The alternative decay modes for these quarks (i.e. $c^+_{y,b} \rightarrow (\nu_e + D^+)$ or $(\nu_\mu + F^+)$) may in fact be suppressed (or absent) if the charmed mesons (D^+, F^+) are heavy compared to the charmed quarks (hence the question mark on these modes).

Consider now decays of red quarks. There are two cases:

Case I: $m_q > m_V$. In this case, at least the charged red quarks (n^-_r and λ^-_r) decay predominantly into charged color-gluons plus neutrinos via the chain ($q^-_{red} \rightarrow V^-_\rho + (q^0_{yel})$ virtual $\rightarrow V^-_\rho + \nu$) with rates comparable to those of the yellow and blue quark-decays; the dominant decay modes of the red quarks thus are:

$$\begin{aligned} p^0_{red} &\rightarrow (n^0_y + \gamma);\ (V^-_\rho + \pi^+ + \nu_e) \quad, \\ n^-_{red} &\rightarrow (V^-_\rho + \nu_e);\ (V^-_\rho + \pi^+ + \pi^- + \nu_e), \\ \lambda^-_{red} &\rightarrow (V^-_\rho + \nu_\mu);\ (V^-_\rho + \text{Mesons} + \nu_\mu), \\ c^0_{red} &\rightarrow (\lambda^0_y + \gamma);\ p^0_r + \pi^0, \lambda^-_r + \pi^+ \quad. \end{aligned} \tag{5}$$

We obtain τ(red-quarks) $\approx 10^{-11}$ to 10^{-12} secs just like τ(yellow and blue quarks) with $m_q \approx 2$ GeV. The radiative decay modes of the neutral red quarks p^0_r and c^0_r (arising through V^+_ρ-W^+_L-mixing) are found to be the dominant decay modes if Q-values for these decays are $\gtrsim$ 10 MeV.

Case II: $m_q < m_V$: In this case, although the neutral red quarks p_r^o and c_r^o still may decay radiatively as in Case I with lifetimes $\lesssim 10^{-11}$ to 10^{-12} secs., the charged red quarks may decay (e.g. $n_r^- \to e^- + \pi^o$, $\lambda_r^- \to \mu^- + \eta$) only by utilizing <u>both</u> $V_\rho^+ - W_L^+$ and $W_L^+ - X^+$-mixings. As a consequence of the double-mixing needed, these quarks (for the present case of $m_q < m_V$) are found to be rather long-lived ($\tau(n_r^-) \approx \tau(\lambda_r^-) \approx 10^{-12}$ sec. $(m_{W_L}/m_V)^4$ with $m_q \approx 2$ GeV).

Turning now to the origin of the SPEAR (μe)-events, pair production of the yellow and blue quarks (even though present) can not be responsible for these events, since these quarks decay predominantly semileptonically into ($\nu_{e,\mu}$ + mesons). On the other hand, if quarks are heavier than the color-gluons (case I), pair-production of charged red quarks ($n_r^- n_r^+$) and ($\lambda_r^- \lambda_r^+$), followed by their sequential decays as shown below can give rise to anomalous (μe)-pairs as seen at SPEAR:

$$\begin{array}{lcl} e^- + e^+ \to n^-_{red} & + & n^+_{red} \\ \quad\downarrow & & \quad\downarrow \\ V_\rho^- + \nu_e & & V_\rho^+ + \bar{\nu}_e \\ \hookrightarrow e^- + \bar{\nu}_e & & \hookrightarrow \mu^+ + \nu_\mu \end{array} \qquad (6)$$

The (μe)-pairs thus arising would appear within the present statistics like three-body leptonic decay of the parent quarks.

Now define $\rho(s) \equiv |f_{qq\gamma}(s)|^2$, where $f_{qq\gamma}(s)$ is the on mass-shell-quark electromagnetic form factor. It follows that $\sigma(e^-e^+ \to q_i\bar{q}_i)/\sigma(e^-e^+ \to \mu^-\mu^+) = R(q_i\bar{q}_i)\rho(s)$, where $R(q_i\bar{q}_i)$ for any given quark-pair is given by Eq. (2). (Note that strictly within the parton model hypothesis we may interpret $\rho(s)$ as the fraction of all

$q_i\bar{q}_i$-parton-pairs created, which "survive" as real particles in the final state. We refer to $\rho(s)$ as the "quark-survival factor"). Noting[11] that asymptotically $R(n_r^- n_r^+) + R(\lambda_r^- \lambda_r^+) = 2/9$ (see Eq. (2)) and that the branching ratio of the ($\mu\nu$) as well as ($e\nu$)-decay modes of $V_\rho^\pm$ is ≈ (20 to 30)% (as discussed before), the net contribution to R of the ($\mu^\pm e^\mp$) signals arising from real quark-production and decays is given by $R_{q\bar{q}}(\mu^+ e^-) = R_{q\bar{q}}(\mu^- e^+) = (2/9)(1/5 \text{ to } 1/3)^2 \rho(s) = \rho(s)(0.9 \text{ to } 2.5)\%$.

Direct production of charged color-gluon pairs followed by their two-body leptonic decays[12] will also contribute to the signature (μe)-events. Noting that $R(V_\rho^+ V_\rho^-) + R(V_{K*}^+ V_{K*}^-) = 1/8$, the net contribution to $R(\mu^\pm e^\mp)$ from color-gluon pair-production is given by $R_{V\bar{V}}(\mu^+ e^-) = R_{V\bar{V}}(\mu^- e^+) = (1/8)(1/5 \text{ to } 1/3)^2 \rho'(s) = \rho'(s)\ (0.5 \text{ to } 1.4)\%$, where $\rho'(s)$ is the "color-gluon-survival factor."

The observed[4] SPEAR (μe)-signal corresponds to a true-signal of $R(\mu^+ e^-) = R(\mu^- e^+) \approx (1 \text{ to } 3)\%$. Comparing with the estimate given above, this can be attributed to quark-decays only provided the square of the quark-electromagnetic form-factor is of order unity ($\rho(s) \times \equiv |f_{qq\gamma}(s)|^2 \sim 1/2$ to 1). Not knowing theoretically the precise nature of the quark-electromagnetic factor, we will proceed in this note with the assumption[13] **that** it is hard (i.e. $f_{qq\gamma}(s) \sim (1/\sqrt{2})$ to 1 and slowly varying at SPEAR energies), unlike the form factors of composite systems such as the nucleon or the pion. There then exist a host of strong experimental predictions of our hypothesis that the SPEAR dilepton events arise from quark-decays <u>and</u> of the assumption that quark-form factors are hard as mentioned above. These we list below:

(i) Jet-like distribution of Hadrons: First, within the quark-hypothesis for the μe-events, a large fraction $\rho(s)$ ($\approx \frac{1}{2}$ to 1) of the total hadronic annihilation events must contain a real-quark and an antiquark emerging with equal and opposite momenta. (The different kinds of quark-pairs would, of course, be produced in proportion to their contribution to the R-parameter (See Eq. (2))). Since quarks decay relatively rapidly predominantly into (neutrino + known mesons) (see Eqs. (4) and (5)); these $q\bar{q}$-pairs would give rise to the final state hadrons (mesons) emerging in the form of two jets opposite to each other with a distribution characteristic of spin-$\frac{1}{2}$-parentage. Such jet structure is indeed observed[5,14] experimentally at SPEAR. Since quarks carry charges, 0, ± 1; within our hypothesis we expect[15] some of the jets to carry a net charge ± 1 and some to be neutral. The ratio of charged to neutral jets should reach a value (18/9)/(12/9) = 3/2 (See Eq. (2)) above the charm-quark-threshold but below threshold for new flavors (if there exist any).

(ii) Energy Crisis: Furthermore, these events must be associated with missing neutral energy and momentum carried away by the neutrinos, which may explain the so-called "energy crisis" and the depletion of charged energy[16] observed in $e^- e^+$-annihilation.

(iii) Non-appearance of Charmed Particles: If the charmed particles D and F are relatively heavy compared to the charmed quarks; the charmed quarks, rather than decaying into (ν+D) or (ν+F), would decay preferentially or entirely into (uncharmed quarks + pions) (see (4) and (5)). In that case, production of real charmed quark-pairs above threshold will lead to an increase in R = 4/3 $\rho(s)$; but this increase will be reflected in the production of pions and kaons rather than charmed mesons consistent with the nonappearance[17] of charmed particles in the continuing SPEAR search.

(iv) V + A-Coupling: Given that the W_L gauge mesons couple to V-A-currents, it is easy to see that the amplitude for the transition $n_r^- \to V_\rho^- + (n_y^0)$ virtual $\to V_\rho^- + \nu_e$ must be of the V+A-form $\propto \nu_e \gamma_\mu (1-\gamma_5) n_r^- V_{\rho\mu}^+$ so far as the basic model is concerned.[18]

(v) Semileptonic Signals: In addition to pure leptonic signals (as in (6)), there must exist semileptonic signals such as $e^- e^+ \to \mu^\pm e^\mp + \pi^+ + \pi^-$ + missing momentum, which arise from semileptonic decay modes of either the color gluons (produced in reaction (6)) or directly of the charged red quarks. We estimate the strength of such semileptonic signals to be ≈ 2 to 10% of the pure leptonic signals. Note by contrast, semileptonic signals as above can not arise[19] via pair-production and decays of heavy leptons. Pair-production of conventional charmed particles ($D\bar{D}$, $F\bar{F}$), if it occurs, can give rise to semileptonic signals, but these signals should preferentially involve K-particles.

(vi) Neutral quark-pair production: One of the most interesting distinctions (within the gauge approach[7]) between the quark and heavy lepton-hypotheses for the SPEAR (μe)-events is that the neutral quark-pairs $p_r^0 \bar{p}_r^0$ and $c_r^0 \bar{c}_r^0$ would each be produced asymptotically with a cross section given by R = (4/9) ρ(s) (See Eq. (2)) which is four times that of the charged pair $n_r^- n_r^+$; neutral heavy leptons on the other hand can not be produced by $e^- e^+$-annihilation. Assuming that p_r^0 decays predominantly radiatively to $(n_y^0 + \gamma)$ (which holds if $m(p_r^0) - m(n_y) \gtrsim 10$ MeV), the production of p_r^0 may be searched for by looking for monoenergetic low-energy (≈ 10 to 50 MeV) γ-rays near threshold for the production of $p_r^0 \bar{p}_r^0$-pairs which should coincide with the threshold for the (μe)-events. Note we expect $m(p_r) \approx m(n_r)$.

(vii) Anomalous leptons in Hadronic Collisions: Finally we expect charged red quarks (and charged color gluons) to be produced in pairs by p+p-collisions with cross sections $\gtrsim 10^{-31}$ cm^2 at Fermilab and ISR energies; their decays should give rise to anomalous dileptons (e^-e^+, $\mu^-\mu^+$ and $\mu^\pm e^\mp$) as well as single leptons; these may in fact be responsible for the "excess" direct leptons[20] observed in such collisions. Note that there is a sharp distinction between the "strong" production of quarks (and gluons) on the one hand and the weak and electromagnetic production of heavy leptons[21] on the other by such collisions.

To conclude, the quark hypothesis for the SPEAR (μe)-events leads to several intriguing consequences; in particular it explains the jet-like-distribution of hadrons[14] and possibly also the depletion of charged-energy as well as the nonappearance of the charmed particles in e^-e^+-annihilation. We emphasize once again that the SPEAR (μe)-events may arise from quark-decays only provided there exist[3] the color-gluons (or similar color vector-mesons) below 2 GeV. We therefore urge a search[22] for narrow neutral and charged color-gluons in the mass region below 2 GeV. Finally, to distinguish between the quark versus heavy lepton-hypotheses for the (μe)-events, we especially urge searches for (i) anomalous semileptonic (μe)-signals in e^-e^+-annihilation, (ii) monoenergetic low energy γ-rays near threshold for the production of (μe)-events and (iii) anomalous leptons in hadronic collisions. Further kinematic distinctions between the two hypotheses are under study.

We thank M. L. Perl, H. Rubinstein, G. A. Snow, J. Sucher and C. H. Woo for several helpful discussions.

REFERENCES AND FOOTNOTES

1. J. C. Pati and Abdus Salam, Phys. Rev. Lett. 31, 661 (1973); Phys. Rev. D10, 275 (1974).

2. For recent estimates and detailed discussions, see J. C. Pati, S. Sakakibara and Abdus Salam (Trieste Preprint IC/75/93, To appear) and also W. R. Franklin Nucl. Phys. B91, 160 (1975).

3. If no such light colored vector mesons exist, the SLAC events would not relevant for quark-decays. One would then need to institute a different type of search for quark-decays (see second paper of Ref. 2).

4. M. L. Perl et al., Phys. Rev. Lett. 35, 1489 (1975).

5. G. Hanson et al. Phys. Rev. Lett. 35, 1609 (1975).

6. The discussions are not altered if we introduce new flavors and heavy leptons through the mirror set, the symmetry being $SU(4)^4$, which leads to a vector-like-theory. (See e.g. J. C. Pati and Abdus Salam, Phys. Lett. 58B, 333 (1975) and R. N. Mohapatra and J. C. Pati, Phys. Rev. D11, 2558 (1975)) or alternatively if the prodigal model is correct (see Ref. 1).

7. J. C. Pati and Abdus Salam, Phys. Rev. Lett. 36, 11 (1976); Univ. of Md. Tech. Rep. 76-061 (To be published); G. Rajesekharan and P. Roy (TIFR preprint TH/75-38 and TH/75-42).

8. The ($\pi e \nu$) and ($K e \nu$)-modes are forbidden to $O(\alpha)$ and $O(G_F)$ respectively in the matrix element compared to the allowed modes. See J. C. Pati and Abdus Salam, Univ. of Md. Tech. Rep. 76-061 for details.

9. Following asymptotic freedom or light cone analysis.

10. Suppression of Fig. 2 is subject to the assumption that the emission of a composite particle (e.g. pion from the quark-line is associated with a form-factor

depending upon the loop-momentum k (e.g. $f_{qq\pi}(k^2) \sim m^2/(k^2-m^2)$; $m \sim m_q$ or m_V), which, if present, makes Fig. 2 over-convergent. Details of these arguments are given in the first paper of Ref. 2.

11. We expect a small difference (≈ 200 to 300 MeV) between the thresholds for pair-production of n and λ-quarks.
12. Note that SPEAR data (Ref. 4) is consistent with the (μe)-events arising partly from three <u>and</u> partly from two-body decays of the parent particles.
13. This question is under study.
14. It appears to us that the familiar parton model considerations do not provide a convincing explanation of the jet structure, unless real quark-pairs are produced as suggested here.
15. We thank G. A. Snow for urging this test.
16. See invited talks by R. Schwitters and F. Gilman, Proceedings of the International Symposium on Lepton and Photon Interactions, Stanford (August, 21, 1975).
17. This does not of course preclude single (and pair)-production of charmed particles by neutrino (and hadronic) interactions. It is conceivable that e^-e^+-annihilation is much more efficient in producing real $q\bar{q}$-pairs than hadron-hadron collisions or neutrino-interactions simply because the "recombination" of quarks into hadrons is more likely in the multiquark environment of a disrupting hadron. This requires further study.
18. This may alter with the "Prodigal" model (Ref.1).
19. Assuming <u>separate</u> muon and electron number conservations and neglecting decay modes with branching ratios $O(\alpha^2)$.

20. See for example L. Lederman, Proceedings of the International Symposium on Lepton and Photon Interactions, Stanford (August 21, 1975).

21. For heavy leptons, we expect $\sigma(L^0\bar{L}^0) \lesssim 10^{-38}$ cm^2 and $\sigma(L^+L^-) \approx 10^{-34}$ to 10^{-35} cm^2 for $m_L \approx 2$ GeV.

22. The question of consistency of light colored vector mesons ($m_V \approx 1$ to 2 GeV) with existing experiments and the means of searching for these narrow mesons (e.g. via e^-e^+-annihilation and photo-production) are discussed in detail in a forthcoming preprint by J. C. Pati, J. Sucher and C. H. Woo.

SPONTANEOUS BREAKING OF SUPERSYMMETRY IN NON-ABELIAN GAUGE THEORIES

A. A. Slavnov

Steclov Mathematical Institute

Moscow, U. S. S. R.

The Higgs mechanism gives at present the only reliable method of spontaneous symmetry breaking in quantum field theory. However, straightforward application of this mechanism to supersymmetric theories meets certain difficulties due to specific form of the supersymmetric potential. Fayet and Illipoulos[1] succeeded to solve the problem for the case of the supersymmetric theory invariant with respect to Abelian gauge group (see also[2]). Unfortunately this method is not directly applicable to the case of semi-simple gauge group which is the most interesting from the practical point of view. In particular it does not allow one to construct a model which is simultaneously infra-red convergent and asymptotically free, for which the supersymmetric Yang-Mills theory is a natural candidate.

Other suggestions on the problem were given in the papers[3,4]. However the models considered there include an additional matter-matter interaction which enters with an independent coupling constant and destroys considerably the original beauty of the supersymmetric gauge theory.

In addition these models contain residual massless scalars not absorbed by the Higgs mechanism.

In the present paper we propose a mechanism of spontaneous supersymmetry breaking, which is in some sense generalization of Fayet-Illiopoulos method but is applicable to the theories invariant with respect to semi-simple gauge group. More precisely the group of invariance of the model should be G(*local*) ⊠ U(1) (global), where G can be semi-simple. Our main observation is that to produce supersymmetry breaking the Abelian subgroup U(1) should not be realized dynamically. Consequently the method works in the case of supersymmetric Yang-Mills theory and in particular allows one to construct asymptotically free and infra-red convergent models.

The paper is organized as follows.
In section 1 we discuss the renormalization procedure for spontaneously broken supersymmetric Abelian theory. Section 2 is devoted to the description of the supersymmetry breaking mechanism for the semi-simple gauge groups. In Section 3 we give an example of a model, based on the proposed mechanism, which is simultaneously infra-red convergent and asymptotically free.

I. RENORMALIZATION OF SPONTANEOUSLY BROKEN SUPERSYMMETRIC ABELIAN THEORIES

The Lagrangian of the supersymmetric quantum electrodynamics given firstly by Wess and Zumino[5] can be written in the form[6]

$$L_s = \frac{1}{8} (\bar{D}D)^2 \{\Phi_+^+ e^{g\psi}\Phi_+ + \Phi_-^+ e^{-g\psi}\Phi_-\} - \frac{M}{2} (\bar{D}D)\{\Phi_-^+\Phi_+ + \Phi_+^+\Phi_-\} + L_\psi^0, \tag{1}$$

where ψ is a gauge supermultiplet with the components $\{A, \chi, F, G, A_\nu, \lambda, D\}$, and $\Phi_\pm$ are chiral superfields describing matter fields

$$\Phi_\pm(\mathbf{x},\theta)=\exp\{\mp\tfrac{1}{4}\bar{\theta}\hat{\partial}\gamma_5\theta\}[A_\pm(x)+\bar{\theta}\psi_\pm(x)+\tfrac{1}{4}\bar{\theta}(1\pm i\gamma_5)\theta F_\pm(x)]. \quad (2)$$

D is a covariant derivative, L^0_ψ is a free Lagrangian for the gauge fields

$$L^0_\psi = -\tfrac{1}{4}\ (\partial_\mu A_\nu-\partial_\nu A_\mu)^2+\tfrac{i}{2}(\bar{\lambda}-i\hat{\partial}\bar{\chi})\hat{\partial}(\lambda+i\hat{\partial}\chi)+\tfrac{1}{2}(D+\Box A)^2. \quad (3)$$

The Lagrangian (1) is manifestly supersymmetric and invariant with respect to generalized gauge transformations

$$\Phi_\pm \to \Omega_\pm\Phi_\pm\ , \qquad e^{g\psi} \to \Omega_- e^{g\psi}\ \Omega_+^{-1}\ , \quad (4)$$

where $\Omega_\pm$ is an arbitrary chiral superfield satisfying the condition $\Omega^+_+ = (\Omega_-)^{-1}$.

To obtain a spontaneously broken solution one should add to the Lagrangian (1) supersymmetric and gauge invariant term $\zeta g^{-1}D$. This leads to a nonvanishing vacuum expectation value of D and performing canonical transformation

$$D\to D-\zeta g^{-1} \quad (5)$$

we obtain the Lagrangian of spontaneously broken theory, which can be written (omitting total divergence) as follows

$$L=L_s-\tfrac{\zeta}{2}\ (\Phi^+_+ e^{g\psi}\Phi_+-\Phi^+_- e^{-g\psi}\Phi_-)_A. \quad (6)$$

where $(\ )_A$ means A-component of a superfield. This Lagrangian is still invariant with respect to the gauge transformations (4) but the supersymmetry transformation is changed:

$$
\begin{aligned}
\delta A &= \bar{E}\gamma_5\chi \quad , \\
\delta \chi &= \tfrac{1}{2}[\gamma_5 F + i\gamma_\mu A_\mu - G - i\gamma_5 \hat{\partial} A]E \quad , \\
\delta F &= \tfrac{1}{2}\bar{E}\gamma_5\lambda - \tfrac{i}{2}\bar{E}\gamma_5\hat{\partial}\chi \quad , \\
\delta G &= -\tfrac{1}{2}\bar{E}\lambda - \tfrac{i}{2}\bar{E}\hat{\partial}\chi \quad , \qquad (7) \\
\delta A_\nu &= \tfrac{1}{2}\bar{E}\gamma_\mu\gamma_\nu\partial_\mu\chi - \tfrac{i}{2}\bar{E}\gamma_\nu\lambda \quad , \\
\delta\lambda &= \tfrac{1}{2}[\gamma_5 D - i\gamma_5\hat{\partial}F - i\hat{\partial}G - i\gamma_\nu i\hat{\partial}A_\nu - \zeta g^{-1}\gamma_5] \\
\delta D &= -i\bar{E}\hat{\partial}\gamma_5\lambda \quad .
\end{aligned}
$$

λ-component is shifted by a constant number, i.e. the super symmetry is spontaneously broken.

In the Zumino-Wess gauge $A=\chi=F=G=0$, the additional term in (6) is nothing but the mass term

$$-\frac{\zeta}{2}(A_+^+A_+ - A_-^+A_-). \qquad (8)$$

The masses of different members of the chiral supermultiplets are no longer equal.

Nevertheless, the renormalization procedure can be performed analogously to the symmetric case[7,8] and no new independant counterterms are needed.

To show that, we note first of all that the Lagrangian (6) can be quantized exactly in the same way as a symmetric one[7,8] because it possesses the same gauge invariance. Introducing the supersymmetric gauge fixing term

$$\frac{1}{4\beta}\,(\bar{D}D)^2\psi_+{}^*P\psi_+, \qquad (9)$$

where P is some polynomial over the d'Alambert operator $\Box$, one easily obtains the generalized Ward identities derived in our paper[7] (eq. 22). These identities express all the diagrams with external unphysical gauge lines in terms of the diagrams without such lines.

The second part of the renormalization procedure, the supersymmetry relations, needs however more careful investigation. We shall show that the renormalized Green function generating functional

$$Z=N^{-1}\int \exp\{i\int [L_R(x)+\tfrac{1}{8}(\bar{D}D)^2(J_\psi \psi)+\tfrac{1}{2}(\bar{D}D)(J_{\Phi\pm}\Phi_\pm)]dx\}d\mu, \quad (10)$$

where

$$L_R=\tfrac{1}{8}(\bar{D}D)^2 Z_2\{\Phi_+^+ e^{g\psi}\Phi_+ + \Phi_-^+ e^{-g\psi}\Phi_-\}+\tfrac{1}{4\beta}(\bar{D}D)^2\psi_+ {}^*P\psi_+ -$$
$$-\tfrac{Z_2}{2}\zeta\{\Phi_+^+ e^{g\psi}\Phi_+ - \Phi_-^+ e^{-g\psi}\Phi_-\}_A - \tfrac{M}{2}(\bar{D}D)\{\Phi_-^+\Phi_+ + \Phi_+^+\Phi_-\}+Z_3 L_\psi^0 \quad (11)$$

leads to the divergence free S -matrix.

The identities associated with the transformations (7) are obtained as usual by making in the integral (10) the change of variables (7) with E depending on x and putting

$$\frac{dZ}{dE}\Big|_{E=0}=0$$

We write down these identities in terms of one particle irreducible Green functions generated by the functional

$$\Gamma(R_\psi, R_\Phi)=W-\int\{\tfrac{1}{8}(\bar{D}D)^2(J_\psi R_\psi)+\tfrac{1}{2}(\bar{D}D)(J_{\Phi\pm}R_{\Phi\pm})\}dx, \quad (12)$$

where

$$W(J_\psi, J_\Phi)=i\ln Z(J_\psi, J_\Phi);\; R_{\psi,\Phi}=\frac{\delta W}{\delta J_{\psi,\Phi}},\; J_{\psi,\Phi}=-\frac{\delta\Gamma}{\delta R_{\psi,\Phi}} \quad (13)$$

Their explicit form is

$$\int\{(\frac{\delta\Gamma}{\delta R_{A_\nu}(X)}\frac{1}{2}\gamma_\nu+\frac{\delta\Gamma}{\delta R_D(X)}\hat{\partial}\gamma_5+i\zeta g^{-1}\gamma_5]R_\lambda(x)+\frac{i}{2}\frac{\delta\Gamma}{\delta R_{\bar{\lambda}}(x)}[\gamma_5 R_D(x)+$$

$$+\hat{\partial}R_{A_\nu}(x)\gamma_\nu-\zeta g^{-1}\gamma_5]-i\frac{\delta\Gamma}{\delta R_{A\pm}(x)}R_{\psi\pm}(x)+\frac{\delta\Gamma}{\delta R_{F\pm(x)}}\hat{\partial}R_{\psi\pm}(x)+ \qquad (14)$$

$$+\frac{1\mp i\gamma_5}{2}(R_{F\pm}(x)-i\hat{\partial}R_{A\pm}(x))\frac{\delta\Gamma}{\delta R_{\bar{\psi}\pm}(x)}\}dx = 0 \quad .$$

(Index c means charge conjugation, we put sources of unphysical gauge components equal to 0).

Differentiating eq. (14) one obtains

$$\frac{i}{2}\frac{\delta^2\Gamma}{\delta R_{\bar{\lambda}}(x)\delta R_\lambda(y)} -\hat{\partial}\frac{\delta^2\Gamma}{\delta R_D(x)\delta R_D(y)}-\frac{i}{2}\zeta g^{-1}\int\frac{\delta^3\Gamma}{\delta R_{\bar{\lambda}}(Z)\delta R_\lambda(y)\delta R_D(x)}dZ= \qquad (15)$$

The last term is absent in the corresponding identity for the symmetric case. But this term is superficially convergent, therefore if the constant Z_3 is fixed by demanding $\Gamma_{DD}(P)$ to be finite, the Green function $\Gamma_{\bar{\lambda}\lambda}(P)$ is also finite (note however that contrary to the symmetric case $\Gamma_{\bar{\lambda}\lambda}(P)$ contains not only a $\hat{P}\Gamma^{(1)}_{\bar{\lambda}\lambda}(P^2)$ term but also a finite term $\Gamma_{\bar{\lambda}\lambda}{}^2(P^2)$).

Analogously one can show that $\Gamma_{A_\nu A_\mu}$ is finite, and three point vertices $\Gamma_{A\pm\lambda\psi\pm}$, $\Gamma_{A_\mu\psi\pm\psi\pm}$, etc.. may differ only by finite terms. Therefore if the constant Z_2 is chosen to make $\Gamma_{A\pm\lambda\psi\pm}$ finite all other three-point functions are finite also. One the other hand electromagnetic three-point functions $\Gamma_{A_\mu\psi\pm\psi\pm}$, etc.. are related to the matter field Green functions by the usual Ward identities. By the usual arguments it follows that the corresponding wave function renormalizations are finite. This result is in fact quite expectable because the Lagrangian (6) differs from L_s essentially by the mass term. But there exists a general statement[9,10,11] that such terms do not affect logarithmic divergences.

The identities (14) allow one to make a stronger statement that also no independent mass renormalization arises.

Differentiating eq. (14) with respect to $A_\pm$, $\psi_\pm$ we obtain

$$-i\frac{\delta^2\Gamma}{\delta R_{A\pm}(x)\delta R_{A\pm}(y)} + \frac{1\mp i\gamma_5}{2}\left(-i\hat{\partial}\frac{\delta^2\Gamma}{\delta R_{\bar{\psi}\pm}(x)\delta R_{\psi\pm}(y)}\right) + \tag{16}$$

$$+\hat{\partial}\frac{\delta^2\Gamma}{\delta R_{A\pm}(x)\delta R_{F\pm}(y)} - \frac{1}{2}\zeta g^{-1}\gamma_5\int\frac{\delta^3\Gamma}{\delta R_{\bar{\lambda}}(z)\delta R_{A\pm}(x)\delta R_{\psi\pm}(y)}dz = 0$$

The last term which is absent in the symmetric case is finite due to our choice of Z_2. Therefore $\Gamma_{A\pm A\pm}(P^2) = \text{const} < \infty$, etc. Finally the relations between $\Gamma_{\bar{\psi}\pm\psi\pm}$ and $\Gamma_{A\pm F\pm}$ are affected also by finite terms, and no independent mass renormalization arises (the absence of mass renormalization in the supersymmetric case was shown in the paper[12]).

This completes the renormalization program for the Abelian gauge theory with spontaneously broken supersymmetry.

II. SPONTANEOUS BREAKING OF SUPERSYMMETRY IN THE CASE OF SEMI-SIMPLE GAUGE GROUP

In this section we shall show that the mechanism described above can be generalized to include the Yang-Mills theories which possess in addition to the local SU(N) invariance global U(1) invariance. The last one is not realized dynamically so in the theory there is only one independent coupling constant.

To illustrate our idea we note firstly that the Fayet-Illiopoulos trick works as well in the case of more complicated group containing Abelian subgroup. For example if the matter fields $\Phi_{\pm}$ are doublets, transforming according to the representation of SU(2) ⊗ U(1) group

$$\Phi_{\pm} \rightarrow \Omega^{1}_{\pm}\Omega_{\pm}\Phi_{\pm} \quad , \tag{17}$$

where

$$(\Omega_{+})^{+} = (\Omega_{-})^{-1}, \; (\Omega^{1}_{+})^{+} = (\Omega^{1}_{-})^{-1}, \; \det \Omega_{\pm} = 1,$$

then the gauge invariant Lagrangian producing spontaneous breakdown of supersymmetry is

$$L=\frac{1}{8}(\bar{D}D)^{2}\{\Phi^{+}_{+}e^{g_{1}\psi_{1}+g\psi}\Phi_{+}+\Phi^{+}_{-}e^{-g_{1}\psi_{1}-g\psi}\Phi\} + \zeta g_{1}^{-1}D_{1} + \ldots , \tag{18}$$

where ψ_1 is an Abelian gauge superfield and ψ is a matrix-superfield transforming as

$$e^{g_{1}\psi_{1}} \rightarrow \Omega^{1}_{-}\, e^{g_{1}\psi_{1}}(\Omega^{1}_{+})^{-1} \quad , \quad e^{g\psi} \rightarrow \Omega_{-}e^{g\psi}\Omega_{+}^{-1} \quad , \tag{19}$$

... denotes the free Lagrangians for the fields ψ and ψ_1 and possible supersymmetric terms.

Performing the canonical transformation $D_1 \rightarrow D_1 - \zeta g_1^{-1}$ one can proceed further in complete analogy with the discussion of the previous section.

The generalized Ward identities derived in[7,8] remain valid and supersymmetry relations are changed only by finite terms.

The Lagrangian (18) written in terms of the shifted fields has a definite limit when $g_1 \rightarrow 0$. The limiting Lagrangian is invariant under local SU(2) transformation and in fact is nothing but the Yang-Mills Lagrangian plus some additional renormalizable interactions of scalars and spinors. This Lagrangian possesses also U(1) invariance, but the local U(1) invariance is lost.

Omitting nonessential additive constant one can write.

$$L=\frac{1}{8}(\bar{D}D)^2\{\Phi_+^+e^{g\psi}\Phi_+ + \Phi_-^+e^{-g\psi}\Phi_-\}-\frac{\zeta}{2}\{\Phi_+^+e^{g\psi}\Phi_+ - \Phi_-^+e^{-g\psi}\Phi_-\}A^+ \dots \quad (20)$$

The additional term is again nothing but the mass term for the scalar fields.

The supersymmetry identities (14) include explicitly the factor ζg_1^{-1} which tends to infinity when $g_1 \to 0$. However this factor is multiplied by $\delta\Gamma/\delta R_\lambda$ which is proportional to g_1. Therefore the supersymmetry relations (14) also have well-defined limit.

One can consider the Lagrangian (18) as an intermediate regularization of the theory defined by the Lagrangian (20). All the discussions of the previous section concerning the renormalization constants are directly applicable to the Lagrangian (18). Therefore we conclude that in exact analogy with the Abelian case the symmetric counterterms are sufficient to remove all ultraviolet divergences. The renormalized Lagrangian can be written as follows:

$$L_R=\frac{1}{8}(\bar{D}D)^2 Z_2\{\Phi_+^+e^{\tilde{g}\psi}\Phi_+ + \Phi_-^+e^{-\tilde{g}\psi}\Phi\}- \\ -\frac{Z_2\zeta}{2}\{\Phi_+^+e^{\tilde{g}\psi}\Phi_+ - \Phi_-^+e^{-\tilde{g}\psi}\Phi_-\}_A + \tilde{Z}_2 L_\psi^0(\tilde{g}) , \quad (21)$$

where $\tilde{g}=Z_2^{-1}Z_1 g=\tilde{Z}_2\tilde{Z}_1^{-1}g$, Z_1 is a three-point vertex re normalization constant and $\tilde{Z}_2$, $\tilde{Z}_1$ are the corresponding constants for the Yang-Mills field.

Summarizing we can say that if one uses for the regularization of the Lagrangian (21) the supersymmetric expression (18), one has a theory which is invariant under the "broken supersymmetry" transformation (7) for any $g_1 \neq 0$ and satisfies the "broken supersymmetry relations" (14) for any g_1 (including $g_1 = 0$). However there does not exist the limiting Lagrangian invariant with respect to limiting transformations (7) (in terms of unshifted fields the Lagrangian becomes singular, and in terms of shifted fields the transformation (7) becomes singular).

This situation is quite common for the quantum theory, where the renormalized Lagrangian simply does not exist and the symmetry manifests itself in the existence of generalized Ward identities or in the invariance of the regularized Lagrangian. In our case the limiting invariant Lagrangian does not exist even at the classical level. The last point has no physical significance and is in fact due to the special formulation of the model we used above. To show that we present an alternative formulation which does not use the limiting procedure and allows to write down the invariant classical Lagrangian.

Such a Lagrangian can be written in the form

$$\begin{aligned} L=&\frac{1}{8}(\bar{D}D)^2\{\Phi_+^+ e^{g\psi+\psi_1}\Phi_+ + \Phi_-^+ \, e^{-g\psi-\psi_1}\Phi_-\}- \\ &-\frac{1}{4}(\partial_\mu A_\nu^1-\partial_\mu A_\mu^1)(\partial_\mu \tilde{A}_\mu^1-\partial_\nu \tilde{A}_\mu)+\frac{1}{2}(\tilde{\lambda}_1-i\partial\tilde{\chi}_1)\partial(\tilde{\lambda}_1+i\partial\tilde{\chi}_1)+ \\ &+\frac{1}{2}(D_1+\Box A_1)(\tilde{D}_1+\Box\tilde{A}_1)+\zeta(\tilde{D}_1+\Box\tilde{A}_1) \quad . \end{aligned} \tag{22}$$

... means the free Lagrangian for the Yang-Mills field ψ and also possible mass terms. ψ_1 and $\tilde{\psi}_1$ are Abelian superfields, transforming as

$$e^{\psi_1} \to \Omega^1_- e^{\psi_1}(\Omega^1_+)^{-1} \ , \quad e^{\tilde{\psi}_1} \to \Omega^1_- e^{\tilde{\psi}_1}(\Omega^1_+)^{-1} \quad . \tag{23}$$

The fields ψ_1 and $\tilde{\psi}_1$ are auxiliary fields which can be eliminated from the Lagrangian (22) providing the Lagrangian of the Yang-Mills theory. The field $\tilde{\psi}_1$ is in fact a Lagrange multiplier: the equations arising after variation over $\tilde{\psi}_1$ are the free field equation for the field ψ_1 .

The Lagrangian (22) is manifestly supersymmetric and gauge invariant. The supersymmetry is spontaneously broken: canonical transformation $D_1 \to D_1 - \zeta$ eliminates linear term and produce a mass form

$$L_m = -\frac{\xi}{2}\{\Phi^+_+ e^{g\psi+\psi_1}\Phi_+ - \Phi^+_- e^{-g\psi-\psi_1}\Phi_-\}_A \quad . \tag{24}$$

The renormalized Lagrangian in terms of the shifted fields is

$$L_R = \frac{1}{8}(\bar{D}D)^2 Z_2\{\Phi^+_+ e^{\tilde{g}\psi+\psi_1}\Phi_+ + \Phi^+_- e^{-\tilde{g}\psi+\psi_1}\Phi_-\} -$$
$$-\frac{\zeta}{2}Z_2\{\Phi^+_+ e^{\tilde{g}\psi+\psi_1}\Phi_+ - \Phi^+_- e^{-\tilde{g}\psi-\psi_1}\Phi_-\}_A + \ldots \quad . \tag{25}$$

We leave to the reader to repeat the arguments given above to show that the corresponding Green functions satisfy generalized Ward identities and "broken supersymmetry relations" analogous to eq. (14).

III. ASYMPTOTICALLY FREE AND INFRA-RED CONVERGENT MODEL

In this section we illustrate the possibilities opened by the mechanism described in Sec. 2 for constructing the model which is simultaneously infra-red convergent and asymptotically free. This model certainly does not pretend to any relation to the experiment and is just a methodical example.

The model is fixed by the Lagrangian

$$L=\frac{1}{16}(\bar{D}D)^2 Tr\{\Omega_+^+ e^{g\psi}\ \Omega_+(1+T_3)\}+\frac{\zeta}{2}Tr\{\Omega_+^+ e^{g\psi}\ \Omega_+(1+T_3)\}_A+\ldots\ . \tag{26}$$

Here we find it more convenient to use a matrix chiral superfield

$$\Omega_+=\exp\{-\frac{1}{4}\bar{\Theta}\hat{\partial}\gamma_5\Theta\}[A_+ +\vec{T}\vec{A}_+ +\bar{\Theta}(\psi_+ +\vec{T}\vec{\psi}_+)+\frac{1}{4}\bar{\Theta}(1+i\gamma_5)\Theta(F_+ +\vec{T}\vec{F}_+)], \tag{27}$$

other notations are the same as before.

Due to the presence of the projection operator $(1+T_3)/2$ Ω_+ is equivalent to the complex doublet Φ_+ with the components

$$\tilde{A}_+=A_+ +A_+^3\ ,\ \tilde{A}_{1+}=A_+^1-iA_+^2\ ,\ \text{etc.}$$

The physical content of the model is most easily analyzed in the Zumino-Wess gauge after elimination of auxiliary fields. In this gauge the Lagrangian can be written as follows:

$$\begin{aligned} L=&\partial_\mu\tilde{A}_+^+\partial_\mu\tilde{A}_+ +\partial_\mu A_{1+}^+\partial_\mu\tilde{A}_{1+} +i\bar{\tilde{\psi}}_+\hat{\partial}\tilde{\psi}_+ +i\bar{\tilde{\psi}}_{1+}\hat{\partial}\tilde{\psi}_{1+}+\tilde{F}_+^+\tilde{F}_+ +\tilde{F}_{1+}^+\tilde{F}_{1+}+\\ &+g^2\vec{A}_\mu^2(\tilde{A}_+^+\tilde{A}_+ +\tilde{A}_{1+}^+\tilde{A}_{1+})-\frac{g^2}{2}(\tilde{A}_+^+\tilde{A}_+ -\tilde{A}_{1+}^+\tilde{A}_{1+})^2-2g^2\tilde{A}_+^+\tilde{A}_+\tilde{A}_{1+}^+\tilde{A}_{1+}+\\ &+\zeta(\tilde{A}_+^+\tilde{A}_+ +\tilde{A}_{1+}^+\tilde{A}_{1+})+\frac{i}{2}\bar{\lambda}^a\hat{\partial}\lambda^a-\bar{\lambda}_3(\tilde{A}_+^+\psi_+ +\tilde{\psi}_+^c\tilde{A}_+)-\\ &-(\bar{\lambda}_1+i\bar{\lambda}_2)(\tilde{\psi}_{1+}\tilde{A}_+^+ +\tilde{\psi}_{1+}^c A_+)+\ \ldots\ . \end{aligned} \tag{28}$$

... denotes interaction terms which are irrelevant for the present discussion.

For positive ζ the Lagrangian (28) evidently

corresponds to the unstable theory, due to the wrong sign of the mass term. That means the gauge invariance is also spontaneously broken and one should translate the fields $\tilde{A}$. Substituting $\tilde{A}$ by $\tilde{A} + \alpha$ we obtain the stability condition

$$-g^2\alpha^3+\xi\alpha=0 \quad . \tag{29}$$

Now the spectrum of the model contains massive vector triplet $\vec{A}_\mu$, two massive complex spinors $\lambda^3_+ + \tilde{\psi}+$ and $\lambda^1_- + i\lambda^2_- + \tilde{\psi}^1_+$, and massive Hermitian scalar $\tilde{A}^+_+ + \tilde{A}_+$, all with the mass ζ. In addition there are two-component massless spinor $\lambda^1_+ + i\lambda^2_+$ and three Goldstone scalars which are absorbed into the longitudinal part of the vector field A_μ.

The model is infrared convergent. On the other hand the only independent coupling constant is the Yang-Mills constant g. The theory is known to be asymptotically free.

It is not difficult to construct other asymptotically free models introducing for example also interaction with the left-handed matter fields and additional supersymmetric mass terms. At present, however, we have not succeeded to invent a parity conserving asymptotically free model.

REFERENCES

1. P. Fayet, J. Illiopoulos, Phys. Let. 51B, 461, (1974).
2. P. Fayet, Nuclear Phys. B90, 104, (1975), Preprint de l'Ecole Normale Superieure. PTENS 7515.
3. S. Browne, L. O'Raifeartaigh, T. Sherry, Preprint Dublin Institute for Advanced Studies DIAS-TP-75-16.
4. P. Fayet, Preprint de l'Ecole Normale Superieure PTENS 75/1.
5. J. Wess, B. Zumino, Nuclear Phys. B78, 1, (1974).
6. A. Salam, J. Strathdee, Preprint ICTP IC/74/42.
7. A. A. Slavnov, Preprint E-2-8308, JINR, 1974 and Theor. and Math. Phys. 23, 3, (1975).
8. A. A. Slavnov, Preprint E2-8443, JINR, (1974).
9. S. Weinberg, Phys. Rev. D, 8, 3497, (1973).
10. J. C. Collins, A. J. MacFarlane, Phys. Rev. D, 10, 1201, (1974).
11. I. V. Tiutin, B. L. Voronov, JETP Let. 21, 396, (1975).
12. S. Ferrara, O. Piguet, Preprint TH 1995-CERN, (1975).

SIGNATURES OF VECTORLIKE WEAK CURRENTS*

Peter Minkowski**

California Institute of Technology

Pasadena, California 91125

ABSTRACT

The general framework of a vectorlike gauge theory of leptons, quarks, and gauge bosons is sketched and particular characteristic features involving the virtual excitation of heavy fermions which are entering the charged weak current right- and left-handedly are discussed and confronted with the non-leptonic decays of strange particles.

INTRODUCTION

I would like to discuss the weak currents within a definite scheme of quarks and leptons characterized by their quantum numbers with respect to the gauge group $SU3^{c} \times SU2^{W} \times U1$:

*Work supported in part by the Energy Research and Development Administration under Contract E-(11-1)-68.

**Present address: Institute of Theoretical Physics, University of Bern, Switzerland

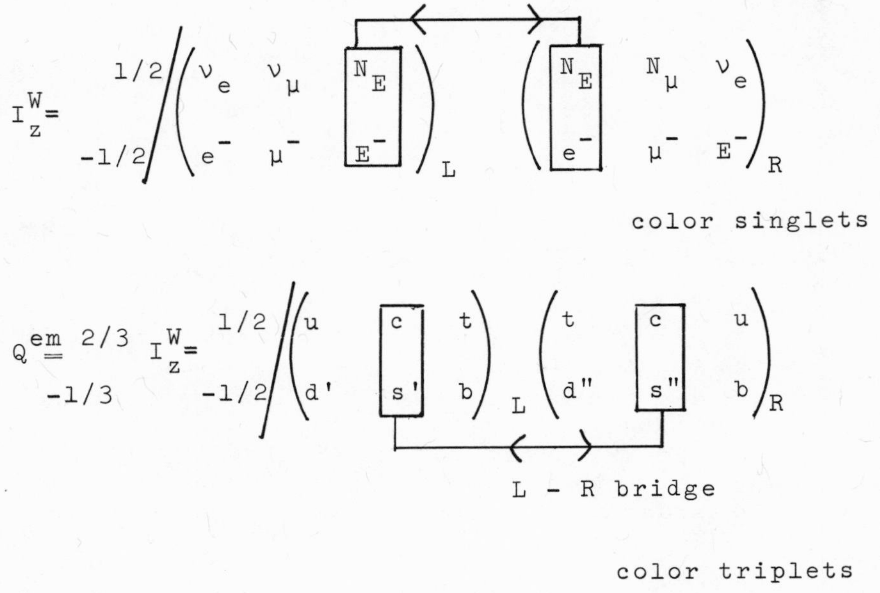

$$I_z^W = \begin{array}{c} 1/2 \\[2em] -1/2 \end{array} \left(\begin{array}{ccc} \nu_e & \nu_\mu & \boxed{\begin{array}{c} N_E \\ E^- \end{array}} \end{array} \right)_L \left(\begin{array}{cccc} \boxed{\begin{array}{c} N_E \\ e^- \end{array}} & N_\mu & \nu_e \\ & \mu^- & E^- \end{array} \right)_R$$

color singlets

$$Q \stackrel{em}{=} \begin{array}{c} 2/3 \\[2em] -1/3 \end{array} \quad I_z^W = \begin{array}{c} 1/2 \\[2em] -1/2 \end{array} \left(\begin{array}{ccc} u & \boxed{\begin{array}{c} c \\ s' \end{array}} & t \\ d' & & b \end{array} \right)_L \left(\begin{array}{ccc} t & \boxed{\begin{array}{c} c \\ s'' \end{array}} & u \\ d'' & & b \end{array} \right)_R$$

L - R bridge

color triplets

FIGURE 1

$$\left(\begin{array}{l} d' = \cos\theta_c \ d + \sin\theta_c \ s \\ s' = -\sin\theta_c \ d + \cos\theta_c \ s \end{array} \right)_L \Bigg/ \left(\begin{array}{l} d'' \simeq d \\ s'' \simeq s e^{i\delta} \end{array} \right)_R$$

θ_c : Cabibbo angle δ : CP violating phase

Although certain specific aspects of the above scheme
may not correspond to the real world, they should be
understood as characteristic phenomena which "mutatis
mutandis" do occur within a unified vectorlike gauge
theory.   Nevertheless the particular interplay of left-
handed and right-handed couplings of the charged weak
boson to $\bar{c} \ \gamma_\mu \ \frac{1+\gamma_5}{2} \ s'$ and $\bar{c} \ \gamma_\mu \ \frac{1-\gamma_5}{2} \ s''$ is proposed to
account for the dominant nonleptonic decays of strange
particles and thus, we feel is a definite guideline for

the construction of the weak (vectorlike) currents*.

The weak gauge group $SU2^W \times U1$ first considered by Salam, Ward and Weinberg[1] is generated by the smallest algebra containing the charged weak currents and the electromagnetic current. Correspondingly, there are four intermediary bosons denoted by W^+, W^-, γ, and Z. This minimal weak gauge group is, of course, also to be considered illustrative in the above sense, at least for the present.

The color exchange forces within $SU3^c$ mediated by eight vector gluons form the substrate of a candidate theory of the strong interactions (QCD)[2]. The relevant questions of color confinement, of the spectrum of physical states within QCD, of the quark masses, and of scale breaking effects in deep inelastic electron (muon) scattering, are under extensive investigation[3] and several contributions to this conference report on the present status of these studies[4].

The universality of electric charge $(Q(p)=Q(e^+))$ and of the weak coupling constants $(G_\mu \cos\theta_c = G_\beta)$ relating properties of quarks <u>and</u> leptons strongly motivates the quest for a unifying simple gauge group G, subject to a hierarchy of spontaneous symmetry breakings characterized by a chain of stability subgroups[5] as e.g.,

---

*It should be stressed at this point that any L-R bridge mediated by a heavy quark h: $(h,s')_L$ $(h,s'')_R$ coupled to one and the same vector boson, produces an effect which cannot be distinguished from the one proposed here in the phenomenology of nonleptonic weak decays of strange particles.

$$G \downarrow SU4^c \times SU2^W \downarrow SU3^c \times SU2^W \times U1 \downarrow SU3^c \times U1^{e.m.}$$

FIGURE 2

As the gauge group is enlarged, three challenging questions arise.

a) Is baryon (and/or lepton)-number conserved?

b) Is the chiral structure of the weak currents and the breakdown of discrete symmetries like P, C, and T compatible with the absence of axial current anomalies[6]?

c) What is the character and the multitude of scalar (Higgs) fields driving the hierarchy of symmetry breaking?

If the parity violation exhibited by the weak currents is to be sought in the spontaneous symmetry breaking only, a vectorlike gauge theory is a valid answer to question b)[6,7,8] which we adopt as a basic feature of the theory*.

New flavors of quarks and leptons besides the u, d, s, c-quarks and ν_e, e^-, ν_μ, μ^- -leptons[9] are necessary ingredients of a realistic vectorlike scheme; they are also imperative in order to prevent proton decay[10] to be induced by the superheavy gauge bosons which unify the

---

*The idea of a vectorlike theory was revived through the suggestion of the right-handed current $\bar{c}\,\gamma_\mu(1-\gamma_5/2)d$ universally coupled to the charged weak boson by a A. de Rujúla, H. Georgi and S. L. Glashow[7].

interactions of quarks and leptons.

A hierarchy of broken symmetries G_1 $G_2 \subset \ldots G_i \subset \ldots G$ characterized by respective mass scales $M_1 \ll M_2 \ll M_i \subset \ldots M$ of representative gauge bosons implies that on the level of G_i say, the lower symmetry group G_{i-1} acts as an exact symmetry group up to corrections of order M_{i-1}/M_i. In this sense the fermions in the scheme of Fig. 1 belong to the lowest level of symmetry, i.e., $SU3^c \times U1^{e.m.}$ and their masses m do not exhibit any further symmetry, whereas $G_2 = SU3^c \times SU2^W \times U1$ is characterized by $M_2 = M_W$ (or M_Z) [or equivalently by $M_2' = \left(\frac{\alpha\pi}{\sqrt{2}\, G_F}\right)^{\frac{1}{2}} \simeq 37.3$ GeV^1].

The physics at energies of the order of M or smaller is largely determined by the pattern of symmetry breaking.

The latter can be studied within the framework of explicit scalar-and pseudoscalar (Higgs) fields in its entire generality[11] or restricting oneself to supersymmetric schemes where the scalar- and pseudoscalar fields form together with spin one-half fermions and/or vector bosons specific supermultiplets[12].

The general discussion suggests, supersymmetry demands, the existence of (heavy) fermions with masses comparable to M_1, M_2, ...M beyond the scheme of Fig. 1, if there is to be a hierarchy of interactions in the first place.

The resulting uncertainities as far as the light fermions are concerned are best exemplified by considering the neutral lepton N_μ. It acquires a mass through a lepton number violating (Majorana) mass term. If $(N_E)_R$ would obtain its mass from the same mechanism, neutrinoless double β decay would be induced according to the diagram in Fig. 3[13].

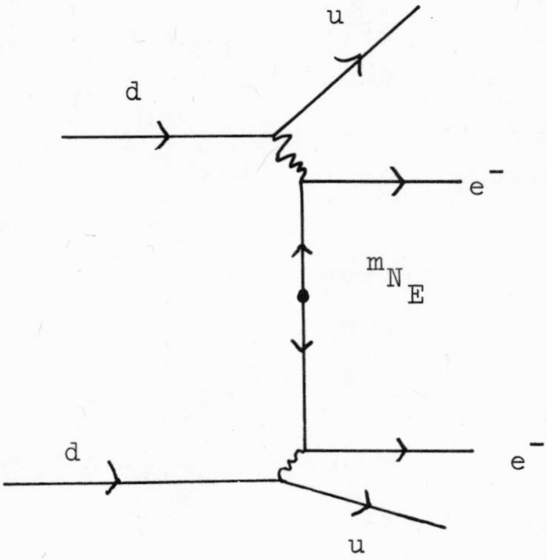

FIGURE 3

In order that the above process be compatible with the observed rate of double β decay (e.g. $Te^{130} \to Xe^{130}$, $Te^{128} \to Xe^{128}$, $Se^{82} \to Kr^{82}$) an upper bound $m_{N_E} > 10^4$ GeV must be imposed. Of course, a vanishing (or sufficiently small) Majorana mass term, as was chosen in the present scheme, does not give rise to any (appreciable) neutrinoless double β decay. If we take seriously, however, the above bound for any Majorana mass term of a neutral lepton, we are forced to a scheme involving at least eight flavors of light quarks and leptons[14] or a deviation from the gauge group $SU2_W \times U1$ as the minimal weak gauge group. The violation of lepton number by a Majorana mass term may still occur, but not within the light fermions and eventually within a larger gauge group with a correspondingly large mass scale.

2. RADIATIVE CORRECTIONS TO THE FERMION MASS MATRIX

A first signature of vectorlike weak currents can

be recognized in the radiative corrections to the mass matrix of the basic fermions. Here the word "radiative" comprises contributions from all gauge bosons relevant to a particular level of symmetry. We will only be interested in the effects on the level of $SU2^W \times U1$ and $W^\pm$ in particular. As is well known, the radiative corrections are not finite, and in general lead to an arbitrary redefinition of the mass matrix[10], except for the presence of so-called zero order symmetries[10,15] which render the corrections calculable. However, such symmetries do not seem to arise in realistic models, at least it has not been demonstrated, to my knowledge, that they do occur in a natural way.

We can nevertheless study the algebraic structure of radiative corrections to the fermion mass matrix following essentially V. Weisskopf[16] in evaluating fermion self energy parts up to a (common) logarithmically divergent multiplier. The latter will be calculable from the hitherto obscure mechanism which governs mass relations due to approximate symmetries.

a) <u>The $(\nu_e N_E)$, $(e^- E^-)$ system</u>

We evaluate the diagram of Fig. 4.

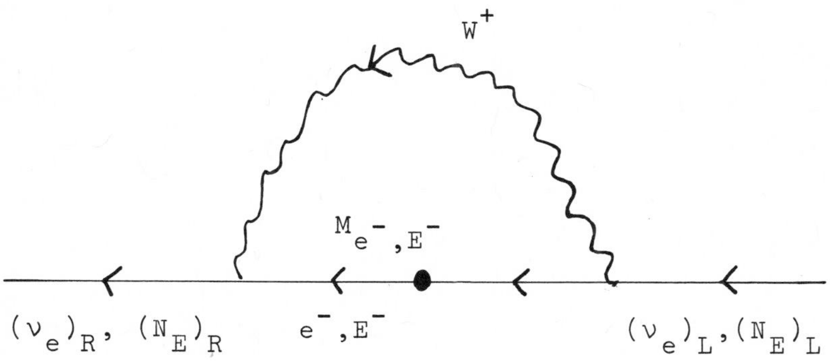

FIGURE 4

Denoting the 2 × 2 mass matrix for the neutral leptons by

$$\left((\nu_e)^+_R (N_E)^+_R\right) \quad m^{(e)} \begin{pmatrix} (\nu_e)_L \\ (N_E)_L \end{pmatrix} + \text{h.c.} \quad ,$$

one finds

$$m^{(e)} = \begin{pmatrix} 0 & r m_E \\ r m_e & m_{N_E} \end{pmatrix} \quad ,$$

$$r = \frac{3\alpha}{4\pi \sin^2\theta_W} \log \frac{\Lambda}{m_W} \simeq \frac{\alpha}{\pi} \log \left(\frac{\Lambda}{m_W}\right) \quad . \tag{2.1}$$

The mass squares of ν_1 and N_1 which are appropriate linear combinations of ν_e and N_E are the eigenvalues of the matrix

$$m^{(e)} + m^{(e)} = \left\{\begin{matrix} r^2 m_e^2 & r m_e m_N \\ r m_e m_N & r^2 m_E^2 + m_{N_E}^2 \end{matrix}\right\} \quad . \tag{2.2}$$

It follows that the neutrino (ν_1) acquires a mass of the order of

$$m_{\nu_e} \simeq r^2 \left(\frac{m_e m_E}{m_{N_E}}\right) \quad . \tag{2.3}$$

For $\log \frac{\Lambda}{m_W} \simeq 1, m_E \simeq m_{N_E}$ one finds the admittedly very uncertain estimate

$$m_{\nu_e} \sim 2 - 3 \text{ eV}. \tag{2.4}$$

b) <u>The $(\nu_\mu)_L$–$(N_\mu)_R$ system</u>

Here we evaluate the diagram of Fig. 5.

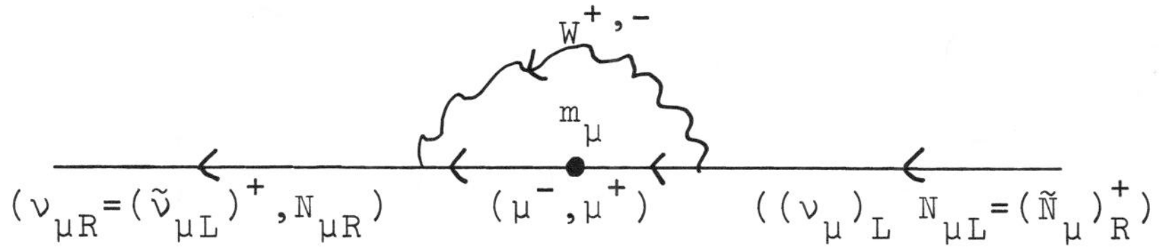

FIGURE 5

$$(N_{\mu L}) = (\tilde{N}_\mu)^+_R \text{ means } (N_{\mu L})^{\dot\alpha} = \varepsilon^{\dot\alpha\dot\rho}(N_{\mu R})^+_{\dot\rho} \quad ,$$

with

$$\varepsilon^{21} = -\varepsilon^{12} = 1, \quad \varepsilon^{\dot\alpha\dot\rho} = -\varepsilon^{\dot\rho\dot\alpha}$$

Denoting the 2 × 2 (Majorana) mass matrix for the $\nu_{\mu L}$, $(N_\mu)_R$ system by

$$\frac{1}{2}\,(\tilde{\nu}_{\mu L},\ N^+_{\mu R})\ m^{(\mu)} \begin{pmatrix} (\nu_\mu)_L \\ (\tilde{N}_{\mu R})^+ \end{pmatrix} + \text{h.c.} \quad ,$$

one finds similarly to case (a):

$$m^{(\mu)} = \begin{pmatrix} 0 & rm_\mu \\ rm_\mu & m_N \end{pmatrix} \quad . \tag{2.5}$$

Again the muon neutrino mixes with $\tilde{N}^+_{\mu R}$, (the anti-particle of $(N_{\mu R})$) and acquires a mass

$$m_{\nu_\mu} \simeq r^2 \frac{m_\mu^{\ 2}}{m_{N_\mu}} \quad . \tag{2.6}$$

The equivalent estimate to eq. (2.4) yields for $m_{N_\mu} \simeq 2$ GeV

$$m_{\nu_\mu} \simeq 20\text{-}30 \text{ eV}^{17} \quad . \qquad (2.7)$$

Since lepton number is broken, the degeneracy in mass of the two Majorana particles composing $(\nu_e)_{L+R}$ will not be exact. One is dealing with a system of three nearly degenerate Majorana fermions $(\nu_{e_1}, \nu_{e_2}, \nu_\mu)$

Thus one may ask whether there exist weak angles or mixing terms of the (neutral) lepton mass matrix, leptonic analogs to the Cabibbo angle. These mixings if indeed present, induce neutrino beam oscillations first discussed by Pontecorvo[18] similar to the K^o-$\bar{K}^o$ oscillations, induced by the $|\Delta S| = 2$ mixing terms in the mass matrix for the K^o-$\bar{K}^o$ system.

The correlation length for a neutrino beam with definite momentum $p \gg m_\nu$ is

$$L_P \simeq 4\pi \frac{p}{|m_{\nu_1}^2 - m_{\nu_2}^2|} \simeq \frac{p/(1 \text{ MeV})}{4|m_{\nu_1}^2 - m_{\nu_2}^2|/(1 \text{ eV})^2} (10\text{m}), \qquad (2.8)$$

whereas the amplitude of the oscillations is proportional to the strength of the mixing*[18a]. For $p = 2$MeV, $|m_{\nu_1} - m_{\nu_2}| = 1\% \cdot m_{\nu_1}$, $m_{\nu_1} = 2$ eV: $L_p = 250$ m.)

It has been stated that the neutrino beam oscillations are not a necessary consequence of vectorlike weak

---

*A complete discussion of lepton mass corrections in terms of a cutoff parameter similar to ours has been given by Mr. Gell-Mann and J. G. Stephenson, Ref. 18a.

currents.

c) Left- and right-handed Cabibbo angles

Starting from unmixed states (d",s") but a nonvanishing (left-handed) Cabibbo angle $\theta_L (=\theta_c \simeq 0.2)$ we evaluate the diagram of Fig. 6

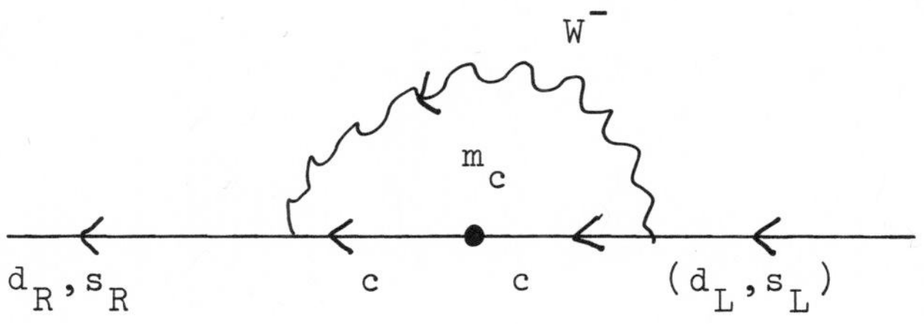

induced by the L-R bridge as indicated in Fig. 1. Using the same notation as in a) and b)

$$(d_R^+, s_R^+)\ m^{(d)} \begin{pmatrix} d_L \\ s_L \end{pmatrix} + \text{h.c.}$$

one obtains

$$m^{(d)} = \begin{pmatrix} m_d & 0 \\ -\sin\theta_L r m_c & \tilde{m}_s \end{pmatrix} ,$$

$$\tilde{m}_s = m_s + \cos\theta_L\ r\ m_c \quad . \tag{2.9}$$

The triangular matrix in eq. (2.9) is decomposed as follows:

$$m^{(d)} = U_R(\theta_R)\rho_D\ U_L^{-1}(\Delta\ \theta_L) , \tag{2.10}$$

where $\rho_D = \begin{pmatrix} m'd & 0 \\ 0 & m's \end{pmatrix}$ is the diagonal matrix with the corrected masses of d and s as eigenvalues and

$$U(\theta)=\begin{pmatrix} \cos\theta & \sin\theta \\ -\sin\theta & \cos\theta \end{pmatrix} ,$$

$$m^{(d)}m^{(d)+} = U_R(\theta_R)\ \rho_D{}^2\ U_R^{-1}\ (\theta_R) ,$$

$$m^{(d)+}m^{(d)} = U_L(\Delta\theta_L)\ \rho_D{}^2\ U_L^{-1}\ (\Delta\theta_L) . \qquad (2.11)$$

The eigenstates of the mass matrix $(d_1)_{L,R}$, $(s_1)_{L,R}$ are related to the eigenstates coupled to the charged weak current

$$(d_L{}'\ s_L{}'),\ (d_R{}'',\ s_R{}'')$$

by

$$(d_L{}',\ s_L{}') = U_L(\theta_L+\Delta\theta_L)\ (d_{1_L},\ s_{1_L}) ,$$

$$(d_R{}'',\ s_R{}'') = U_R(\theta_R)\ (d_{1_R},\ s_{1_R}) . \qquad (2.12)$$

One finds to lowest order:

$$\theta_R=-(\sin\theta_c)r\ \frac{m_d m_c}{m_s{}^2-m_d{}^2} \simeq -\ \sin\theta_c\ r\frac{m_d m_c}{m_s{}^2} ,$$

$$\Delta\theta_L=-(\sin\theta_c)r\ \frac{m_s m_c}{m_s{}^2-m_d{}^2} \simeq -\ \sin\theta_c\ r\frac{m_c}{m_s} . \qquad (2.13)$$

For $r\simeq \frac{\alpha}{\pi}$; $\frac{m_d m_c}{m_s{}^2} \simeq 1$ one finds

$$|\theta_R|\simeq 0.5 \cdot 10^{-3} . \qquad (2.14)$$

Although the above analysis seems straightforward it is puzzling in one respect: If the Cabibbo angle (θ_c) is turned off it is <u>not</u> generated through the above mechanism ($\Delta\theta_c \propto \sin\theta_c$); thus a natural scale for mixing angles is established by $|\theta_R| \simeq 0.5 \cdot 10^{-3}$ and it looks like θ_c is substantially enhanced relative to these crude expectations for angles like θ_R.

This fact was taken seriously by Glashow[19] in his lecture at this conference in which he chose to switch the roles of d''_R and s''_R or equivalently, to start out from an angle $\theta_R \sim 90°$. In this case the (lefthanded) Cabibbo angle does become calculable (modulo r).

The induced θ_R of the order as given in eq.(2.14) does not allow, always within $SU2^W \times U1$, to separate s_R'' and d_R'' and make s_R'' part of an $SU2^W$ doublet and d_R'' part of an $SU2^W$ singlet because of the generation of a dangerous large strangeness changing neutral current with a component proportional to

$$\theta_R \; \bar{s} \; \gamma_\mu \; \frac{1-\gamma_5}{2} \; d \qquad ,$$

which is not sufficiently suppressed by θ_R. This critique applies to the model based on the hierarchy $E_7 \supset \ldots \supset SU3^c \times SU2^W \times U1$ as discussed by Gürsey at this conference[21].

3. RADIATIVE DECAYS OF LEPTONS AND QUARKS

The left-right bridges $(E^- N_E)_L \leftrightarrow (e^- N_E)_R$ for leptons and $(cs')_L \leftrightarrow (cs)_R$ for quarks occur with different charge assignment in the models of Prentki and Zumino[22] and of Georgi and Glashow[23]. They give rise to magnetic and electric dipole transitions according to the diagrams of Fig. 7 of the type

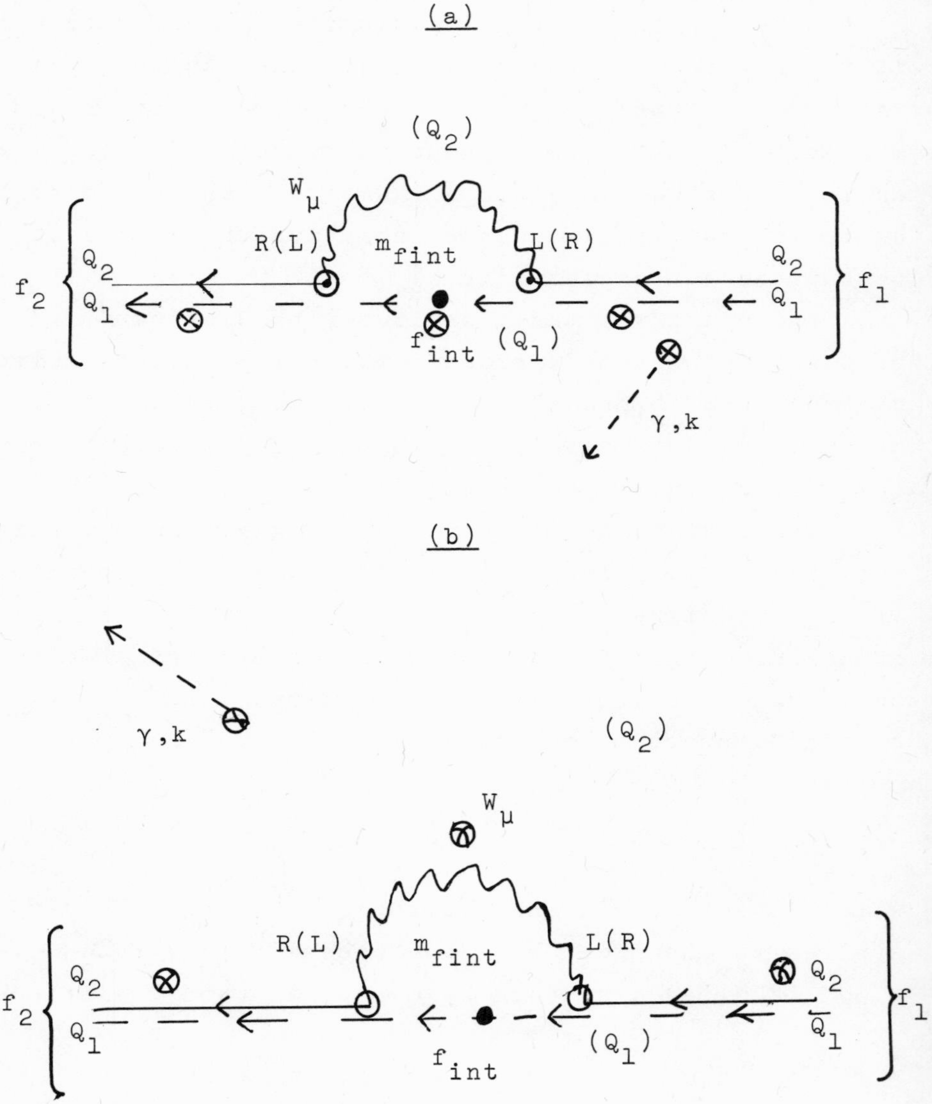
(a)
(Q2)
Wμ
R(L)
mfint
L(R)
Q2
Q1
f2
Q2
Q1
f1
fint
(Q1)
γ,k
(b)
γ,k
(Q2)
Wμ
R(L)
mfint
L(R)
Q2
Q1
f2
Q2
Q1
f1
fint
(Q1)

FIGURE 7

A wavy (massive) W-line denotes the contributions from the (gauge dependent) W-propagator and from the corresponding Goldstone boson. The calculation can be divided according to the two fluid concept in Fig. 7 in the radiation off the intermediary fermion with charge Q_1 and off the intermediary vector boson with charge Q_2 depicted in Fig. 7 a) and b).

One is evaluating generalized weak contributions to the anomalous magnetic moment of the electron (muon)[24]. One obtains

$$m=2(Q_2-Q_1)\left(\frac{e\ G_F}{\sqrt{2}\cdot 4\pi^2}\right)m_{f_{int}}\ \bar{u}_2\left(\frac{1}{2}\ \sigma_{\mu\nu}\ F^{\mu\nu}\right)\frac{1\pm\gamma_5}{2}\ u_1 \quad ,$$

$$F_{\mu\nu} = i[\varepsilon_\mu k_\nu - \varepsilon_\nu k_\mu], \qquad \varepsilon:\ \text{Polarization vector of the photon,} \tag{3.1}$$

+: L → R transition, −: R → L transition.

If f_1 also couples to a light partner f_1': $(f_1,f_1')\rightarrow(E^-_R,\nu_{e_R})$ for the leptonic case, $\rightarrow (s_L,u_L)\sin\theta_c$ for hadrons we have

$$\frac{\Gamma(f_1 \rightarrow f_2+\gamma)_-}{\Gamma(f_1 \rightarrow f_1'e^-\bar{\nu}_e)} = 6(Q_2-Q_1)^2\left(\frac{\alpha}{\pi}\right)\left(\frac{m_{int}}{m_{f_1}}\right)^2 \tag{3.2}$$

for m'_{f_1}, $m_e \ll m_{f_1}$; m_{f_1}, $m_{fint} \ll m_W$.

In the case of $E^-\rightarrow (e^-)_R+\gamma$, eq. (3.2) gives

$$\frac{\Gamma(E^- \rightarrow e^-+\gamma)}{\Gamma(E^- \rightarrow \gamma_{eR}+e^-_L+(\bar{\nu}_e)_R)} = 6\frac{\alpha}{\pi}\left(\frac{m_{N_E}}{m_{E^-}}\right)^2 . \tag{3.3}$$

We note the new mass scale m_{N_E} which governs the radiative decay $E^- \to e^- \gamma$. This decay is enhanced relative to the estimates which would involve α and the kinematic decay parameters only if $m_{N_E} > m_{E^-}$.

One may envisage the possibility that N_E has not been observed for a given stage of experiments because of its large mass, and yet would induce an unexpected enhancement of the above radiative decay.

For $m_{E^-} = 2$ GeV and $m_{N_E} = 5(10)$ GeV the branching ratio for the radiative decay of E^- is 4.5% (15%).

The above analysis has been applied to the study of the radiative decay of hyperons through the subprocess $s \to d+\gamma$, mediated by the charmed quark over the L-R bridge $(cs')_L(cs'')_R$ according to Fig. 8[25]. It has been argued by Ahmed and Ross, that the above subprocess dominates the decay $\Sigma^+ \to p+\gamma$.

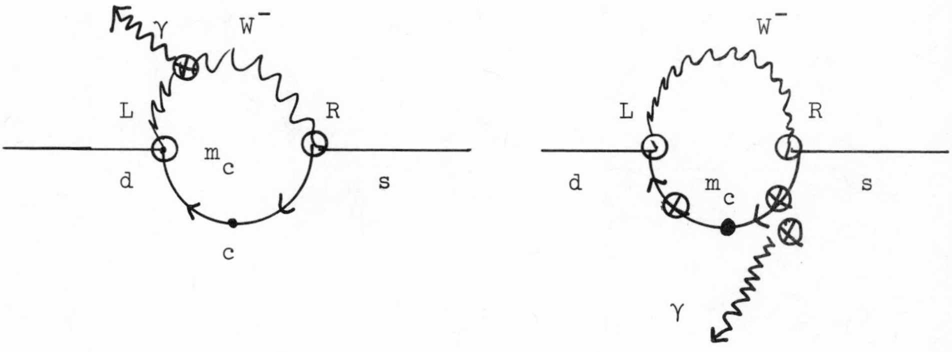

FIGURE 8

The transition amplitude for this decay is of the general form

$$m_{\Sigma^+ \to p\gamma} = \bar{u}_p \frac{1}{2} \sigma_{\mu\nu} F^{\mu\nu} \frac{1-\gamma_5}{2} u_\Sigma A_L$$

$$+ \bar{u}_p \frac{1}{2} \sigma_{\mu\nu} F^{\mu\nu} \frac{1+\gamma_5}{2} u_\Sigma A_R \quad . \tag{3.4}$$

The subprocess of Fig. 8 if dominating the decay $\Sigma^+ \to p\gamma$ leads to the following amplitudes for $s_R'' \simeq s_R$:

$$A_R \simeq 0, A_L \simeq -\sin\theta_c \sigma \frac{5G_F m_p^2}{\sqrt{2}\, 6\pi^2} \frac{m_c}{m_p} \frac{e}{m_p} \quad ,$$

$$\alpha_{\Sigma^+ p\gamma} = \frac{|A_L|^2 - |A_R|^2}{|A_L|^2 + |A_R|^2} \simeq +1 \quad , \tag{3.5}$$

where σ is defined by the matrix element of the tensor current $\bar{d}\, \sigma_{\mu\nu}\, s$:

$$\langle p|\bar{d}\, \sigma_{\mu\nu} s|\Sigma^+\rangle = \sigma\, \bar{u}_p\, \sigma_{\mu\nu}\, u_\Sigma \quad , \tag{3.6}$$

with $\sigma \simeq -1/3$ in the nonrelativistic SU6 limit.

Using this value for σ and $m_c \sim 1.5$ GeV one finds

$$|A_L| \simeq 0.6 \cdot 10^{-7}\, e/m_p,\; A_R = 0 \quad . \tag{3.7}$$

For the choice $s_R'' \simeq d_R$ [7,25] one obtains

$$A_L = 0,\; A_R = \cos\theta_c \sigma \frac{5G_F m_p^2}{\sqrt{2}\, 6\pi^2} \frac{m_c}{m_p} \frac{e}{m_p} \simeq 3.0 \cdot 10^{-7}\, e/m_p \quad ,$$

$$\alpha_{\Sigma^+ p\gamma} = -1 \quad . \tag{3.8}$$

The measurement of the asymmetry parameter[26]

$$\frac{d\Gamma_{\Sigma p\gamma}}{d\cos\theta} = \frac{1}{2}\,\Gamma_{\Sigma p\gamma}\,(1 + \alpha_{\Sigma p\gamma}\cos\theta\,[\vec{p},\,\vec{s}_{\Sigma}])\ ,$$

$\vec{p}$: proton momentum, $\vec{s}_{\Sigma}$ = spin direction of a hypothetical 100% polarized Σ and the branching ratio for the radiative decay give

$$\Gamma_{\Sigma p\gamma}/\Gamma^{\Sigma}_{tot} = (1.24 \pm 0.18)\ 10^{-3}\ ,$$

$$\alpha_{\Sigma p\gamma} = -1.03\ {}^{+\ 0.5}_{-(0.4)}\ . \tag{3.9}$$

For $\alpha_{\Sigma p\gamma} = -1$ one then finds

$$A_L = 0,\ |A_R| = 1.15 \cdot 10^{-7}\ e/m_p\ .$$

One may then try to argue that the $\bar{c}\,\gamma_\mu \frac{1-\gamma_5}{2}\,d$ righthanded current dominates the radiative hyperon decays through the radiative transition $s_L \to d_R + \gamma$. This mechanism implies that all asymmetry parameters for the radiative transition $\Xi^\circ \to \Lambda\gamma$, $\Lambda \to n\,\gamma$ and $\Xi^- \to \Sigma^-\gamma$ are near -1. A. Zee, in his talk presented at this conference, mentioned further experimental work on the radiative decay $\Xi^\circ \to \Lambda\gamma$ by Overseth et al., with the result

$$\alpha_{\Xi^\circ \to \Lambda\gamma} = 0.2 \pm 0.3\ .$$

We conclude from the above estimates combined with the analysis of the nonleptonic weak decays to which we now turn that the mechanism of the radiative decays

$s_L \to d_R + \gamma$ inside the baryons has to be rejected as a dominating decay process even though it would explain the large negative asymmetry of the decay $\Sigma^+ \to p\gamma$.

4. NONLEPTONIC DECAYS OF STRANGE PARTICLES

Let us first compare the two decays $\Sigma^+ \to p\pi^\circ$ and $\Sigma^+ \to p\gamma$.

a) $\Sigma^+ \to p\gamma$.

The amplitudes $A_{L,R}$ in eq. (3.4) are proportional to the helicity amplitudes

$$T_{\Sigma^+\to p\gamma} = d^{1/2}_{-1/2\ \ 1/2}(\theta)\ T^\gamma_R$$

$$\lambda_p - \lambda_\gamma \quad \lambda_\Sigma$$

$$+ d^{1/2}_{1/2\ \ 1/2}(\theta)\ T^\gamma_L \quad ,$$

$$T^\gamma_R = -c\ A_R,\ T^\gamma_L = c\ A_L,\ c = 4\sqrt{2}\ m_\Sigma k,\ k = \frac{m^2_\Sigma - m^2_p}{2m_\Sigma} \quad ,$$

$$\Gamma_{\Sigma^+\to p\gamma} = \frac{k}{16\pi\ m^2_\Sigma}\left(|T^\gamma_R|^2 + |T^\gamma_L|^2\right) \quad ,$$

$$\alpha_{\Sigma^+\to p\gamma} = \frac{|T^\gamma_L|^2 - |T^\gamma_R|^2}{|T^\gamma_L|^2 + |T^\gamma_R|^2}\ (= -\ 1.03^{+0.5}_{-0.4}) \quad . \qquad (4.1)$$

In the forward or backward direction of the proton momentum relative to the spin of Σ, T^γ_L and T^γ_R describe the following processes as shown in Fig. 9.

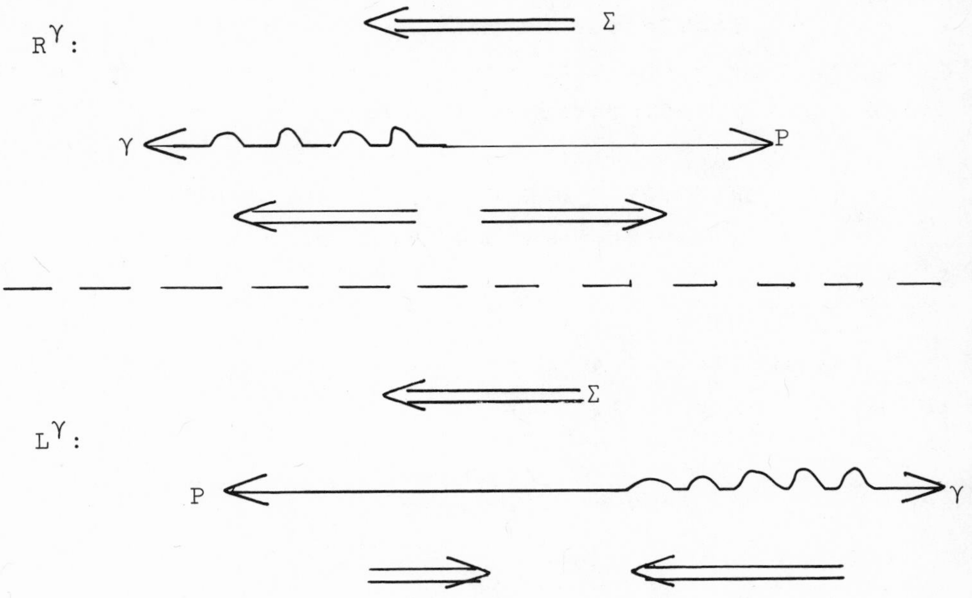

FIGURE 9

b) $\Sigma^+ \to p\pi°$.

As in a), we describe the decay $\Sigma^+ \to p\pi°$ by the helicity amplitudes

$$T_{\Sigma^+\to p\pi°} = d^{1/2}_{-1/2\ \ 1/2}(\theta)\, T_L^{\pi°} + d^{1/2}_{1/2\ \ 1/2}(\theta)\, T_R^{\pi°}$$

$$\lambda_p - \lambda_{\pi°}$$
$$(=0)$$

$$\Gamma_{\Sigma^+\to p\pi°} = \frac{P}{16\pi\, m_\Sigma^2}\left(|T_L^{\pi°}|^2 + |T_R^{\pi°}|^2\right) \quad ,$$

$$\alpha_{\Sigma^+\to p\pi°} = \frac{|T_R|^2 - |T_L|^2}{|T_R|^2 + |T_L|^2}\ (= -\ 0.98 \pm 0.02)\ . \qquad (4.2)$$

The relations to the Lorentz invariant amplitudes are given by

$$T_{\Sigma^+\to p\pi^\circ} = \bar{u}_p\,(A+\gamma_5 B)_{\Sigma^+ p\pi^\circ}\, u_\Sigma$$

$$= \psi_p^+ (f_1 + f_2\, \vec{e}_p \vec{\sigma})_{\Sigma^+ p\pi^\circ}\, \psi_\Sigma \quad ,$$

$$\begin{pmatrix} f_1^{\pi^\circ} \\ f_2^{\pi^\circ} \end{pmatrix} = [(m_\Sigma \pm m_p)^2 - m_\pi^2]^{1/2} \begin{pmatrix} A^{\pi^\circ} \\ B^{\pi^\circ} \end{pmatrix} \quad ,$$

$$T_L^{\pi^\circ} = (f_1 - f_2)^{\pi^\circ}, \quad T_R^{\pi^\circ} = (f_1 + f_2)^{\pi^\circ} \quad . \qquad (4.3)$$

As in the radiative Σ^+ decay for $\Sigma^+ \to p\pi^\circ$ the helicity amplitudes $T_L^{\pi^\circ}$, $T_R^{\pi^\circ}$ describe the subprocesses for proton momenta parallel or antiparallel to the spin direction of Σ^+ as given in Fig. 10.

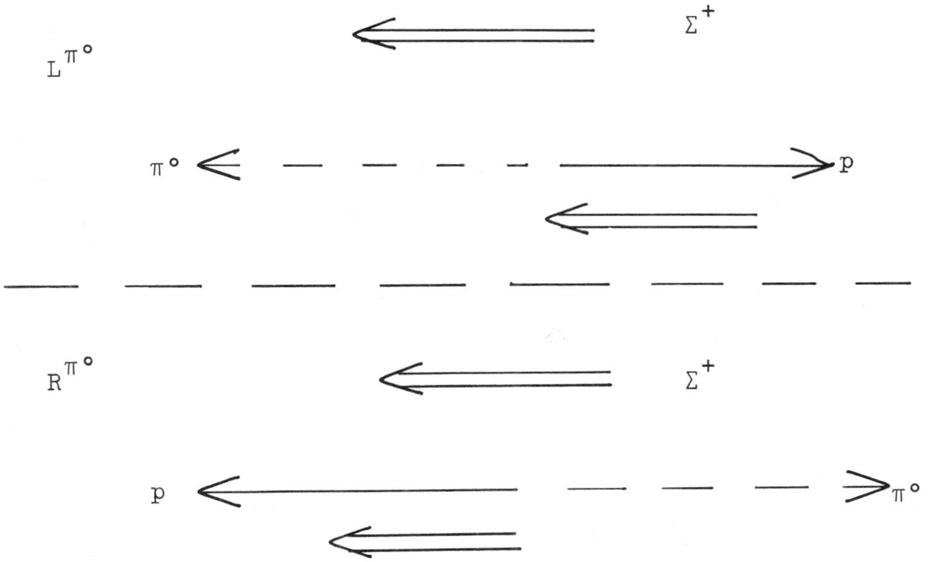

FIGURE 10

If we adopt $\alpha_{\Sigma p\gamma} \simeq -1$, for the sake of definiteness (a new measurement of this parameter with improved precision would be, I think, of crucial significance) we are led to a striking parallel between π° and γ emission in the transition $\Sigma^+ \to p$. Together with $\alpha_{\Sigma^+ p\pi^\circ} = -0.98 \pm 0.02$ this corresponds to

$$|T_L^{\pi^\circ}| \simeq 1.33 \cdot 10^{-6}\ m_p, |T_R^{\gamma}/T_L^{\pi^\circ}| \simeq 0.05 \ .$$

Within quantum chromodynamics there exists another parallel between the subprocesses

$$s \to d + \gamma \leftrightarrow s \to d + \text{glue}.$$

The latter process corresponds to the emission of a vector gluon in the transition $s \to d$. It proceeds according to the diagram in Fig. 11 (to be compared with the process in Fig. 8).

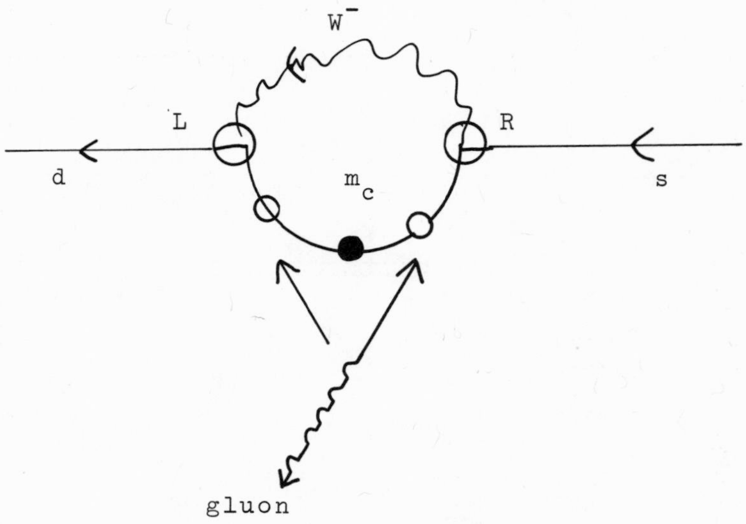

FIGURE 11

For $s_R'' \simeq s_R$ one obtains[27]

$$T_{s\to d+glue} = \sin\theta_c \frac{G_F}{\sqrt{2}} K \, m_c \, \bar{u}_d \, \sigma_{\mu\nu} \, G^{\mu\nu} \, \frac{1-\gamma_5}{2} \, u_s \, ,$$

$$K = \frac{g}{4\pi^2} + O(g^3), \quad G_{\mu\nu} = \sum_A G^A_{\mu\nu} \frac{\chi^A}{2} \, , \qquad (4.4)$$

$$\chi^A : \quad 8 \text{ matrices of } SU3^{(c)} .$$

In eq. (4.4) K denotes a parameter of the strong interactions, given within QCD as a function of the coupling constant g (and the quark masses).

$$L_q = \bar{q} \, (i\gamma^\mu \partial_\mu - g \, \gamma^\mu V_\mu) q \quad ,$$

$$V_\mu = \sum_A V^A_\mu \frac{\chi^A}{2}, \quad G_{\mu\nu} = \partial_\nu V_\mu - \partial_\mu V_\nu + i \, g | V_\nu , \, V_\mu | \, .$$

In the sense of the short distance expansion[28] eq. (4.4) corresponds to an induced effective Hamiltonian density

$$H^{II}_{|\Delta S|=1} = \frac{G_F}{\sqrt{2}} K_M \, m_c \left(- \sin\theta_c \right) \left\{ \begin{array}{l} \bar{d}_L \, \sigma_{\mu\nu} \, G^{\mu\nu} \, s_R \\ \\ + \bar{s}_R \, \sigma_{\mu\nu} \, G^{\mu\nu} \, d_L \end{array} \right\}_M . \quad (4.5)$$

The subscript M in eq. (4.5) means that the form factors of the local operators $\bar{d}_L \, \sigma_{\mu\nu} \, G_{\mu\nu} \, s_R$, $\bar{s}_R \, \sigma_{\mu\nu} \, G_{\mu\nu} \, d_L$ have to be normalized to a given value at momentum transfers characterized uniquely by M. We note that $H^{II}_{|\Delta S|=1}$ only induces transitions with $\Delta I = 1/2$.

H^I stands for the directly induced current-current Hamiltonian density

$$H^{I} = \frac{4G_F}{\sqrt{2}} [(\bar{u}d')_L + (\bar{c}s')_L + (cs'')_R + \ldots]$$

$$[(\bar{d}'u)_L + (\bar{s}'c)_L + (\bar{s}''c)_R + \ldots]. \qquad (4.6)$$

Defining the octet operators (with respect to the u, d, s flavor SU3)

$$\text{parity conserving:} \quad S_i = Km_c\, \bar{q}\, \sigma_{\mu\nu}\, G^{\mu\nu} \frac{\lambda_i}{2} q\,,$$

$$C_n = +1$$

$$\text{parity violating:} \quad P_i = (-i)\, Km_c \bar{q}\, \sigma_{\mu\nu}\, G^{\mu\nu} \frac{\lambda_i}{2} \gamma_5\, q,$$

$$C_n = + \qquad (4.7)$$

λ_i: eight SU3 matrices.

One can expand $H^{II}_{|\Delta S|=1}$ in the form:

$$H^{II}_{|\Delta S|=1} = \frac{4G_F}{\sqrt{2}} \sin\theta_c (S_6(C_n=+) + P_7(C_n=+)) \quad . \qquad (4.8)$$

$H^{I}_{|\Delta s|=1}$ contains the contribution from the product of left-handed currents

$$H^{I\,LL}_{|\Delta S|=1} = \frac{4G_F}{\sqrt{2}} \sin\theta_c \cos\theta_c \left\{ \begin{array}{ll} (\bar{s}u)_L & (\bar{u}d)_L \\ +(\bar{d}u)_L & (\bar{u}s)_L \end{array} \right\}$$

$$\sim S_6' \, (C_n=+) + P_6' \, (C_n=-)\,, \qquad (4.9)$$

(octet part only)

and for $s''_R \simeq s_R$ the characteristic product of left-handed and right-handed currents involving pairs of c-quark and $\bar{c}$-antiquark.

$$H^{I\ LR}_{|\Delta S=1|} = -\frac{4G_F}{\sqrt{2}} \sin\theta_c \begin{Bmatrix} (\bar{d}c)_L & (\bar{c}s)_R \\ (\bar{s}c)_R & (\bar{c}d)_L \end{Bmatrix}$$

$$\sim S''_6\ (C_n=+) + P''_7\ (C_n=+). \qquad (4.10)$$

The combination of S_6 and P_7 (i.e., equal positive natural charge conjugation parity of the scalar- and pseudoscalar octets forming $H^{II}_{|\Delta S|=1}$) in eq. (4.8) and (4.10) as opposed to the combination S'_6 and P'_6 (i.e., opposite natural charge conjugation parity for the scalar (+) and pseudoscalar (-) octets in $H^{LL}_{|\Delta S|=1}$) which occur in the reduction of the product of left-handed currents in eq. (4.9) is a most definite signature of a L-R bridge (as specified in Fig. 1).

Let us first discuss the problem of whether $H^{II}_{|\Delta S|=1}$ can dominate $H^{I\ LL}_{|\Delta S|=1}$ in nonleptonic weak transitions.

A crude but simple estimate of the relevant decay amplitudes can be obtained by calculating the ratio of two hypothetical decays

$$\Gamma[s \to d + \text{glue}]/\Gamma[s \to \bar{u}\ u\ d] = \rho\ ,$$

neglecting the strong final state interactions which do not allow colored quanta to become separated. One obtains

$$\rho \simeq \frac{8}{3}\left(\frac{\kappa_M}{\pi}\right)\left(\frac{m_c}{m_s}\right)^2_M \quad ,$$

$$M \simeq 1\text{-}3 \text{ GeV}, \quad \kappa = g^2/4\pi \ . \qquad (4.11)$$

Typical estimates of κ_M evaluated in the range M = 1-3 GeV yield a value $\kappa \sim 1/3^3$. Thus for $m_c/m_s \simeq 16$ one obtains

$$\rho \simeq 80. \qquad (4.12)$$

The enhancement of strange particle decays induced by $H^{II}_{|\Delta S|=1}$ is due, in our view, to the large mass $m_c(m_h)$ which exceeds the mass scales which characterize these processes kinematically. Only if left- and right-handed currents are coupled to the same vector boson can the above generalized radiative transitions occur to lowest order in G_F[24].

What we mean by "mutatis mutandis" can be illustrate considering the effect of a L-R bridge involving a vector boson W' different from $W^{\pm}$ and a heavy quark flavor h different from c yielding an induced Hamiltonian density very similar to $H^{II}_{|\Delta S|=1}$ in eq. (4.5),

$$H^{II}_{|\Delta S|=1} \rightarrow \frac{G'_F}{\sqrt{2}} K_M \, m_h \, f[\theta_c,]\times[\bar{d}_L \, \sigma_{\mu\nu} \, G^{\mu\nu} \, s_R + \text{h.c.}],$$

$$G'_F = G_F \frac{m_{W'}{}^2}{m_N{}^2} f[\theta_c,\ldots] \quad ; \qquad (4.13)$$

$f[\theta_c,\ldots]$: appropriate function of the Cabibbo angle and other eventual weak interaction angles.

The properties of strange particle decays cannot distinguish between the different mechanisms leading to eq. (4.5) or (4.13) except for the strength of the $\Delta I=1/2$ enhancement as confronted with the rather uncertain theoretical evaluation of parameters like ρ in eq. (4.11). It will be very interesting to evaluate ρ in the bag model to obtain a better feeling for the crude estimate of eq. (4.12).

Let us now compare the characteristic features of a dominating $H^{II}_{|\Delta S|=1}$ as opposed to a dominating $H^{I\ LL}_{|\Delta S|=1}$

1) The enhancement of ($\Delta I=1/2$) matrix elements of $H^{II}_{|\Delta S|=1}$ relative to ($\Delta I=3/2$) transition elements is due to (m_c/m_s); the hypothesis of octet enhancement for $H^{I\ LL}_{|\Delta S|=1}$ is not convincing[29].

2) The decay $K_s \to 2\pi$ induced by $H^{II}_{|\Delta S|=1}$ (P_7) is not inhibited by SU3 forbiddenness contrary to the case of $H^{I\ LL}_{|\Delta S|=1}$ (P_6)[30].

3) $K \to 3\pi$ and $K \to 2\pi$ relations. Because only d_L is involved in $H^{II}_{|\Delta S|=1}$ and $H^{I\ LL}_{|\Delta S|=1}$ the axial charge of the nonstrange chiral SU2 × SU2 group acts in the following way:

$$\left[Q_3^{(5)},\ H^{I,\ II}_{\Delta S=1}\right] = \sigma\left[Q_3,\ H^{I,\ II}_{\Delta S=1}\right] ,$$

where σ is the same for $\Delta I=1/2$ and $\Delta I=3/2$. Then the sum rule due to Bouchiat and Meyer[31] is valid in both cases.

4) $\Sigma^+ \to p\gamma$: $\alpha_{\Sigma^+ p\gamma}$ is not restricted for transitions

induced by $H^{II}_{|\Delta S|=1}$ (S_6+P_7). For $H^{ILL}_{|\Delta S|=1}$ $(S_6'+P_6')$ $\alpha_{\Sigma^+ p\gamma}=0$ in the limit of U-spin invariance[31a]. This fact is related to the vanishing of $K_s \to 2\pi$ in the SU3 symmetry limit.

5) Hyperon decays. Using the parameterizations of eq. (4.3)

$$<B_2\pi|H_{|\Delta S|=1}|B_1> = \bar{u}_2 \; (A^{21} + \gamma_5 \; B^{21})u_1 \qquad (4.14)$$

and the PCAC reduction

$$<B_2\pi°|H_{|\Delta S|=1}|B_1> \simeq (-\frac{i}{f_\pi}) <B_2|[Q_3^{(5)}, H_{|\Delta S|=1}]|B_1>$$

$$+ \frac{1}{f_\pi} q_\pi^\mu \int d^4x \; e^{+iq_\pi} \times (B_2|T(J_{\mu 3}^{(5)}(x) \; H^{(0)}_{|\Delta S|=1}|B_1> \qquad (4.15)$$

($f\pi \; g_{\pi N} \simeq g_A \; m_N$ Goldberger and Treiman relation), one obtains within the baryon pole model the following soft pion approximations to the amplitudes A^{21}, B^{21}

$$A^{(\pi°)} \simeq -\frac{1}{2f_\pi} <B_2|H^{P.C.}_{\Delta S=1}|B_1> = -\frac{1}{2f_\pi} h^{P.C.}_{21} \quad ,$$

$$B^{(\pi°)} = -\frac{1}{2f_\pi} <B_2|H^{P.V.}_{\Delta S=1}|B_1> + \frac{1}{\delta m} <B_2|[g^{\pi°}, h^{P.C.}]|B_1>, \qquad (4.16)$$

$g^{\pi°}$: coupling matrix of $\pi°$ to $(\bar{B}B)$.

In eq. (4.16) δm is left as a free parameter related to but not equal to the mass splitting in the baryon octet. Since one is extrapolating the above amplitudes to $q_\pi \simeq 0$ it is not correct to retain the physical masses of the baryons and hence $\delta_m \neq m_\Sigma - m_N$ even in the approximation $m_\Lambda = m_\Sigma$. We do not admit K^* dominated s-wave amplitudes in the soft pion limit[32]. For P_6' the p-wave commutator term $<B_2|H^{P.V.}_{\Delta S=1}|B_1>$ vanishes in the limit of SU3 invariance for the same $C_n(P_6')$=-property causing $K_s \to 2\pi$ and $\alpha_{\Sigma^+ p\gamma} = 0$.

Although we can account for all the hyperon decay amplitudes to within the errors of their determination choosing

$$(d/f)[H^{P.C.}] = -0.3 \quad ,$$
$$(d/f)[H^{P.V.}] = 1.8 \quad ,$$

$$\delta m \simeq 75 \text{ MeV} \quad ,$$
$$d\,[H^{P.V.}]/d[H^{P.C.}] \simeq 22.5 \quad . \qquad (4.17)$$

The large p-wave in $\Sigma^+ \to n\pi^+$ remains unexplained. It is this fact which is relevant for the almost maximal asymmetry parameter $\alpha_{\Sigma^+ \to p\pi^\circ} \simeq -1$, which in turn appears to be at least similar to $\alpha_{\Sigma^+ \to p\gamma}$.

5. CP-VIOLATION AS RELATED TO THE $\Delta I = 1/2$ DOMINATING INTERACTION.

The system of left-handed charged currents $(\bar{u}d')_L$, $(\bar{c}s')_L$ when extended to include the rignt-handed $(\bar{c}s)_R$ current can show a CP-violating phase*[34],

---

*This was first remarked by R. N. Mohapatra[7] in conjunction with the $(\bar{c}d)_R$-current. For left-handed charged currents only an extension to six-quark flavors with the charged current couplings $(\bar{u}d')_L$, $(\bar{c}s')_L$, $(\bar{t}b)_L$ can generate a CP-violating phase[33].

$$s''_R = s_R^{(0)} \; e^{+i\delta} \quad ,$$

$$s'_L = \cos\theta_c \; s_L^{(0)} - \sin\theta_c \; d_L^{(0)} \quad ,$$

$$d'_L = \cos\theta_c \; d_L^{(0)} + \sin\theta_c \; s_L^{(0)} \quad . \tag{5.1}$$

One can just as well define $s''_R = s_R$ which gives

$$s''_R = s_R , \; s'_L = \cos\theta_c \; s_L \; e^{-i\delta} - \sin\theta_c \; d_L \quad ,$$

$$d'_L = \cos\theta_c \; d_L + \sin\theta_c \; s_L \; e^{-i\delta} . \tag{5.2}$$

$H^{II}_{|\Delta S|=1}$ as given in eq. (4.5) is part of a Hamiltonia density H^{II} with $|\Delta S|=1$ and $|\Delta S|=0$ components

$$H^{II} = \frac{G_F}{\sqrt{2}} K \; m_c \left\{ \begin{array}{l} \bar{s}'_L \; \sigma_{\mu\nu} \; G_{\mu\nu} \; s_R \\ + \bar{s}_R \; \sigma_{\mu\nu} \; G_{\mu\nu} \; s'_L \end{array} \right\} \tag{5.3}$$

$$= - \frac{G_F}{\sqrt{2}} \sin\theta_c \; (S_6 + P_7)$$

$$+ \frac{G_F}{\sqrt{2}} \cos\theta_c \; [\cos\delta \; S_{\bar{s}s} + \sin\delta \; P_{\bar{s}s}]$$

(↓ $P_{\bar{s}s}$: CP-violating term,)

$$S_{\bar{s}s} = K \; m_c \; (\bar{s} \; \sigma_{\mu\nu} \; G_{\mu\nu} \; s) \quad ,$$

$$P_{\bar{s}s} = (-ik \; m_c)(\bar{s} \; \sigma_{\mu\nu} \; G_{\mu\nu} \; \gamma_5 \; s) \quad .$$

$H^{I\;LL}_{|\Delta S|=1}$ now inherits the phase δ:

$$H^{I\ LL}_{|\Delta S|=1} = \frac{4G_F}{\sqrt{2}} \sin\theta_c \cos\theta_c \left\{ \begin{array}{l} e^{+i\delta} (\bar{s}u)_L (\bar{u}d)_L \\ +e^{-i\delta} (\bar{d}u)_L (\bar{u}s)_L \end{array} \right\} . \tag{5.4}$$

Thus if we consider only effects due to the $|\Delta S|=1$ interactions, CP violation only becomes observable if amplitudes induced by H^{II} interfere with amplitudes induced by $H^{I\ LL}$. If, however, H^{II} dominates the non-leptonic decays (as compared to H^{I}) as we proposed in the last chapter, this effect may be secondary to the $|\Delta S|=2$ effects governing the mass - and decay matrix relevant to the K^o-$\bar{K}^o$ system.

Using the K^o, $\bar{K}^o$ basis states $|K^o\rangle = |\bar{d}s\rangle$, $\bar{K}^o = |\bar{d}s\rangle$ with d and s defined in eq. (5.2), we define the mass- and decay matrix of the K^o-$\bar{K}^o$ system

$$m = \begin{pmatrix} M & \mu \\ \mu^* & M \end{pmatrix} - \frac{i}{2} \begin{pmatrix} \Gamma & -\tilde{\Gamma} \\ -\Gamma^* & \Gamma \end{pmatrix} ,$$

$$\mu = |\mu| e^{-2i\theta} , \ \tilde{\Gamma} = |\tilde{\Gamma}| e^{-2i\tau} . \tag{5.5}$$

Since the $|\Delta S|=1$ decays induced by H^{II} have been defined such as not to depend on δ, μ and Γ will in this basis be complex and cannot be redefined to real values.

The width parameters Γ, $\tilde{\Gamma}$ are given by

$$\Gamma = \frac{1}{2m_K} \int d\Omega_{(n)} \; <\overset{(-)}{K^\circ}|H_{|\Delta S|=1}|n><a|H_{|\Delta S|=1}|\overset{(-)}{K^\circ}> \; ,$$

$$\tilde{\Gamma} = \frac{1}{2m_K} \int d\Omega_n \; <K^\circ|H_{|\Delta S|=1}|n><u|H_{|\Delta S|=1}|\bar{K}^\circ> \; . \qquad (5.6)$$

To lowest order in θ, τ one finds

$$2\Gamma = \gamma_s + \gamma_L \simeq \gamma_s \; ,$$

$$2\tilde{\Gamma} = \gamma_s - \gamma_L \simeq \gamma_s \; ,$$

$$2|\mu| = m_L - m_s = \Delta m \; . \qquad (5.7)$$

The complex eigenvalues of m are [34a] determined from the equations

$$\chi = m - i\,\frac{\gamma}{2},\; M - m = x,\; |\mu| = x_o (\geq 0) \; ,$$

$$\frac{\Gamma - \gamma}{2} = y \; , |\Gamma| = y_o (\geq 0) \; ,$$

$$x^2 - y^2 = (x^o)^2 - (y^0)^2 ,$$

$$xy = -x^o y^o + 2\, x^o y^o \sin^2(\theta - \tau) \; . \qquad (5.8)$$

For $|\theta-\tau| << 1$ the eigenstates K_s, K_ℓ correspond to

$$m_s = M-|\mu| - \frac{2\, x_o(y_o)^2}{(x_o)^2+(y_o)^2}\sin^2(\theta-\tau) + 0[(\theta-\tau)^4],$$

$$\gamma_s = \Gamma+|\tilde{\Gamma}|+ \frac{4(x_o)^2 y_o}{(x_o)^2+(y_o)^2}\sin^2(\theta-\tau)+0[(\theta-\tau)^4] ,$$

$$m_L = M+|\mu|- \frac{2x_o (y_o)^2}{(x_o)^2+(y_o)^2}\sin^2(\theta-\tau)+0[(\theta-\tau)^4] ,$$

$$\gamma_L = \Gamma-|\tilde{\Gamma}|+ \frac{4(x_o)^2 y_o}{(x_o)^2+(y_o)^2}\sin^2(\theta-\tau)+0[(\theta-\tau)^4] . \tag{5.9}$$

To first order in θ and τ the parameter

$$\varepsilon = \frac{<(2\pi)I=0/H_{weak}/K_\ell>}{<(2\pi)I=0/H_{weak}/K_s>} \simeq \frac{<2\pi|H^{II}_{|\Delta S|=1}|K_\ell>}{<2\pi|H^{II}_{|\Delta S|=1}|K_s>}$$

$$\simeq \frac{1}{(x_o)^2+(y_o)^2}\left[x_o y_o(\theta-\tau)+i[(x_o)^2\theta-(y_o)^2\tau]\right] , \tag{5.10}$$

$$\arg\varepsilon \simeq \text{arc tg}\left[\frac{\theta}{\theta-\tau}\,\frac{2\,\Delta m}{\gamma_s} - \frac{\tau}{\theta-\tau}\,\frac{\gamma_s}{2\,\Delta m}\right]\frac{x_o}{y_o} = \frac{2\,\Delta m}{\gamma_s} .$$

It is puzzling to note that $x_o=y_o$ or $\Delta m = \gamma_s/2$, a relation for which I cannot see any justification except for an accident, one obtains

$$\varepsilon = \frac{\theta-\tau}{2}(1+i) = \arg\varepsilon = 45°, \text{ independent of } \theta,\ \tau. \tag{5.11}$$

In our case

$$\tau = 0\left(|\delta| \left| \frac{<2\pi|H^{I\ LL}|K_s>}{<2\pi|H^{II}|K_s>} \right| \right), \qquad (5.12)$$

and we neglect it in the following relative to θ because of the $\Delta I = 1/2$ enhancement of $<2\pi|H^{II}|K_s>$ relative to $<2\pi|H^{I\ LL}|K_s>$.

Thus the CP-violation in our scheme due to the relation of δ to the $\Delta I = 1/2|\Delta S|=1$ dominating Hamiltonian, is restricted to the mass matrix of the K^0-$\bar{K}^0$ system as in a superweak theory[34b].

We assume that the main contribution to μ results from the short distance part of the $|\Delta S| = 2$ amplitudes according to the diagrams in Fig. 12:

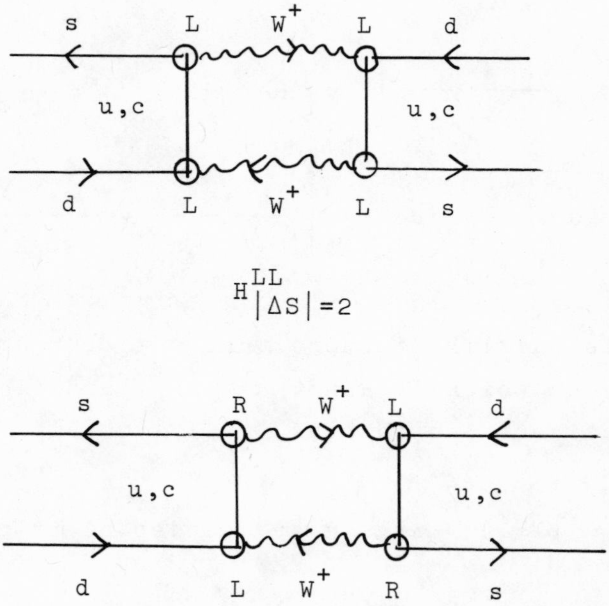

$H^{LR}_{|\Delta S|=2}$

FIGURE 12

The corresponding induced effective $|\Delta S| = 2$ Hamiltonian

$$H^{LL}_{|\Delta S|=2} = \frac{(G_F)^2\, m_c^{\,2}}{\sqrt{2}\,\pi^2} \sin^2\theta_c \cos^2\theta_c \times$$

$$\times \left\{ e^{2i\delta} \left(\bar{s}\, \gamma_\mu\, \frac{1+\gamma_5}{2} d \right) \left(\bar{s}\, \gamma^\mu\, \frac{1+\gamma_5}{2} d \right) \right\} + \text{h.c.} \quad ,$$

$$H^{LR}_{|\Delta S|=2} = - \frac{(G_F)^2\, m_c^{\,2}}{\sqrt{2}\,\pi^2} \sin^2\theta_c \left[\log \frac{m_W^{\,2}}{m_c^{\,2}} - \frac{3}{2} \right] \times$$

$$\times \left\{ \bar{s}\, \gamma_\mu\, \gamma_\nu\, \frac{1+\gamma_5}{2}\, d\ \bar{s}\, \gamma_\nu\, \gamma_\mu\, \frac{1+\gamma_5}{2} d \right\} + \text{h.c.} \quad , \tag{5.13}$$

with

$$\mu = \mu_{LR} + \mu_{LL}\, e^{-2i\delta} = (\mu) e^{-2i\theta} \quad ,$$

$$\mu_{LR} = \frac{1}{2m_K} \langle K^0 | H^{LR}_{|\Delta S|=2} | \bar{K}^0 \rangle \quad ,$$

$$\mu_{LL} e^{-2i\delta} = \frac{1}{2m_K} \langle K^0 | H^{LL}_{|\Delta S|=2} | \bar{K}^0 \rangle \quad . \tag{5.14}$$

Contrary to the situation for $|\Delta S|=1$, $H^{LR}_{|\Delta S|=2}$ does not dominate over $H^{LL}_{|\Delta S|=2}$ by a power of $\frac{m_c}{m_s}$.

The insertion of the vacuum state in the evaluation of $H_{|\Delta S|=2}$ yields μ_{LL} and μ_{LR} both positive; however, one has every right to doubt the validity of the vacuum insertion especially if refined questions like $\frac{\mu_{LL}}{\mu_{LR}} \gtrless 0$ are posed. We note the relations between

$$\theta, \delta, \frac{\gamma_s}{2\Delta m}, \frac{\mu_{LL}}{\mu_{LR}} (> -1),$$

$$|\varepsilon| = |\theta| \left(1+\left(\frac{\gamma_s}{2\Delta m}\right)^2\right)^{-1/2} \simeq \frac{|\theta|}{\sqrt{2}},$$

$$|\delta| = \left(1+\frac{\mu_{LL}}{\mu_{LR}}\right)\left(1+\left(\frac{\gamma_s}{2\Delta m}\right)^2\right)^{1/2}, |\varepsilon| \simeq \left(1+\frac{\mu_{LL}}{\mu_{LR}}\right)\sqrt{2}\,|\varepsilon|. \quad (5.15)$$

The term $\frac{G_F}{\sqrt{2}} \cos\theta_c \sin\delta\, P_{\bar{s}s}$ in $H^{II}_{|\Delta S|=0}$, induces an electric dipole moment for Σ^+. The latter can be estimated through the parity violating amplitude $(A_R - A_L)$ in eq. (3.4) for the radiative decay $\Sigma^+ \to p\gamma$.

Neglecting $\bar{q}q$-pairs it follows

$$\langle p\gamma | P_7 | \Sigma^+ \rangle = 1/2 \langle \Sigma^+, \gamma | P_{\bar{s}s} | \Sigma^+ \rangle .$$

Thus one obtains for the electric dipole moment of Σ^+,

$$\mathcal{D}_{\Sigma^+} \simeq \text{cotg}\,\theta_c\, |\delta|\,|A_R - A_L|,$$

and for $\alpha_{\Sigma p\gamma} = 1$,

$$\mathcal{D}_{\Sigma^+} \simeq (3.2) \cdot 10^{-23} (\text{e cm}) \left(1+\frac{\mu_{LL}}{\mu_{LR}}\right). \quad (5.16)$$

In order to estimate the electric dipole moment of the neutron one needs to know the suppression factor

$$\varepsilon_s = \frac{\langle n\gamma | P_{\bar{s}s} | n \rangle}{\langle \Sigma^+\gamma | P_{\bar{s}s} | \Sigma^+ \rangle} \quad \mathcal{D}_n = \varepsilon_s \cdot \mathcal{D}_{\Sigma^+} .$$

6. THE NEUTRAL CURRENT

If the gauge group $SU2^{weak} \times U1$ forms indeed the lowest level of a hierarchy of weak gauge groups, the neutral current in the scheme of Fig. 1 is a vector current apart from a straightforward modification in the neutral lepton sector. This is not a necessary feature of vectorlike theories, rather it follows from the fact that all fermions belong to a doublet of $SU2^{weak}$.

Nevertheless the fact discussed in section 2 that radiative corrections to the fermion mass matrix generate a righthanded Cabibbo angle ($\theta_R = O(\alpha/\pi \sin\theta_c \log \Lambda^2/m_W^2)$) can only be reconciled with the smallness of the K_L-K_s mass difference and the absence of strangeness changing neutral currents to a high accuracy if (s_L, d_L) as well as (s_R, d_R) are assigned to identical representations of $SU2^{weak}$.

Furthermore, if the large y -anomaly discussed by A. Mann[36] at this conference shall be interpreted as evidence for a (delayed) threshold involving the new piece of the charged current $(\bar{b}u)_R$ [37], the vectorial character of the neutral current is a consequence.

The analysis of the Caltech FNAL neutral current experiment[38] is not yet completed, as it has been realized that a potentially significant deviation from scaling in antineutrino charged current scattering seems to manifest itself, eventually confirming the results of the Harvard-Pennsylvania-Wisconsin collaboration[36,37].

ACKNOWLEDGEMENT

It is a pleasure to thank H. Fritzsch, R. P. Feynman, M. Gell-Mann, S. P. Rosen, R. D. Tripp, P. Ramond, Y. Frishman, B. Barish, F. Sciulli, F. Merritt, S. Pakvasa, Y. Hara, R. M. Barnett, and A. Zee for many elucidating discussions, questions and objections to the ideas presented here.

REFERENCES

1. A. Salam and J. C. Ward, Phys. Letters 13, 168 (1964); S. Weinberg, Phys. Rev. Letters 19, 1264 (1967).
2. Y. Nambu, Preludes in Theoretical Physics, North Holland (1966); H. Fritzsch and M. Gell-Mann, Proceedings of the XVI Int. Conf. on High Energy Physics, Chicago, 1972, Vol. 2; S. Weinberg, Phys. Rev. Letters 31, 494 (1973).
3. K. G. Wilson, Phys. Rev. D10, 2445 (1974); J. Kogut and L. Susskind, Phys. Rev. D11, 395 (1975);K. G. Wilson, Phys. Reports, to be published; K. Johnson, Acta Polonica B6, 865 (1975); J. M. Cornwall and G. Tiktopoulos, Phys. Rev. Letters 35, 338 (1975); UCLA preprint 75/TEP/21, also contributions to this conference. H. D. Politzer, Phys. Reports 14C 130 (1974); H. Fritzsch and P. Minkowski, Nucl, Phys. B76, 365 (1974).
4. M. Gell-Mann, K. G. Wilson, H. D. Politzer, J. C. Pati, K. Johnson, J. M. Cornwall and T. Appelquist: contributions to this conference.
5. J. C. Pati and A. Salam, Phys. Rev. D8, 1240 (1973); H. Georgi and S. L. Glashow, Phys. Rev. Letters 32, 438 (1974); H. Fritzsch and P. Minkowski, Annals of Phys. 93, 193 (1975); F. Gürsey, P. Ramond and P. Sikivie, Yale preprint (1975); F. Gürsey, contribution to this conference.
6. H. Georgi and S. L. Glashow, Phys. Rev. D6, 429 (1972).
7. A. de Rùjula, H. Georgi, and S. L. Glashow, Phys. Rev. Letters 35, 69 (1975); R. N. Mohapatra, Phys. Rev. D6, 2023 (1972).
8. H. Fritzsch, M. Gell-Mann and P. Minkowski, Phys. Letters 59B, 256 (1975); F. A. Wilczek, A. Zee, R. L. Kingsley and S. B. Treiman, Phys. Rev. D12, 2768

(1975); S. Pakvasa, W. Simmons, and S. F. Tuan, Phys. Rev. Letters 35, 702 (1975); A. de Rùjula, H. Georgi, and S. L. Glashow, Phys. Rev. D12, 3589 (1975); for earlier considerations of vectorlike gauge theories see also: R. N. Mohapatra and J. C. Pati, Phys. Rev. D11, 2558 (1975).

9. R. M. Barnett, Phys. Rev. D11, 3246 (1975); H. Harari, Phys. Letters 57B, 265 (1975); F. Wilczek, Princeton University preprint (1975).

10. H. Fritzsch and P. Minkowski, Phys. Letters 56B, 69 (1975).

11. See, e.g., S. Weinberg, Phys. Rev. D7, 2887 (1973) and invited talk given at the Conference on Gauge Theories and Modern Field Theory, Northeastern University, Boston (1975); H. Georgi, H. Quinn, and S. Weinberg, Phys. Rev. Letters 33, 451 (1974); E. Gildener and S. Weinberg, Harvard University preprint (1976).

12. See e.g., L. O'Raifeartaigh, Lecture Notes on Supersummetry, Dublin Institute for Advanced Studies (1975); B. Zumino, Invited talk given at the Conference on Gauge Theories and Modern Field Theory, Northeastern University, Boston (1975), CERN preprint TH 2120 (1975); P. G. O. Freund, contribution to this conference.

13. A. Halprin, P. Minkowski, H. Primakoff, and S. P. Rosen, Caltech preprint CALT-68-533 (1976).

14. H. Fritzsch and P. Minkowski, (Caltech preprint CALT-68-503, unpublished).

15. H. Georgi and S. L. Glashow, Phys. Rev. D6, 2977 (1972); A. Zee and S. L. Glashow, contributions to this conference.

16. V. Weiskopf, Phys. Rev. 56, 72 (1939).

17. R. Cowsik and J. M. McClelland, Phys. Rev. Letters 29, 669 (1972); for the problems relating to an eventual "missing mass" in the universe, see e.g., J. P. Ostriker, P. J. E. Peebles and A. Yahil, Astrophys, Journal 193, 21 (1974); R. V. Wagoner, contributions presented at the American Association for the Advancement of Science Annual Meeting, Boston, Mass. 1976.

18. B. Pontecorvo, J. Exp. and Theor. Phys. U.S.S.R., 53, 1717(1967), translation Soviet Phys. JETP 26,986(1968).

18a. M. Gell-Mann and J. B. Stephenson, to be published

19. S. L. Glashow, contribution to this conference.

20. F. Gürsey, contribution to this conference, see also P. Ramond, Caltech preprint CALT-68-540.

21. I am indebted to S. Weinberg for a thorough discussion of this point.

22. J. Prentki and B. Zumino, Nucl. Phys. B47, 95 (1972).

23. H. Georgi and S. L. Glashow, Phys. Rev. Letters 28, 1494 (1972).

24. K. Fujikawa, B. W. Lee and A. I. Sanda, Phys. Rev. D6, 2929 (1972); R. Shrock, Phys. Rev. D9, 743 (1974); S. Pi and J. Smith, Phys. Rev. D9, 1498 (1974); R. Bertlmann, H. Grosse and B. Lautrup, Nucl, Phys. B73; 523 (1974). See also, A. Zee, contribution to this conference; F. Wilczek and A. Zee, Princeton University preprint (1976); K. Fujikawa, DESY preprint (1976); H. Fritzsch and P. Minkowski, Caltech preprint CALT 68-538 (1976).

25. M. Ahmed and G. G. Ross, Phys. Letters 59B, 293 (1975); N. Vasanti, Princeton University preprint (1975), unfortunately the evaluation of the diagrams correspon-ding to the ones in Fig. 7 are numerically incorrect in the above papers.

26. L. K. Gershwin, M. Alston-Garnjost, R. Q. Bangerter, A. Barbaro-Galtieri, T. S. Mast, F. T. Solmitz, and R. D. Tripp, Phys. Rev. 188, 2077 (1969).
27. H. Fritzsch and P. Minkowski, Caltech preprint CALT-68-532 (1975).
28. K. Wilson, Phys. Rev. 179, 1499 (1969).
29. M. K. Gaillard and B. Lee, Phys. Rev. Letters 33, 108 (1974); G. Altarelli and L. Maiani, Phys. Letters 52B, 351 (1974).
30. N. Cabibbo, Phys. Rev. Letters 12, 62 (1964); M. Gell-Mann, Phys. Rev. Letters 12, 155 (1964).
31. C. Bouchiat and Ph. Meyer, Phys. Letters 25B, 282 (1967); M. A. B. Bég and A. Zee, Phys. Rev. D8, 1460 (1973); E. Golowich and B. Holstein, Phys. Letters 35, 83 (1975).
31a. Y. Hara, Phys. Rev. Letters 12, 378 (1964).
32. For the contrary point of view, see e.g., G. Branco and R. N. Mohapatra, City College of New York preprint CCNY-HEP-76/1 (1976); J. Sakarai, Phys. Rev. 156, 1508 (1967); M. Gronau, Phys. Rev. Letters 28, 188 (1972); For a review of earlier work, see e.g., R. E. Marshak, Riazuddin, and C. P. Ryan, "Theory of Weak Interactions in Particle Physics," John Wiley and Sons (1969); also, P. C. McNamee and M. D. Scadron, Univ. of Arizona preprint (1976).
33. M. Kobayashi and K. Maskawa, Progr. Theor. Physics 49, 652 (1973); L. Maiani, University of Rome preprint (1975);
34. H. Fritzsch and P. Minkowski, Caltech preprint CALT-68-537 (1976).
34a. For a review of the CP-violation in the kaon system see e.g., J. S. Bell and J. Steinberger, Weak Ingeractions of Kaons, Proceedings of the Int. Conf. on Elementary Particles, Oxford, 1965.

34b. L. Wolfenstein, Phys. Rev. Letters 13, 180 (1964).

35. M. K. Gaillard and B. Lee, Phys. Rev. D10, 897 (1974); See also, F. A. Wilczek, A. Zee, R. L. Kingsley and S. B. Treiman and A. de Rujùla, H. Georgi and S. L. Glashow in Ref. 8.

36. A. Mann, contribution to this conference.

37. See e.g., A. de Rujùla, contribution to this conference; R. M. Barnett, Harvard University preprint (1976).

38. F. Sciulli, contribution to this conference.

PARTICIPANTS

Carl H. Albright
Northern Illinois University

Thomas Appelquist
Yale University

Marshall Baker
University of Washington

William Bardeen
University of Wisconsin

Vernon Barger
University of Wisconsin

Isac Bars
Yale University

P. N. Bogolubov, P.N.
Ins. for Nuclear Research
Academy of Science U.S.S.R.

Richard Brandt
New York University

Laurie Brown
Northwestern University

Arthur Broyles
University of Florida

Nina Byers
University of California

Robert Cahn
University of Washington

Peter Carruthers
Los Alamos Scientific Lab.

George Chapline
Lawrence Livermore Lab.
University of California

M. Chen
Massachusetts Institute of Technology

Fred Cooper
Los Alamos Scientific Lab.

John Cornwall
University of California--Los Angeles

Michael Creutz
Brookhaven National Lab.

Richard Dalitz
Oxford University

Stanley Deans
University of South Fla.

Alvaro de Rújula
Harvard University

P. A. M. Dirac
Florida State University

Max Dresden
State University of New York--Stony Brook

Loyal Durand
Institute for Advanced Study--Princeton

Glennys Farrar
California Institute of Technology

Gordon Feldman
John Hopkins University

Paul Fishbane
University of Virginia

Peter Freund
University of Chicago

Harald Fritzsch
California Institute of Technology

Frederick Gilman
Stanford University

Sheldon Glashow
Harvard University

Alfred Goldhaber
State University
of New York Stonybrook

Ahmad Golestaneh
Fermi National Accelerator
Laboratory

Christian Le Monnier
de Gouville
Center for Theoretical
Studies
University of Miami

O. W. Greenberg
University of Maryland

Feza Gürsey
Yale University

Alan Guth
Columbia University

C. R. Hagen
University of Rochester

Leopold Halpern
Florida State University

M. Y. Han
Duke University

Joseph Hubbard
Center for Theoretical
Studies
University of Miami

Muhammad Islam
University of Connecticut

Ken Johnson
Massachusetts Institute of
Technology

Lorella Jones
University of Illinois

Gabriel Karl
University of Guelph

T. E. Kalogeropoulos
Syracuse University

Abraham Klein
University of Pennsylvania

Behram Kursunoglu
Center for Theoretical
Studies
University of Miami

Willis Lamb
University of Arizona

Benjamin Lee
Fermi National Accelerator
Laboratory

Don Lichtenberg
Indiana University

K. T. Mahanthappa
University of Colorado

Alfred Mann
University of Pennsylvania

André Martin
CERN

Caren Ter Martirosiyan
Institute for Theoretical
and Experimental Physics
Moscow, U.S.S.R.

V. A. Matveev
JINR, Moscow, U.S.S.R.
and Fermi Laboratory

Meinhard Mayer
University of California-
Irvine

Barry McCoy
SUNY, Stonybrook

Sydney Meshkov
National Bureau of Standards

Peter Minkowski
California Institute of Technology

J. W. Moffat
University of Toronto

Yoichiro Nambu
University of Chicago

Yuval Ne'eman
Tel-Aviv University

André Neveu
Institute for Advanced Studies
Princeton

Roger Newton
Indiana University

Kazuhiko Nishijima
University of Chicago

Richard Norton
University of California - Los Angeles

Horst Oberlack
Max Planck Institut for Physics and Astronomy
Munich, West Germany

Reinhard Oehme
University of Chicago

Lars Onsager
Center for Theoretical Studies
University of Miami

Heinz Pagels
Rockefeller University

Sandip Pakvasa
University of Hawaii

Michael Parkinson
University of Florida

Jogesh C. Pati
University of Maryland

R. D. Peccei
Stanford University

Arnold Perlmutter
Center for Theoretical Studies
University of Miami

H. David Politzer
Harvard University

P. G. Price
University of California - Berkeley

Pierre Ramond
California Institute of Technology

Rudolf Rodenberg
III Physikalisches Institut der Technischen Hochschule
Aachen, West Germany

Fritz Rohrlich
Syracuse University

S. P. Rosen
ERDA

Ronald Ross
University of California - Berkeley

V. I. Savrin
Institute for High Energy Physics
Serpukhov, U.S.S.R.

Howard Schnitzer
Brandeis University

Frank Sciulli
California Institute of Technology

Gordon Shaw
University of California Irvine

Dennis Silverman
University of California - Irvine

Alberto Sirlin
New York University

L. Slavnov
Institute of Math. of the Academy of Sciences
Moscow, U.S.S.R.

L. Soloviev
Serpukhov Institute for Theoretical Studies
Serpukhov, U.S.S.R.

George Soukup
Center for Theoretical Studies
University of Miami

Joseph Sucher
University of Maryland

Katsumi Tanaka
Ohio State University

William Tanenbaum
Stanford Linear Accelerator Center

John R. Taylor
University of Colorado

Vigdor Teplitz
Va. Polytechnic Institute and State University

George Tiktopoulos
University of California - Los Angeles

Yukio Tomozawa
University of Michigan

T. L. Trueman
Brookhaven National Lab.

V. S. Vladimirov
Institute of Math. of the Academy of Sciences
Moscow, U.S.S.R.

Kameshwar C. Wali
Syracuse University

Geoffrey West
Los Alamos Scientific Lab.

Ken Wilson
Cornell University

Lincoln Wolfenstein
Carnegie-Mellon University

T. T. Wu
Harvard University

G. B. Yodh
University of Maryland

Fredrick Zachariasen
California Institute of Technology

Anthony Zee
Princeton University

Daniel Zwanziger
New York University

SUBJECT INDEX